The Society of Fire Protection Engineers Series

Series Editor

Chris Jelenewicz, Society of Fire Protection Engineers, Gaithersburg, USA

The Society of Fire Protection Engineers Series provides rapid dissemination of the most recent and advanced work in fire protection engineering, fire science, and the social/human dimensions of fire.

The Series publishes outstanding, high-level research monographs, professional volumes, contributed collections, and textbooks.

Pat Brock · Virginia Charter

Fire Protection Hydraulics and Water Supply Analysis

Fourth Edition

SFPE
Engineering A Fire Safe World

Springer

Pat Brock
Oklahoma State University
Stillwater, OK, USA

Virginia Charter
Oklahoma State University
Stillwater, OK, USA

ISSN 2731-3638 ISSN 2731-3646 (electronic)
The Society of Fire Protection Engineers Series
ISBN 978-3-032-09876-4 ISBN 978-3-032-09877-1 (eBook)
https://doi.org/10.1007/978-3-032-09877-1

This Springer imprint is published by the registered company Springer Nature Switzerland AG
The registered company address is: Gewerbestrasse 11, 6330 Cham, Switzerland

Preface

When Pat Brock, who was not only my teacher and mentor, asked me to update the book I now use in the classroom, I was honored. I first met Pat in Fall of 2001 as a Freshman at Oklahoma State University in Fire Protection and Safety Engineering Technology. While at school, he was always one of my favorite teachers and has continued to be a mentor and colleague as I have progressed through my career. When I returned to OSU to teach, I had the privilege of teaching the class that uses this book. It is one of my favorite classes to teach and I enjoy working with the students to learn both the theory and application of this book. Pat is now enjoying retirement but is always willing to listen and provide guidance.

Floyd Luinstra, who is now also retired from being the lab manager for our Fire Protection and Safety Engineering Technology lab, should be recognized for his advice, photography, and general support.

As with the previous editions, the students of the Fire Protection and Safety Engineering Technology program at Oklahoma State University have, and continue to be, both mine and Pat's inspiration. This textbook was always for them and will continue to be. I would also like to especially thank two students, Caleb Pierce and Nick Schopfer (who have now graduated), for their additional assistance in getting equations formatted correctly. Though I hope that this textbook continues to reach beyond the classroom so that others can find the joy and inspiration of being a fire protection professional and working to protect people, property, and the environment.

Stillwater, USA

Pat Brock
Virginia Charter

Competing Interests The authors have no competing interests to declare that are relevant to the content of this manuscript.

Introduction

With all of the scientific advancements that have occurred in recent decades, it is surprising that the most popular fire extinguishing agent continues to be water. At the same time, water is becoming less abundant and more valuable. Therefore, the conservation and efficient use of our water supplies is more important than ever. Efficient use of water for fire protection purposes involves being able to establish how much water is needed, determining the energy required to get the water to the fire in proper quantities to achieve fire control or extinguishment, and analyzing the water distribution systems for inefficient or deficient design. This text was created to address these issues.

This is the Fourth Edition of this textbook. The First Edition was published in 1990, with the Second Edition being published in 2000, and the Third Edition was published in 2012. This edition of the text has been updated to reflect changes that have occurred in those codes and standards that have impact upon water-based fire protection systems and equipment, particularly the technical design changes that have occurred in NFPA 13.

This text was always intended primarily for use in the classroom at colleges and universities where fire protection is being taught. However, in its more than three and a half decades of existence, it has also become a valuable resource to the practicing fire protection specialists and fire protection engineers. It has become part of the professional libraries of fire marshals, insurance specialists, fire system designers, and fire protection engineering consultants all across the country. This is due, in part, because the text is not only used in undergraduate classrooms, but also as the basis of continuing education classes taught both in the traditional format and online.

As with previous editions, practical application in real-world situations takes precedence over theory, and the text is intended to be useful, not only in the classroom, but also on jobsites around the world.

Contents

List of Tables

Chapter 1
Water as an Extinguishing Agent

Abstract This chapter provides an introduction to water as one of the most important components of fire protection systems due to its effectiveness, relative availability, and low cost. Water extinguishes fires through cooling and through excluding oxygen by means of the volumetric expansion as the liquid changes to the gaseous state. The properties of water are discussed and presented along with that water, for all practical purposes, can be considered incompressible. Finally, common conversions between units are provided.

Water is the most important single component of fire protection. Therefore, for credibility as a fire protection specialist, a person must possess a knowledge of the behavior and physical characteristics of water and water supply systems. Water is so important as an extinguishing agent because of its effectiveness, general availability, and low cost. Although water is not effective in every situation, water is an effective extinguishing agent over a broader spectrum of hazards than any other agent. With the development of water-mist extinguishing systems, the range of hazards over which water may have application continues to increase. With water-mist technology, very small quantities of water prove effective in fire extinguishment in situations where primarily gaseous agents have been used in the past.

Either through accident or experiment, man has known that putting water on a fire will put it out or at least retard the fire's growth. Following is a simplified equation illustrating the combustion process:

$$\text{Heat of Ignition} + \text{Oxygen} + \text{Fuel} \Rightarrow \Delta\text{H(heat)} + \text{Products of Combustion}$$

By examining this relationship, one can conclude that fire control will result through elimination of any one of the basic components required in the combustion process. Elimination of the fuel source, the oxygen source, or the ignition source will result in fire control. Also, there is every indication that the fire can be controlled by chemically inhibiting the combustion chain reaction.

Why is water so important as an extinguishing agent? Water is an efficient agent because it works at least two ways: First, by cooling, water helps to eliminate the

P. Brock and V. Charter, *Fire Protection Hydraulics and Water Supply Analysis*, The Society of Fire Protection Engineers Series,
https://doi.org/10.1007/978-3-032-09877-1_1

ignition source. In changing water from a liquid at 32 °F (0 °C) to steam, 1190 Btu per pound (768.3 W/kg) of water is absorbed. Second, in this transition to steam, water undergoes a volume expansion by a factor of about 1700. Therefore, 1 cubic foot (1 L) of water can create 1700 cubic feet (48.14 m^3) of steam. The dilution of the air by the steam results in displacement of some of the oxygen molecules by water vapor molecules. This effectively reduces the amount of oxygen available to support the combustion process and tends to smother the fire.

The preceding discussion establishes the effectiveness of water as an extinguishing agent. Effectiveness alone, however, does not promote the widespread use of water in fire protection. On the other hand, if effectiveness is considered along with the availability and cost, no other agent is comparable. Except for the most arid regions of the world, water is available. It is not necessary that water be fresh or pure to work as an extinguishing agent. Even though saltwater is highly corrosive to many metals, saltwater supply systems are being used for fire protection purposes. Non-potable water is also used in fire protection systems. However, economics usually dictates that a single water system provides both public consumption and fire protection supplies.

Water is by far the least expensive of agents. Historically, property owners have not been required to pay directly for water used in extinguishing fires in their buildings. In more recent times, many water utility companies have established monthly or annual rates, to be paid by property owners whose buildings are protected by automatic sprinkler systems.

There is a real cost associated with the production of potable water, especially if changing saltwater to fresh water. These costs, however, are usually defrayed through rates paid by domestic and commercial consumers. Even for industries that drill their own wells or provide their own water supplies by other means, water is still relatively inexpensive compared to other agents, such as dry chemicals, carbon dioxide, or other gaseous agents.

Although water is generally available, everyone is aware that water shortages are becoming increasingly commonplace. This is usually not a result of increasing fire protection needs. On the contrary, because of aggressive enforcement of modern building codes and requirements for sprinklers in many buildings, the total water supply requirement for fire protection has been reduced in many communities. The shortages are most often a result of a combination of population increases, environmental factors, and extreme weather events. An increase in the number of people or industrial and commercial concerns will always result in greater water consumption. Increased consumption coupled with extended dry weather conditions or extreme weather events such as polar vortexes or wildland urban interface fires can leave little water to meet the fire protection needs of a city.

The fire protection specialist must have a good understanding of hydraulics. Adequate fire protection is often equated to adequate water supplies, and problems like water shortages affect the amount of water available for fire protection. In analyzing water supply systems, the fire protection specialist must be prepared to answer the following questions:

Table 1.1 Variation of water density and specific gravity with temperature

Temperature		Weight density		Specific gravity
°F	°C	(lb/ft^3)	(kg/L)	
32.0	0.0	62.416	0.9998	0.9998
39.2	4.00	62.425	1.0	1.0
50.0	10.0	62.408	0.9997	0.9997
60.0	15.56	62.366	0.9991	0.9991
70.0	21.11	62.3	0.9980	0.9980
80.0	26.67	62.217	0.9967	0.9967
90.0	32.22	62.118	0.9951	0.9951
100.0	37.78	61.998	0.9932	0.9932
150.0	65.56	61.203	0.9804	0.9804
200.0	93.33	60.135	0.9633	0.9633

- *How many gallons per minute are needed for adequate protection?*
- *At what pressures must the water be available?*
- *Are there methods to transform deficient water supply systems into systems adequate for fire protection purposes?*
- *How do you test a water system for adequacy?*
- *What is the relationship between pipe arrangements and the materials from which piping is made and the amount of water available?*
- *Is the water system reliable?*

A study of engineering fluid dynamics in its purest form would involve a study of the physical laws and principles that might apply to any fluid. For the reasons presented in the previous section, the fluid of primary importance to the fire protection specialist is water. This text is therefore primarily devoted to discussion of the behavior exhibited by water **at rest** and **in motion**. Such discussion must include study of the relevant physical properties of water.

Water, of course, has the chemical formula H_2O and has a molecular weight of approximately 18 g per mole. Of more practical importance, the weight density or specific weight **(w)** of liquid water is 62.4 lb./ft^3 (1.0 kg/L). This density will vary slightly with temperature and pressure (see Table 1.1).

The changes in density are sufficiently small over the temperature and pressure ranges commonly encountered; therefore, deviations from 62.4 lb./ft^3 (1.0 kg/L) can be ignored for practical application. The concentration of impurities in the water will also affect the density. For example, ocean salt water has a density of 64.1 lb./ft^3 (1.027 kg/L). For purposes of practical application, water can be considered incompressible because of the insignificant change in density under a broad range of pressures. However, the decrease in density when water is both cooled and heated indicates that the water is actually expanding when it is heated above, or cooled below, 39.2 °F (4 °C). It is this volume expansion that causes the rupturing of pipes containing water when piping is subjected to freezing temperatures. It also explains

Table 1.2 Specific gravity of common substances

Substance	Specific gravity at 68 °F (20 °C)
Water	0.998
Salt water	1.027
Commercial solvent	0.717
Carbon tetrachloride	1.582
Medium lubrication oil	0.891
Medium fuel oil	0.854
Heavy fuel oil	0.908
Regular gasoline	0.724
Turpentine	0.862
Ethyl fuel	0.789
Benzene	0.879
Glycerin	1.262
Light machinery oil	0.907
Air	0.0012
Ammonia	0.0007
Carbon dioxide	0.0018
Methane	0.0017
Oxygen	0.0013

why sprinkler systems installed with piping in a gridded configuration and without pressure relief valves can come apart during summer months as ceiling temperatures increase.

Another important relative measurement is specific gravity. The specific gravity (**S_g**) of a substance may be defined generally as the ratio of the weight density of the substance to the weight density of another substance. Or,

$$S_g = \frac{W_a}{W_b} \tag{1.1}$$

As most commonly used, specific gravity refers to the ratio of the weight density of a substance to the weight density of water at 39.2 °F (4 °C). Thus, the specific gravity of water is 1.0 at 39.2 °F (4 °C). Care must be taken to include the effect of pressure and temperature when calculating specific gravities. Particularly in the cases of gases, the weight density varies significantly with pressure and temperature. Table 1.2 contains values of the specific gravities of some common substances.

Example 1.1

If a particular grade of crude oil has a weight density of 52 lb./ft^3 (0.833 kg/L) at 39.2 °F (4 °C), what is the specific gravity of the oil?

Solution:

$$S_g = \frac{W_a}{W_b}$$

$$S_g = \frac{W_{oil}}{W_{H_2O}}$$

$$S_g = \frac{52\,\text{lb./ft}^3}{62.4\,\text{lb/ft}^3}$$

$$S_g = 0.833$$

In S.I. Units, the Solution is:

$$S_g = (0.833\,\text{kg/L})/(1.0\,\text{kg/L}) = 0.833.$$

Example 1.2

A particular foam plastic has a specific gravity of 0.008. What is its specific weight?

Solution:

$$S_g = \frac{W_a}{W_b}$$

Then:

$$w_a = (w_b)(S_q)$$

And:

$$w_{plastic} = (62.4\,\text{lb./ft}^3)(0.008)$$
$$w_{plastic} = 0.4992\,\text{lb./ft}^3.$$

In S.I. Units, the Solution is:

Table 1.3 Important conversion factors

1 ft = 12 in = 0.3048 m
1 in = 2.54 cm
1 ft^3=1728 in^3=28,315 cc
1 ft^3=7.48 gal = 28.315 L
1 gal = 231 in^3=3785.4 cc
1 gal = 0.1337 ft^3=3.7854 L
1 ft^3/s = 449 gpm = 28.315 L/s
1 lb = 0.4536 kg
1 BTU = 3.415 Watts
1 gal of water weighs 8.3453 lb. and 3.7854 kg 1 psi = 0.0689 bar = 6894.757 Pa
1 bar = 10^5 Pa = 100 Kpa
1 bar = 1.0197 kg/cm^2=14.5 psi

$$W_{\text{plastic}} = 0.008 \times 1\,\text{kg/L} = 0.008\,\text{kg/L}.$$

The boiling point and freezing point of water are 212 °F (100 °C) and 32 °F (0 °C) respectively. These values apply exactly at atmospheric pressure at sea level only. As might be expected, the boiling and freezing points are affected by pressure. However, it has been pointed out that a rather unique feature is that water undergoes a volumetric expansion both upon freezing and when changing into the vapor state. The 1700:1 expansion ratio during vaporization was previously pointed out. The expansion of water as it freezes causes undrained automatic sprinkler piping in improperly heated buildings to break and underground piping not buried deep enough to rupture. The maximum density and minimum volume of water occurs at 39.2 °F (4 °C).

Table 1.3 contains some additional relationships and conversion factors that will be useful in following discussions.

Important Terms and Definitions

These terms are used in future chapters and should be a part of the vocabulary of the fire protection specialist. The following definitions are those compiled by Henke in his fluid mechanics text [1].

Fluid—Any substance that can flow; a substance that has definite mass and volume at constant temperature and pressure, but no definite shape, and with the inability to sustain shear stresses.

Fluid Mechanics—A branch of physics dealing with the behavior of fluids, particularly with respect to their reaction to forces applied to them.

Force—In the general terms of physics, force is that which causes motion and is illustrated mathematically by

$$F = (m)(a) \tag{1.2}$$

where:

F = the sum of all forces acting on an object,

m = the mass of the object,

a = the acceleration of the object.

Since:

$$m = \frac{w}{g} \tag{1.3}$$

where:

W = weight in pounds (kg), and

g = acceleration due to gravity in ft/s^2 (m/s^2).

It follows that:

$$F = \frac{(W)(a)}{g} \tag{1.4}$$

And where the units of acceleration (**a**) are also in ft/s^2 (m/s^2) the units of force are given in pounds (kg). And when the acceleration (**a**) is due only to gravity (**a = g**),

$$F = W \tag{1.5}$$

Therefore, for many practical applications, force and weight are synonymous.

Example 1.3

A 1 × 1 × 2 foot (0.3 × 0.3 × 0.6 m) rectangular metal ingot weighs 235 pounds (106 kg). If it rests upon the ground, how much force does it exert upon the ground?

$$F = W$$

The dimensions for the ingot, in this case, are irrelevant and $F = 235$ pounds (106 kg).

Example 1.4

What force is exerted by the water on the bottom of a full rectangular storage tank that has a base 30 × 40 ft (9.144 m × 12.192 m) and is 10 ft (3.048 m) high?

Solution:

$$F = W$$

And:

$$W = (w)(V)$$

where:

V = Volume

So:

$$F = \left(62.4\,\text{lb.}/\text{ft}^3\right)(30\,\text{ft} \times 40\,\text{ft} \times 10\,\text{ft})$$

$$F = \left(62.4\,\text{lb.}/\text{ft}^3\right)\left(12{,}000\,\text{ft}^3\right)$$

$$F = 748{,}800\,\text{lb.}$$

In S.I. Units, the Solution is:

$$F = \left(1000\ \text{kg/m}^3\right)(9.144\,\text{m} \times 12.192\,\text{m} \times 3.048\,\text{m})$$
$$F = 339,802\,\text{kg.}$$

Hydraulics—The branch of fluid mechanics dealing with the mechanical properties of liquids (in this text, water) and the application of these properties in engineering.

Hydrokinetics (Hydrodynamics)—A branch of hydraulics having to do with liquids (water) in motion, particularly in relation to forces created by or applied to the liquid in motion.

Hydrostatics—A branch of hydraulics dealing with the properties of liquids (water) at rest, particularly in relation to pressures resulting from or applied to the static liquid.

Liquid—A fluid having a definite volume, unlike gases, which will expand to fill the vessels containing them.

Pressure—A force per unit of area. Mathematically, this is represented as

$$P = \frac{F}{A} \tag{1.6}$$

P = pressure in pounds/unit of area (kg/cm^2 or bars),

F = force in pounds (kg),

A = area, usually in square inches (square centimeters).

Commonly, pressure is talked about in units of pounds (kg). This is technically incorrect and should be in *pounds per unit of area (kg/cm^2 or bars)*. In fire protection, pressure is most often dealt with in *units of pounds per square inch(psi)*. In Standard International Units, pressure is described using units of bars or kilopascals. One bar is approximately equal to atmospheric pressure. Fire protection pressures are also most often *gauge pressures (psig)*. That is, pressure readings are deviations from atmospheric pressure. Thus, a gauge reading of 100 psi (6.9 bar) would indicate that the pressure being measured is 100 psi (6.9 bar) above atmospheric pressure. Pressures that include atmospheric pressures are referred to as *absolute pressures(psia)*. At sea level, atmospheric pressure is approximately 14.7 psia (1.0 bar). Therefore, 100 psig (6.9 barg) equals 114.7 psia (7.9 bar).

Example 1.5

What pressure, in psi, would be acting on the base of the tank described in Example 1.4?

Solution:

$$P = \frac{F}{A}$$

In this case:

$$A = (30\,\text{ft} \times 40\,\text{ft})\left(144\,\text{in.}^2/\text{ft}^2\right)$$
$$A = 172{,}800\,\text{in.}^2$$

From Example 1.4:

$$F = 748{,}800\,\text{lb.}$$

And finally:

$$P = \frac{748{,}800\,\text{lbs}}{172{,}800\,\text{in}^2}$$

$$P = 4.33\,\text{psi.}$$

In S.I. Units, the Solution is:

$$A = (9.144\,\text{m})(12.192\,\text{m}) \times 10^4\,\text{cm}^2/\text{m}^2 = 1.115 \times 10^6\,\text{cm}^2$$
$$P = (339{,}802\,\text{kg})/\left(1.115 \times 10^6\,\text{cm}^2\right) = 0.305\,\text{kg/cm}^2.$$

Or $\text{P} = \left(0.305\text{kg/cm}^2\right)\left(1\text{bar}/1.0197\text{kg/cm}^2\right) = 0.299\text{bar}.$

Work—The product of the magnitude of the force acting on an object and the distance through which the object moves. This is expressed algebraically as

$$G = (F)(D) \tag{1.7}$$

where:

G = the work in units of foot-pounds (centimeter-grams or meter-kilograms),

F = the force in pounds (grams or kilograms),

D = the distance of displacement in feet (centimeters or meters).

Example 1.6
How much work is involved in transporting 1000 ft^3 of water 1 mile?

Solution:

$$G = (F)(D)$$

And where:

$$F = W$$
$$F = \left(1000\,\text{ft}^3\right)\left(62.4\,\text{lb.}/\text{ft}^3\right)$$

$$F = 62{,}400\,\text{lb.}$$

And.

$$D = (1\,\text{mile})(5280\,\text{ft/mile})$$

Then:

$$G = (62{,}400\,\text{lb.})(5280\,\text{ft})$$
$$G = 3.29472 \times 10^{8}\,\text{ft - lb.}$$

Example 1.7
How much work is involved in transporting 1000 L of water a distance of 1500 m?

Solution:

Again,

$$G = (F)(D)$$

And where:

$$F = W$$
$$F = (1000\,\text{L})(1\,\text{kg/L})$$
$$F = 1000\,\text{kg}$$

Then:

$$G = (1000\,\text{kg})(1500\,\text{m}) = 1.5 \times 10^{6}\,\text{m-kg}$$

Example 1.8
How much work is involved in elevating 1 gallon of water to the top of a storage tank that is 80 ft high?

Solution:

Again,

$$G = (F)(D)$$

And where:

$$F = (1\ \text{gal})(8.3453\ \text{lb./gal})$$
$$F = 8.3453\ \text{lb.}$$

Then:

$$G = (8.3453\ \text{lb.})(80\ \text{ft})$$
$$G = 667.624\ \text{ft - lb.}$$

Example 1.9

How much work is involved in elevating one liter of water to the top of a storage tank that is 30 m high?

Solution:

Again,

$$G = (F)(D)$$

And where:

$$F = W$$
$$F = (1\ \text{L})(1\ \text{kg/L})$$
$$F = 1\ \text{kg}$$

Then:

$$G = (1\ \text{kg})(30\ \text{m}) = 30\ \text{m-kg}.$$

Summary of Important Concepts

1. Water is the most important component of good fire protection because of its effectiveness, availability, and low cost.
2. Water extinguishes fire through cooling and through excluding oxygen by volumetric expansion as the liquid changes to the gaseous state.
3. Water has a weight density of 62.4 lb./ft^3 (1 kg/L), changes to a vapor at 212 °F (100 °C), and freezes at 32 °F (0 °C). Water undergoes a volumetric expansion during vaporization and during freezing.
4. For most practical purposes, water can be considered incompressible.

Exercises

1.1 Compare and contrast fluids, solids, and gases.

1.2 List some significant ways that liquids differ from gases.

1.3 How many gallons (liters) of water could be contained in a 30 × 40 × 12 foot (10 × 12 × 4 m) rectangular storage tank?

1.4 If a spherical storage tank holds 40,000 gallons (120,000 L) of water, what diameter must it have?

1.5 If a cylindrical storage tank has a 40-foot (12 m) diameter circular base, how high must it be to hold 1 million gallons (3 million liters)?

1.6 If water is moving at a rate of 3.2 ft^3/s (90,000 cc/s), how many gallons (liters) per minute is this?

1.7 What is the difference between specific gravity and specific weight?

1.8 If the specific gravity of gold is 19.0, what is the weight of a cubic inch (cubic centimeter) of gold? A cubic foot (cubic meter)?

1.9 If the weight density of air at sea level is 0.08 lb./ft^3 (0.0012 kg/L), how much should the air weigh in a room that measures 20 × 12 × 8 ft (7 × 4 × 2.4 m)?

1.10 A man supporting himself on the heel of his shoe, which has an area of 4 in^2 (100 cm^2), produces a pressure on the ground of 50 psi (3.45 bar). How much does the man weigh?

1.11 If a 180-pound (82 kg) man sits on a 5-pound (3 kg) chair, and each of the 4 chair legs has a circular cross section with a 1-in (2.5 cm) diameter, what pressure is exerted on the floor by the chair?

1.12 A heavy fuel oil has a specific gravity of 0.918. How much will 100 gallons (300 L) of this oil weigh?

1.13 If gasoline has a specific gravity of 0.738, what force will be exerted on the circular bottom of a cylindrical gasoline tank that has a diameter of 8 ft (2.4 m) and is 12 ft(4 m) high?

1.14 What amount of work will be accomplished in transporting 500 gallons (1500 L) of water 1760 ft (1500 m)?

1.15 What amount of work will be accomplished in elevating 40,000 gallons (120,000 L) of water to the top of an elevated tank that is 80 ft (24 m) high?

1.16 The circular clapper on the water side of a dry pipe sprinkler valve has a diameter of 4 in (10 cm). What force will be acting on this clapper due to a water pressure of 72 psi (5 bar)?

1.17 The circular clapper on the air side of a dry pipe sprinkler valve has a diameter of 10 in (25 cm). What force will be acting on this clapper due to an air pressure of 40 psi (3 bar)?

1.18 If 4-in (100 mm) Schedule 40 sprinkler piping has an actual inside diameter of 4.026 in (102.3 mm), how many gallons (liters) of water could be contained in 135 ft (40 m) of this pipe? How much would the water in this pipe weigh?

1.19 If 1-in (25 mm) Schedule 40 sprinkler piping has an actual inside diameter of 1.049 in (26.6 mm), how many gallons (liters) of water will 800 ft (240 m) of this pipe hold? How much would the water in this pipe weigh?

1.20 It is commonly desired to keep water velocities in automatic sprinkler systems below 20 ft/s (fps) (6 m/s). How many gallons per minute (liters per minute) would be represented by a water velocity of 20 fps (6 m/s) in 1-in (25 mm) pipe with an actual inside diameter of 1.049 in (26.6 mm)?

Selected Answers Found in Appendix K.

Reference

1. Russell W. Henke, *Introduction to Fluid Mechanics* (Reading: Addison-Wesley Publishing Co., 1972) pp. 1–9.

Supplemental Readings

Hickey, Harry E. *Hydraulics For Fire Protection.* National Fire Protection Association: Quincy, MA, 1980 pp. 1–6.

Chapter 2
Hydrostatics

Abstract This chapter presents the science behind liquids at rest or hydrostatics. The five hydrostatic principles of hydrostatics are presented and demonstrated with the mathematical relationships including pressure created by a column of water and how the relationship of pressure, height, and density with proper unit conversion is required. Potential energy within a system is also presented to understand the interaction of hydrostatic pressures from elevation and external pressures create potential energy. Common instrumentation to measure hydrostatic pressures is also presented.

As pointed out in Chap. 1, hydrostatics is the branch of hydraulics dealing with fluids at rest. Because water is the fluid of primary importance to the fire protection specialist, this first section is devoted to the features of water in a static condition.

Basic Principles of Hydrostatics

There are five hydrostatic principles that should be initially emphasized. From these principles, practical applications involving water at rest can be promoted. The principles of hydrostatic pressure are as follows:

Principle 1: The pressure at a point in a liquid is constant in every direction.

Principle 2: Pressure applied from external sources on a confined liquid is transmitted in all directions throughout the liquid without reduction in magnitude.

Principle 3: The pressure created by a liquid in an open container is directly proportional to the depth of the liquid.

Principle 4: The pressure created by a liquid in an open container is directly proportional to the density of the liquid.

Principle 5: The shape or volume of a container has no bearing on the pressure created by the liquid.

P. Brock and V. Charter, *Fire Protection Hydraulics and Water Supply Analysis*, The Society of Fire Protection Engineers Series,
https://doi.org/10.1007/978-3-032-09877-1_2

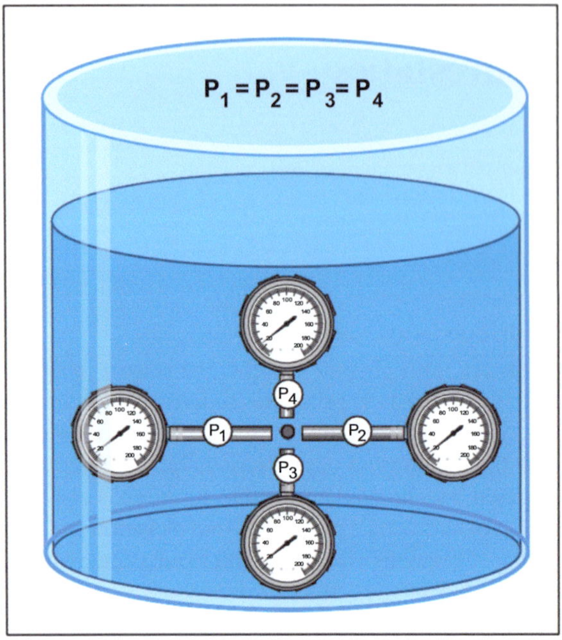

Fig. 2.1 Four gauges taking pressure readings at the same point will register the same pressure in accordance with Principle 1

People sometimes have the idea that pressure always acts in the downward direction. This fallacy is fostered by the understanding that pressure is closely related to weight. However, Principle 1, known as Pascal's Law, points out that pressure acts in all directions and not simply downward [1]. Fig. 2.1 shows four different pressure gauges taking readings at the same point in a system but with different orientations. In accordance with Principle 1 of hydrostatic pressure, all four show the same pressure reading.

Principle 2 has more obvious practical implications than Principle 1. The concept of undiminished pressure transmission throughout a confined liquid can be more readily accepted if the relatively incompressible nature of liquids is understood. It was previously pointed out that for practical applications liquids are incompressible. Unlike gases, which will contract in volume under pressurization, liquids will substantially maintain their original volume when confined and subjected to external pressures. Because the external pressure is not absorbed in a volume contraction, the pressure is simply transmitted throughout the liquid in every direction without reduction in magnitude.

Note that in Fig. 2.2 the pump operating at 0 gpm (0 L/min) against a closed system is developing a pressure of 100 psi (6.89 bar) at its discharge port. Because no water is flowing and the system is closed, all gauges show an increase in pressure of 100 psi (6.89 bar). If the gauges are located at different elevations, the pressures are not the same, but all show the same pressure increase as long as the water is in a static condition. This principle explains why pressure surges (water hammers), due

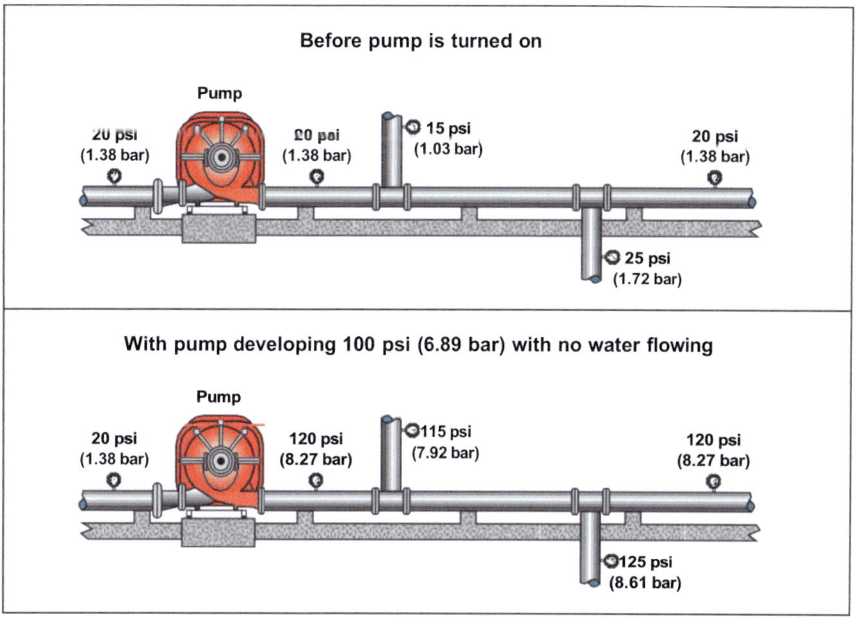

Fig. 2.2 Illustration of Principle 2 of hydrostatics

to a pump turning on or a valve being closed too quickly, can be transmitted over great distances inside a water system.

Principle 3 is a basic and critical part of fire protection hydraulics. The pressure created by a column of liquid is directly proportional to the depth of the liquid. Water, as indicated in Chap. 1, has a specific weight of 62.4 lb./ft^3(1 kg/L). Therefore, 1 cubic foot (liter) of water will weigh 62.4 lb. (1.0 kg). Visualize a cubic foot (1000 cc) of water (Fig. 2.3). From the discussion presented in Chap. 1, it is known that force can often be equated to weight, or the following:

$$F = W = 62.4\,\text{lb.}(1.0\,\text{kg})$$

And then:

$$\frac{F}{A} = \frac{62.4\,\text{lb.}}{144\,\text{in}^2} = 0.433\,\text{psi}$$

In S.I. Units:

$$P = \frac{F}{A} = \frac{1.0\,\text{kg}}{100\,\text{cm}^2} = 0.01\ \text{kg/cm}^2 = 0.0098\,\text{bar.}$$

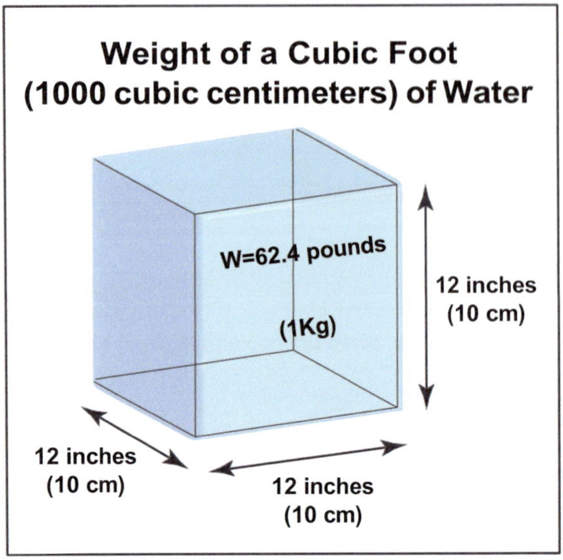

Fig. 2.3 Cubic foot of water weighs 62.4 pounds. One thousand cubic centimeters of water weighs one kilogram

Now, visualize 2 cubic feet (2000 cc) of water (Fig. 2.4). The weight of the water is 124.8 pounds (2 kg), and, in the example, the area at the base remains constant at 144 in^2 (100 cm^2).

$$P = \frac{F}{A} = \frac{124.8\,\text{lb.}}{144\,\text{in}^2} = 0.866\,\text{psi}$$

In S.I. units:

$$P = \frac{F}{A} = \frac{2.0\,\text{kg}}{100\,\text{cm}^2} = 0.02\,\text{kg/cm}^2 = 0.0196\,\text{bar}$$

When the height of the water doubled, the pressure created at the base also doubled.

Principle 4 is closely associated with Principle 3 in that the pressure created by a liquid is also directly proportional to the density. For example, since the density of mercury is 13.6 times greater than water, the pressure created by one foot (30.48 cm) of mercury is 13.6 times greater than the pressure created by one foot (30.48 cm) of water. Both Principle 3 and Principle 4 can be summarized by the following general relationship:

$$P = (w)(h) \tag{2.1}$$

where:

P = pressure (usually in psi or bar),

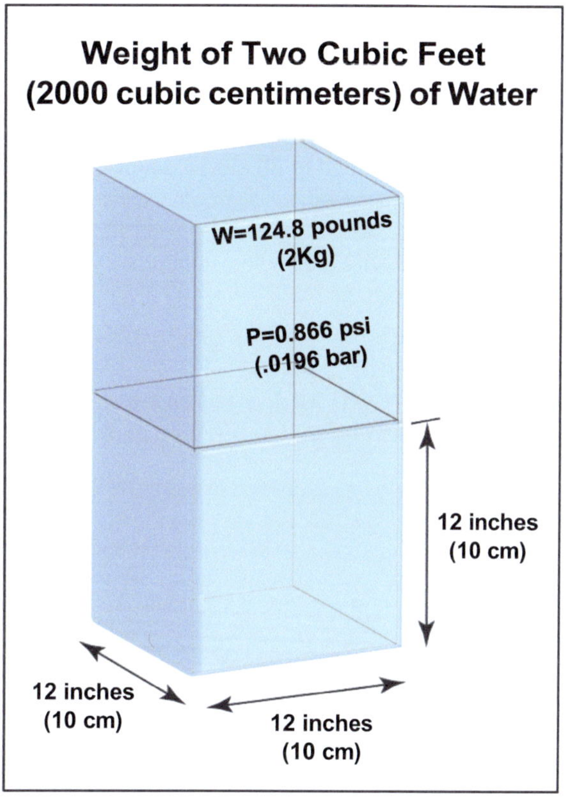

Fig. 2.4 Two cubic feet of water will weigh 124.8 pounds. Two thousand cubic centimeters of water weighs two kilograms

w = specific weight of the liquid in lb./ft^3 (kg/L),

h = height of liquid column in feet (m).

More specifically, for water:

$$P = \left(62.4\,\text{lb.}/\text{ft}^3\right)(\text{h})(\text{ft})$$

$$P = \frac{\left(62.4\,\text{lb.}/\text{ft}^3\right)(\text{h})(\text{ft})}{\left(144\,\text{in}^2/\text{ft}^2\right)}$$

$$P = (0.433\,\text{psi}/\text{ft})(\text{h})(\text{ft}) \tag{2.2}$$

In S.I. units:

$$P = (1.0\ \text{kg/L})(\text{h})(\text{m})$$

$$P = \frac{(1.0\,\text{kg/L})(100\,\text{cm/m})(1.0\,\text{bar})(\text{h})(\text{m})}{(1000\,\text{cc/L})\left(1.0197\ \text{kg/cm}^2\right)}$$

$$P = 0.098(\text{bar/m})\text{h}(\text{m}) \tag{2.2a}$$

Note: Where **h** is in feet (meters), the units of **P** is psi (bar).

This is one of the basic relationships of fire protection hydraulics and should be memorized; Eqs. 2.2 and 2.2a will be a frequent tool of the fire protection specialist. These formulas can be used to determine the following:

- The pressure needed to fill elevated tanks.
- The pressure needed to overcome elevation in fire water systems, including sprinkler systems.
- The back pressure created by standpipe systems in high-rise buildings.

Example 2.1

What pressure will be observed on a gauge at the base of an elevated tank in which the water level is 85 ft (26 m) above the pressure gauge?

Solution:

$$P = (0.433\,\text{psi/ft})(\text{h})(\text{ft})$$
$$P = (0.433\,\text{psi/ft})(85\,\text{ft}) = 36.8\,\text{psi}$$

In S.I. Units, the Solution is:

$$P = (0.098\,\text{bar/m})(\text{h})(\text{m})$$
$$P = (0.098\,\text{bar/m})(26\,\text{m}) = 2.548\,\text{bar.}$$

Example 2.2

What pressure must be available to overcome elevation alone in pumping water to the top of a 160-foot (50-m) high-rise building?

Solution:

$$P = (0.433\,\text{psi/ft})(\text{h})(\text{ft})$$
$$P = (0.433\,\text{psi/ft})(160\,\text{ft}) = 69.3\,\text{psi}$$

In S.I. Units, the Solution is:

$$P = (0.098\,\text{bar/m})(\text{h})(\text{m})$$
$$P = (0.098\,\text{bar/m})(50\text{m}) = 4.9\,\text{bar.}$$

Note that the application of Eqs. 2.2 and 2.2a is limited to fresh water. The equations can be altered to work for any liquid by simply inserting the specific gravity, S_g, of the liquid as follows:

Example 2.3

If an oil storage tank holding oil with a specific gravity of 0.92 is 45 ft (14 m) high, what pressure will exist at the base of the tank when the tank is full?

Solution:

$$P = (W)(h)$$
$$P = \left(S_{g_{oil}}\right)\left(W_{H_2O}\right)(h)$$
$$P = (0.92)\left(62.4\,\text{lb.}/\text{ft}^3\right)\frac{\left(1\,\text{ft}^2\right)(45\,\text{ft})}{\left(144\,\text{in}^2\right)}$$
$$P = 17.9\,\text{psi}$$

In S.I. Units, the Solution is:

$$P \doteq 0.098\,(\text{bar/m})h(\text{m})\text{Sg}$$
$$P = (0.098\,(\text{bar/m})(14\,\text{m})(0.92)$$
$$P = 1.26\,\text{bar.}$$

Example 2.4

What magnitude of force (F) will be required in the situation shown in Fig. 2.5 to maintain a state of equilibrium if the cross-sectional area of Piston A is 4 in^2 (26 cm^2)? The weight of Piston A may be considered negligible; however, Piston B weighs 500 pounds (225 kg) and has a cross-sectional area of 420 in^2 (2700 cm^2). The hydraulic fluid used has a specific gravity of 0.82.

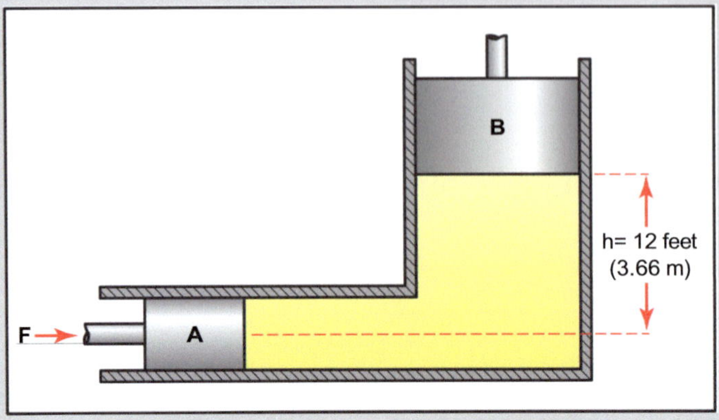

Fig. 2.5 Illustration for Example 2.4

Solution:

The pressure acting on the face of Piston A is due to the combination of pressures created by the weight of the fluid plus the weight of Piston B.

Thus:

$$P_A = (w)(h) + \frac{F_b}{A_b}$$

$$P = (0.82)\left(62.4\,\text{lb.}/\text{ft}^3\right)(12\,\text{ft})\frac{\left(1\,\text{ft}^2\right)}{\left(144\,\text{in}^2\right)} + \frac{500\,\text{lb.}}{420\,\text{in}^2}$$

$$P_A = 5.454\,\text{psi}$$

Then:

$$F = (P_A)(A_A)$$

$$F = (5.454\,\text{psi})\left(4\,\text{in.}^2\right) = 21.82\,\text{lb.}$$

In S.I. Units, the Solution is:

$$P_A = 0.098\,(\text{bar}/\text{m})\text{h}(\text{m})S_g + \frac{F_b}{A_b}$$

$$P_A = (0.098\,\text{bar}/\text{m})(3.66\,\text{m})(0.82)$$

$$\left(1.0197\,\text{kg}/\text{cm}^2\right)/\text{bar} + \frac{225\,\text{kg}}{2700\,\text{cm}^2}$$

$$P_A = 0.383\,\text{kg}/\text{cm}^2.$$

Then:

$$F = (P_A)(A_A)$$
$$F = (0.383\,\text{kg/cm}^2)(26\,\text{cm}^2) = 9.96\,\text{kg}$$

Principle 5 is often the most difficult to understand. Pressure developed is not related to size or shape of the container, only the depth and density of the liquid. Figure 2.6 shows three situations where the pressure readings on the gauges would be identical. Remember, Principle 5 applies to pressure, not force. The weight of the water in each case is different, but all three pressure readings will be the same.

The previous discussion established that the pressure beneath one cubic foot (1000 cc) of water that is one foot (10 cm) high is 0.433 psi (0.098 bar/m). What about a situation where the water is again one foot (10 cm) high but only one square inch (1 cc) in area? (Fig. 2.7).

What is the weight of such a small column of water? It is obvious that 144 (100) such columns exist in a cubic foot (1000 cc) of water. Therefore, the weight of one such column would be:

$$W = \frac{62.4\,\text{lb.}}{144} = 0.433\,\text{lb.}$$

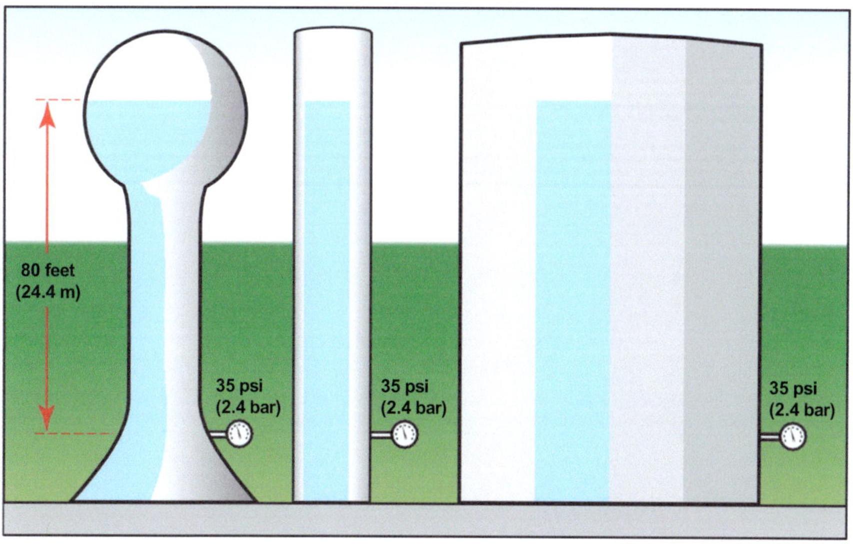

Fig. 2.6 Pressure developed depends only upon the density and height of the liquid, not upon container shape

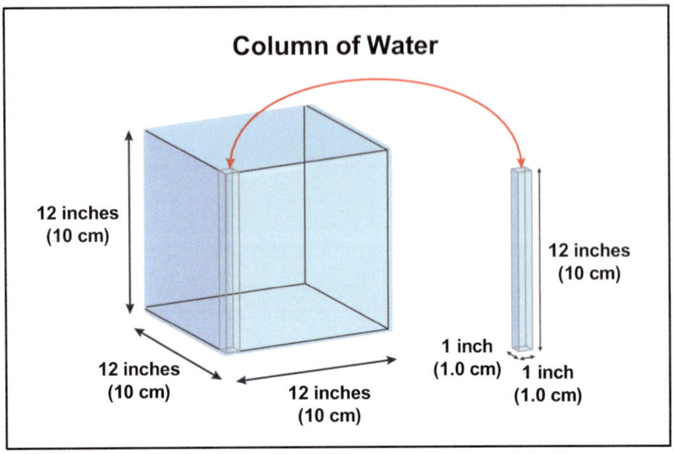

Fig. 2.7 Column of water one foot (10 cm) high with a base of one square inch (one square centimeter)

And now:

$$P = \frac{F}{A} = 0.433 \frac{(\text{lb.})}{\text{in}^2}$$

Or:

$$P = 0.433\,\text{psi}$$

In S.I. Units, the Solution is:

$$W = \frac{1.0\,\text{kg}}{100} = 0.01\,\text{kg}$$

And now:

$$P = \frac{F}{A} = 0.01\,\text{kg}/1.0\,\text{cm}^2$$

Or:

$$P = 0.01\,\text{kg}/\text{cm}^2$$

And in bars:

$$P = \left(0.01\,\mathrm{kg/cm^2}\right)\frac{(1\,\mathrm{bar})}{\left(1.0197\,\mathrm{kg/cm^2}\right)} = 0.0098\,\mathrm{bar}$$

Then, in units of bar/m:

$$P = (0.0098\,\mathrm{bar}/10\,\mathrm{cm})(100\,\mathrm{cm/m}) = 0.098\,\mathrm{bar/m}$$

The conclusion is that the pressure depends only on the depth (and specific weight) and is independent of the shape of the container or the volume of the container.

Head

It is common to hear fire protection specialists refer to water pressure in terms of head. *Head* is pressure expressed in units of feet (meters) of water instead of psi (bar) and can be expressed in the general sense as follows:

$$h = \frac{P}{w} \tag{2.3}$$

where:

h = head in feet (meters),

P = pressure psi (bar),

w = the specific weight in lb./ft^3 (kg/L).

More specifically, for water:

$$h = \frac{P(\mathrm{psi})}{62.4\,\mathrm{lb./ft^3}}$$

$$h = \frac{P(\mathrm{psi})\left(144\,\mathrm{in.^2/ft^2}\right)}{62.4\,\mathrm{lb./ft^3}}$$

$$h = \frac{(144)}{(62.4)}(P)\frac{(\mathrm{lb.})}{\left(\mathrm{in.^2}\right)}\frac{\left(\mathrm{in.^2}\right)}{\left(\mathrm{ft.^2}\right)}\frac{\left(\mathrm{ft.^3}\right)}{(\mathrm{lb.})}$$

$$h = 2.31P(\mathrm{ft/psi}).$$

Note: Where **h** is in feet, the units of **P** are psi.

In S.I. Units, the Solution is:

$$h = \frac{P(\text{ bar })}{1\,\text{kg/L}}$$
$$h = 10.197P$$

Note: Where **P** is in bar, **h** is in meters.

Buoyancy

For any object to remain stationary in a liquid without sinking, the force in the downward direction must be equalized by the force in the upward direction (Fig. 2.8). The force acting down (F_1) will equal the weight density of the material times the volume of the object.

$$F_1 = (w)(V)$$

For the object to remain stationary, (F_2) must also equal $(w)(V)$. In approximately 250 B.C., Archimedes discovered that the upward, or buoyant, force (F_2) is equal to the weight of the liquid displaced [2]. Therefore, whether a material will float or not and how much of the object will protrude above the surface of the liquid depends in

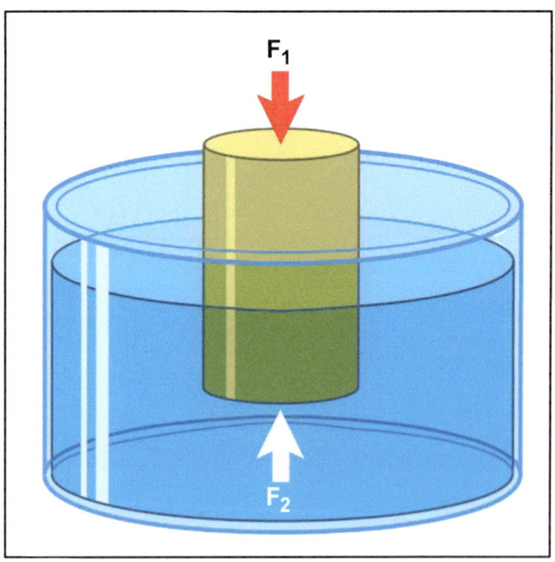

Fig. 2.8 When an object floats in a liquid, F_1 must equal F_2

part upon the relationship between the weight densities of the objects and the liquid. A solid object will float in a liquid only if its weight density is less than or equal to that of the liquid. The proportion of a floating object that sinks below the surface of a liquid depends upon how much of the liquid equals the weight of the object.

Example 2.5

A solid cube of wood, 2 ft (0.6 m) on each side, has a weight density of 48 lb./ft^3 (0.77 kg/L). If it is placed in a container of water, how far above the surface does the cube extend?

Solution:

The buoyant force is equal to the amount of water displaced by the cube, and the weight of the water displaced equals the weight of the cube. Thus,

$$F = (w)(V)$$
$$F = \left(48\,\text{lb./ft}^3\right)(2\,\text{ft})^3$$
$$F = \left(48\,\text{lb./ft}^3\right)\left(8\,\text{ft}^3\right)$$
$$F = 384\,\text{lb.}$$

The volume of water displaced must also weigh 384 pounds. Therefore, the volume of water displaced is

$$V_\text{w} = \frac{384\,\text{lb.}}{62.4\,\text{lb/ft}^3}$$
$$V_\text{w} = 6.154\,\text{ft}^3$$

Then, the volume of the cube protruding above the water surface is the volume of the cube minus the volume of the displaced water.

$$V_\text{c} = 8\,\text{ft}^3 - 6.154\,\text{ft}^3$$
$$V_\text{c} = 1.846\,\text{ft}^3$$

Finally, if it is assumed that the cube floats with its bottom parallel to the water surface, the height above the water that the cube extends is the volume protruding divided by the area of the bottom of the cube, or

$$H = \frac{1.846\,\text{ft}^3}{4\,\text{ft}^2} = 0.46\,\text{ft}$$

In S.I. Units, the Solution is:

$$F = (w)(V)$$
$$F = (0.77\,\text{kg/L})(0.6\,\text{m})^3\left(1000\,\text{L/m}^3\right)$$
$$F = 166.3\,\text{kg}$$

The volume of water displaced must also weigh 166.3 kg. Therefore, the volume of water displaced is

$$V_\text{w} = \left(\frac{166.3\,\text{kg}}{1\,\text{kg/L}}\right)\left(\frac{1\ \text{m}^3}{1000\,\text{L}}\right) = 0.166\,\text{m}^3$$

Then the volume of the cube protruding above the water surface is the volume of the cube minus the volume of the displaced water.

$$V_\text{c} = 0.216\,\text{m}^3 - 0.166\,\text{m}^3 = 0.05\,\text{m}^3$$

Finally, if it is assumed that the cube floats with its bottom parallel to the water surface, the height above the water that the cube extends is the volume protruding divided by the area of the bottom of the cube, or

$$H = \frac{0.05\,\text{m}^3}{(0.6\,\text{m})(0.6\,\text{m})} = 0.139\,\text{m}.$$

Potential Energy

It is important to understand the part that hydrostatic pressure plays in the energy composition of a hydraulic system. Hydrostatic pressure composes the potential energy component of the total energy of a hydraulic system. Potential energy is stored energy. It has the ability to perform work once released. It is easy to visualize the potential of a stretched rubber band or a coiled spring. However, elevated or pressurized water also contains potential energy. Indeed, potential energy has two components in a water system: (1) potential energy due to elevation and (2) potential energy due to external pressure sources, such as pumps.

Because *energy* is the ability to do work, and *work* is the product of force times distance and has units of foot-pounds (meters-kilograms); *potential energy* is the potential or ability to move some unit of weight (force) some distance or:

$$\mathrm{PE} = (W)(h) \tag{2.4}$$

where:

W = some unit of weight,

h = height in feet (meters).

This relationship applies to elevated water as well as elevated solids. Also, we learned in the proceeding section that the pressure in a system can be expressed in head of water by the relationship.

$$h = \frac{P}{w}$$

Therefore the potential energy due to external pressures can be expressed as:

$$\mathrm{PE_p} = (W)\left(\frac{P}{W}\right)$$

And then the total potential energy existing at a point in a system is

$$\mathrm{PE_t} = \mathrm{PE_h} + \mathrm{PE_p}$$

$$\mathrm{PE^t} = (W)(h) + (W)\left(\frac{P}{w}\right).$$

Fig. 2.9 Illustration for Example 2.6

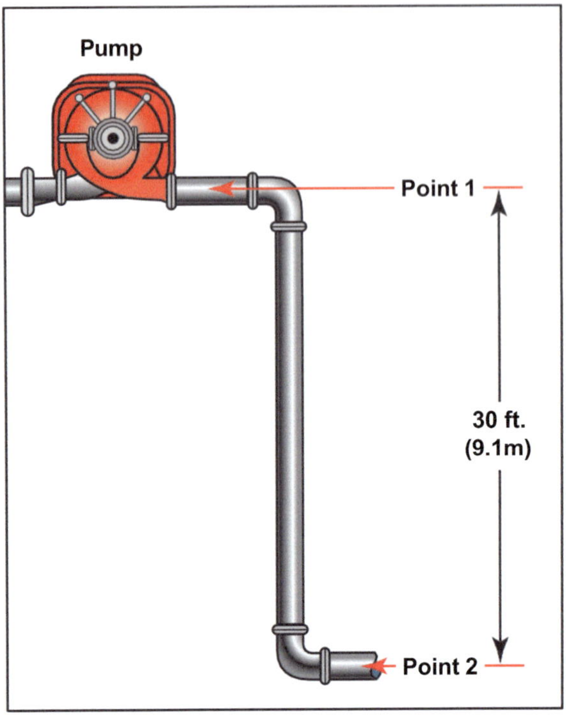

Example 2.6

In the static system shown in Fig. 2.9, determine the total potential energy in units of feet (meters) at Points 1 and 2 if the pump is creating a pressure of 100 psi (6.9 bar) and Point 1 is 30 ft (9.1 m) higher than Point 2.

Solution:

At Point 1:

$$\mathrm{PE}_1 = Wh + W\left(\frac{P}{W}\right)$$
$$\mathrm{PE}_1 = (30\,\mathrm{ft})W + (2.31\,\mathrm{ft/psi})(100\,\mathrm{psi})W$$
$$\mathrm{PE}_1 = (261\,\mathrm{ft})W$$

At Point 2, the 30 ft of elevation has been converted into pressure due to the weight of the water. This pressure is:

$$P = (0.433\,\mathrm{psi/ft})(30\,\mathrm{ft}) = 12.99\,\mathrm{psi}$$

Then, at Point 2:

$$\mathrm{PE}_2 = (0\,\mathrm{ft})W + (2.31\,\mathrm{ft/psi})(100\,\mathrm{psi} + 12.99\,\mathrm{psi})W$$
$$\mathrm{PE}_2 = (261\,\mathrm{ft})W$$

It should be noted that, for any unit of weight W, the potential energy is constant throughout the system.

In S.I. Units, the Solution is:

At Point 1:

$$\mathrm{PE}_1 = Wh + W\left(\frac{P}{W}\right)$$
$$\mathrm{PE}_1 = (9.1\,\mathrm{m})W + (10.197\,\mathrm{m/bar})(6.9\,\mathrm{bar})W$$
$$\mathrm{PE}_1 = (79.46\,\mathrm{m})W$$

At Point 2, the 9.1 m of elevation have been converted into pressure due to the weight of the water. This pressure is:

$$P = (0.098\,\mathrm{bar/m})(9.1\,\mathrm{m}) = 0.892\,\mathrm{bar}$$

Then, at point 2:

$$\mathrm{PE}_2 = (0\,\mathrm{m})W + (10.197\,\mathrm{m/bar})(0.892\,\mathrm{bar} + 6.9\,\mathrm{bar})W$$
$$\mathrm{PE}_2 = (79.46\,\mathrm{m})W.$$

For any unit of weight, W, the potential energy is constant at all points.

Devices for Measuring Hydrostatic Pressures

The fire protection specialist should have an understanding of several devices used to measure hydrostatic pressure (and thus potential energy). These devices are:

- The barometer.
- The piezometer tube.
- The differential manometer.
- The Bourdon gauge.
- Electronic gauges.

Barometers

Barometers are widely used for the measurement of atmospheric pressure. The basic construction of a barometer involves the inversion of a tube of liquid in a container open to the atmosphere (Fig. 2.10). Because the tube is closed on one end, the inversion results in the absence of atmospheric pressure inside the tube on top of the column of liquid. Therefore, the only pressure existing at the base of the liquid column in the tube is a result of the height and density of the liquid. Atmospheric pressure acts on the liquid in the open dish. The pressure created by the column must equal atmospheric pressure for a state of equilibrium to exist. This provides a simple method to measure atmospheric pressure. When using a barometer, measure the height of the column above the surface of the liquid in the open container and use Eq. 2.1 to obtain the pressure in psi or bar. Fluctuations in atmospheric pressure are reflected by changes in the column height. A few calculations will reveal why mercury barometers are more popular than water barometers.

Piezometer Tubes

These devices again use the heights of liquid columns to illustrate the pressures existing in hydraulic systems. These devices typically have application only where small hydrostatic pressures exist, and their range is governed by the density of the liquid within the system. Greater versatility can be obtained by incorporating a U-tube into the piezometer (Fig. 2.11a, b). This allows the use of a high density liquid (e.g. mercury) which permits measurement of greater pressures. The U-tube also allows measurement of vacuum pressures.

Example 2.7

If the piezometer tube in Fig. 2.10a is measuring the pressure in a piping system filled with oil ($S_g = 0.86$), what height in inches (cm) in the tube would indicate a system pressure of 0.3 psi (0.02 bar)?

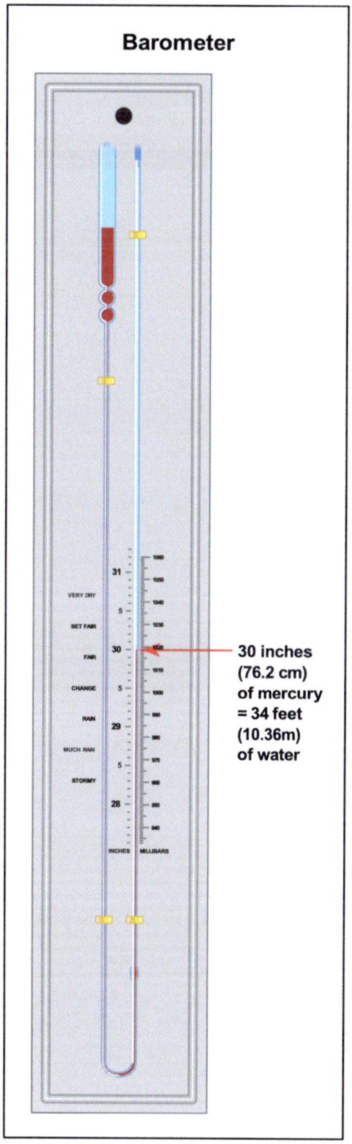

Fig. 2.10 Barometer is used to measure atmospheric pressure. The pressure represented by 30 in. (76.2 cm) of mercury will equate to the pressure created by 34 ft (10.36 m) of water

Solution:

$$P = (W)(h)$$

$$P = \left(S_{g_{\text{oil}}}\right)\left(W_{H_2O}\right)(h)$$

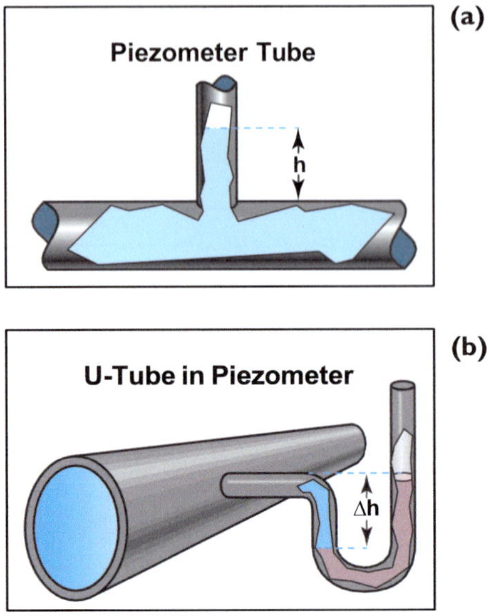

Fig. 2.11 **a** Piezometer tubes illustrate the pressure existing in hydraulic systems **b** Greater versatility can be obtained by incorporating a U-tube into the piezometer

$$P = \frac{P}{\left(S_{g_{oil}}\right)\left(W_{H_2O}\right)}$$

$$h = \frac{0.3\,\text{psi}}{(0.86)\left(62.4\,\text{lb.}/\text{ft}^3\right)\left(1\,\text{ft}^3/1728\,\text{in}^3\right)}$$

$$h = 9.66\,\text{in.}$$

In S.I. Units, the Solution is:

$$P = (w)(h)$$

$$P = \left(S_{g_{oil}}\right)\left(w_{H_2O}\right)(h)$$

$$P = \frac{P}{\left(S_{g_{oil}}\right)\left(w_{H_2O}\right)}$$

$$h = \frac{0.02\,\text{bar}}{(0.86)\left(1\,\text{g}/\text{cm}^3\right)\left(\frac{1\,\text{bar}}{1.0197\,\text{kg}/\text{cm}^2}\right)\left(\frac{1\,\text{kg}}{1000\,\text{g}}\right)}$$

$$h = 23.7\,\text{cm.}$$

Example 2.8

If a U-tube piezometer, illustrated in Fig. 2.10b, is used with mercury in the U-tube, what system pressure will be indicated by a mercury height difference of 3.5 in. (8.9 cm)?

Solution:

$$P = (w)(h)$$
$$P = (13.6)\left(62.4\,\text{lb./ft}^3\right)(3.5\,\text{in.})\left(1\,\text{ft}^3/1728\,\text{in}^3\right)$$
$$P = 1.72\,\text{psi}$$

In S.I. Units, the Solution is:

$$P = (w)(h)$$
$$P = \left(S_{\text{aHg}}\right)(w_{\text{H2O}})(h)$$
$$P = (13.6)\left(1\,\text{g/cm}^3\right)\left(\frac{1\,\text{bar}}{1.0197\,\text{kg/cm}^2}\right)\left(\frac{1\,\text{kg}}{1000\,\text{g}}\right)(8.9\,\text{cm})$$
$$P = 0.119\,\text{bar}.$$

Manometers

Differential manometers are similar to piezometers; however, their primary application is in reflecting the differences in pressures between two points in a system (Fig. 2.12a, b). Although the measured pressure differences are in potential energy between two points, the manometer is used frequently in applications involving liquids in motion. Coupled with devices, such as venturi meters or orifice plate meters, manometers can be an accurate method of establishing liquid velocities.

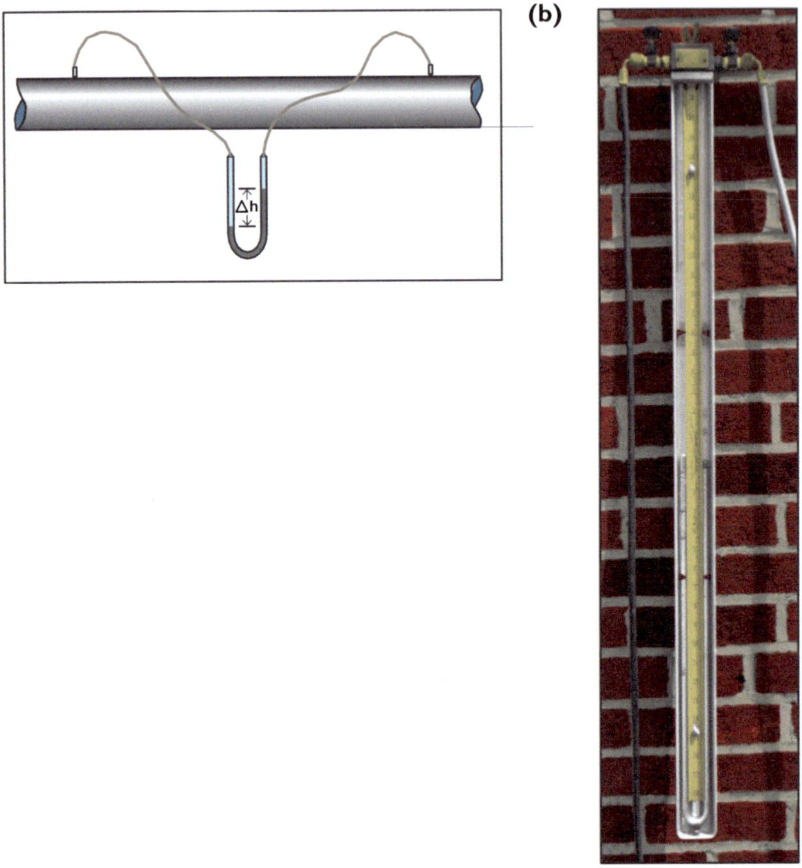

Fig. 2.12 **a** Differential manometer is used frequently in applications involving liquids in motion. **b** Differential manometer measuring pressure differences between two points on a piping system

Example 2.9

If using a differential manometer to measure the pressure difference between two points in a piping system (Fig. 2.12b), what pressure difference in psi (bar) is indicated if h is 8 in. (20.32 cm), and mercury is used in the manometer?

Solution:

$$P = wh$$
$$P = (S_{g\mathrm{Hg}})(W_{\mathrm{H_2O}})h$$
$$P = (13.6)(62.4)(\mathrm{lb./ft^3})(8\,\mathrm{in})(1\,\mathrm{ft^3}/1728\,\mathrm{in^3})$$
$$P = 3.93\,\mathrm{psi}$$

In S.I. Units, the Solution is:

$$P = (W)(h)$$
$$P = (S_{gHg})(W_{H_2O})(h)$$
$$P = (13.6)(1\ \text{g/cm}^3)\left(\frac{1\ \text{bar}}{1.0197\ \text{kg/cm}^2}\right)$$
$$\left(\frac{1\ \text{kg}}{1000\ \text{g}}\right)(20.32\ \text{cm}).$$

Bourdon Gauges

Bourdon gauges are the most common devices used for measuring pressures (Fig. 2.13). It can be seen that the principle employed is that of a coiled spring. Increasing pressure within the coiled tube will have the tendency to straighten the coil. This results, through the tiny gearing, in a deflection of the indicating needle. With proper and regular calibration, the Bourdon gauges give very satisfactory representation of pressures existing in static or non-static systems.

If the pressures represented by Bourdon gauges are hydrostatic pressures, they are a reflection of the potential energy within a system. However, when used with a pitot tube (as discussed in Chap. 3, Hydrokinetics) they are employed to measure total pressures or velocity pressures.

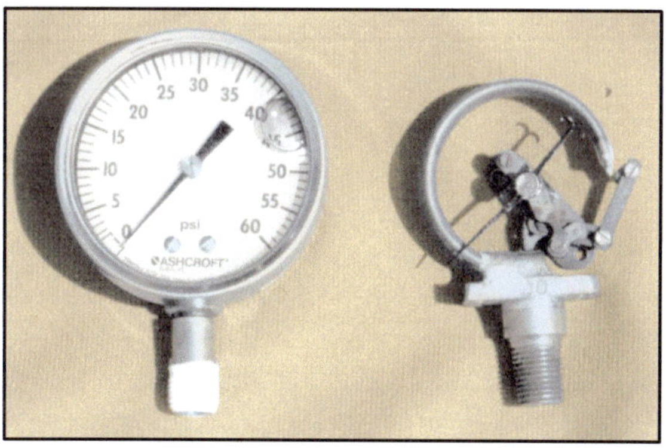

Fig. 2.13 Bourdon gauges are the most common devices used for measuring pressures. This illustrates the inside and outside of a Bourdon gauge

Electronic Pressure Gauges

More recently, electronic pressure gauges have become available (Fig. 2.14). With these pressure gauges, an electronic transponder senses the pressure and transforms it into a digital readout. These gauges also have an application on both static and non-static systems. However, they are probably more susceptible to mechanical damage and damage from getting the exterior wet if not properly cared for.

Summary of Important Concepts

1. Hydrostatics is the study of liquids at rest. There are five basic principles of hydrostatic pressure.
2. The pressure created by a column of water (in psi) is equal to $P = 0.433$ h (h in feet). And, in S.I. units, the pressure (in bar) is calculated from $P = 0.098$ h (h in meters).
3. The head of water in feet (meters) equivalent to a given pressure is $H = 2.31$ P (P in psi), or $H = 10.2$ P (P in bar).
4. The general relationship equating pressure, height, and density is $P = wh$, with proper unit conversion required.

Fig. 2.14 Electronic pressure measuring devices are increasingly popular in modern systems

5. Hydrostatic pressures caused by elevation and external pressure sources (such as pumps) create the potential energy component of the total energy of a hydraulic system.
6. Important instruments for measuring pressures include the differential manometer and the Bourdon gauge.

Exercises

2.1 If an elevated tank containing water is 85 ft (26 m) high, what pressure will exist at the base of the tank if the tank is full?

2.2 If a spherical water storage tank has a diameter of 50 ft (15 m), and the top of the tank is 106 ft (32 m) above the ground, what pressure exists at ground level when the tank is full?

2.3 If a tank of identical height as the tank in Exercise 2.1 is constructed with twice the volume, how does this affect the pressure available at the base of the tank?

2.4 If a tank of identical height as the tank in Exercise 2.2 is constructed with a diameter of 75 ft (23 m), how does this affect the pressure at ground level?

2.5 If a high-rise building is 108 ft (33 m) tall, how much pressure must be available at ground level to overcome elevation and deliver the 100 psi (6.9 bar) standpipe water pressure requirement at the roof?

2.6 If the standpipe in Exercise 2.5 is attached to a city water main and filled with water, what pressure will exist at the roof level if the pressure at ground level is 127 psi (9 bar)?

2.7 A 500,000 barrel crude oil storage tank is 46 ft (14 m) high. If the tank is full and the oil has a specific gravity of 0.89, what pressure will exist at the base of the tank?

2.8 A tank containing 200,000 gallons (750 m^3) of regular gasoline is 32 ft (10 m) high. If the tank is full, what pressure in psi (bar) will exist at the base of the tank?

2.9 In the hydraulic system shown in Fig. 2.15, what force (F_2) must be applied

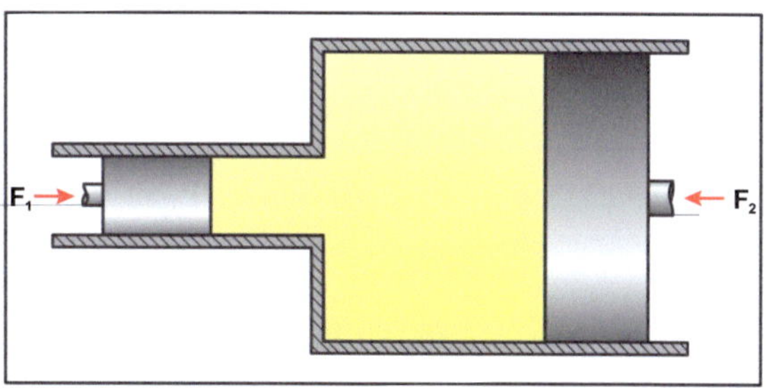

Fig. 2.15 Illustration for Exercise 2.9

Fig. **2.16** Illustration for Exercise 2.12

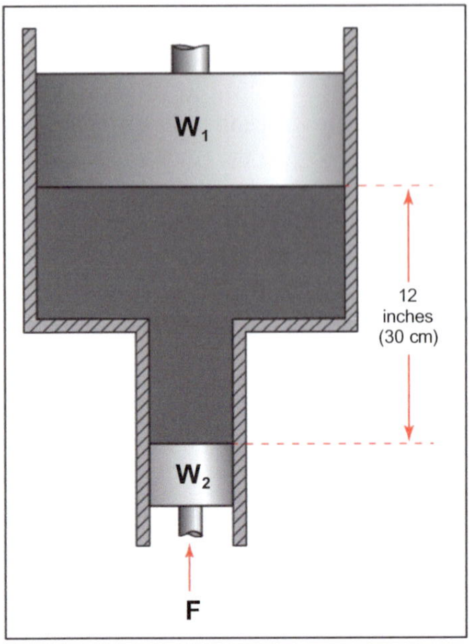

to attain a state of equilibrium if the circular surface upon which F_1 acts has a diameter of 6 in. (15 cm), the circular surface acted upon by F_2 has a diameter of 18 in. (46 cm), and F_1 is 85 pounds (39 kg)?

2.10 In a dry-pipe sprinkler valve, the air seat typically has a surface area on which the air pressure acts six times as large as the water seat upon which the water pressure acts. If the water pressure is 105 psi (7.2 bar), what air pressure is required to attain a state of equilibrium?

2.11 In a multiple-story building, a water column has formed on the top side of a dry-pipe sprinkler valve having a 6:1 air seat to water seat ratio. If the column of water is 42 ft (12.8 m) high, and the city water pressure is 60 psi (4.1 bar), will the dry valve be able to open when the air pressure is evacuated from the system?

2.12 In Fig. 2.16, the weight of the upper piston (W_1) is 85 pounds (39 kg) and has a surface area of 113 in^2 (729 cm^2). The weight (W_2) of the lower piston is 14 pounds (6 kg) and has a surface area of 12.6 in^2 (81 cm^2). If the liquid in the hydraulic system is an oil with a specific gravity of 0.90, what force (F) will be required to attain a state of equilibrium?

2.13 If the industrial fire pump (Fig. 2.17) is developing 270 ft (82 m) of head at the discharge port of the pump (Point A), what pressures in psi (bar) will be indicated by the gauges at Points B, C, and D?

2.14 If a system contains an oil with a specific gravity of 0.87, what height in inches (cm) on a piezometer tube would indicate a pressure of 1.0 psi (0.07 bar)?

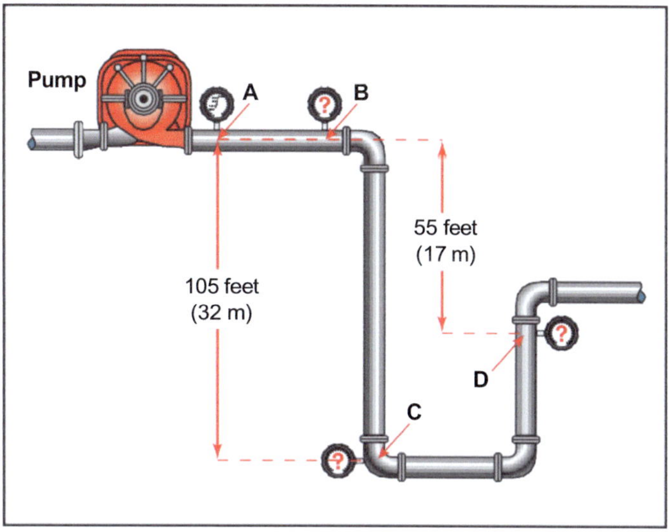

Fig. 2.17 Illustration for Exercise 2.13 and 2.20

2.15 In Fig. 2.18, the liquid contained in the system is water. What is the pressure reading on Gauge P in psi (bar) if W_1 is 1500 pounds (680 kg) and F_1 is 1000 pounds (450 kg)? What force (F_2) would be required to achieve equilibrium? Assume both cylinders have circular cross sections.

2.16 A cargo ship has a volume of 480,000 ft^3 (13,590 m^3). The ship alone weighs 800,000 pounds (360,000 kg). What weight of cargo can the ship carry and still remain afloat? Assume the ship is to be used in ocean water ($w = 64$ lb./ft^3 and 1.026 g/cm^3).

2.17 A barge with a rectangular cross section of 40 ft (12 m) wide by 10 ft (3 m) high is used on the Mississippi River for transporting oil. If the barge is 60 ft (18 m) long, how much does it weigh when empty if 7 ft (2 m) of the barge extends above the water surface? How much of the barge would extend above the water surface if half full of oil (Sg = 0.9)?

2.18 If a man is placed into a completely full container of water and his body causes 24.9 gallons (94 L) of water to run over the top, how much does the man weigh?

2.19 If a solid sphere of some material has a diameter of 15 in. (38 cm) and weighs 2 lb. (0.9 kg), what is the specific weight of the material and what percent of the sphere's volume extends above the surface if placed in a vat of water?

2.20 In Fig. 2.17, if Point C is at ground level, calculate the total potential energy in feet (meters) of water existing at all four gauge locations.

Selected Answers Found in Appendix K.

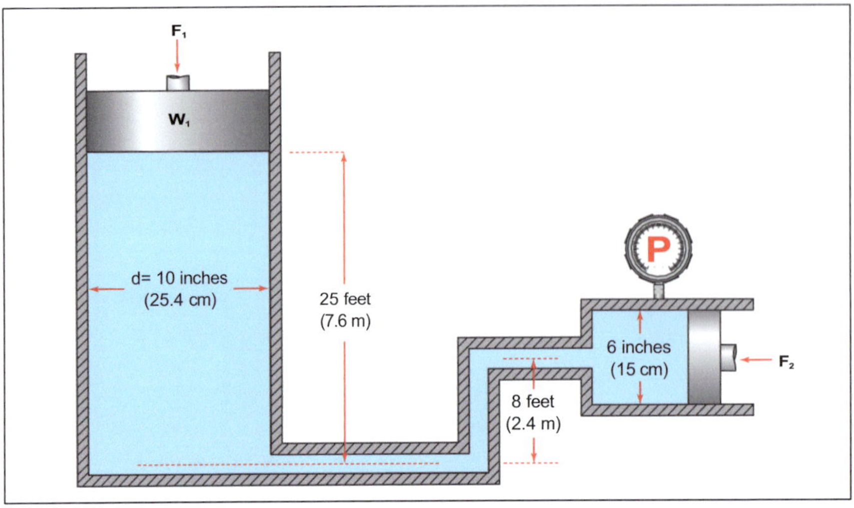

Fig. 2.18 Illustration for Exercise 2.15

References

1. Russell W. Henke, *Introduction to Fluid Mechanics* (Reading: Addison-Wesley Publishing Co., 1972), p. 8.
2. Henke 32.

Supplemental Readings

Hickey, Harry E. *Hydraulics for Fire Protection*. National Fire Protection Association: Quincy, MA, 1980, pp. 14–22.

Chapter 3
Hydrokinetics

Abstract This chapter introduces the concept of the study of water in motion, or hydrokinetics. Information on kinetic energy along with Bernoulli's equation is provided, including the application of Bernoulli's equation. The concept of Conservation of Matter is introduced as well as how flow, velocity, and area interface in this concept mathematically.

Hydrokinetics involves the study of the characteristics and physical properties of water in motion. Water in motion must be understood and dealt with when working with automatic sprinkler systems, standpipe systems, fire department pumpers, or industrial fire pump installations.

The importance of the principles of static water pressure cannot be denied and should not be minimized. The fire protection specialist continually works with water supply situations that mainly involve application of hydrostatic principles, particularly involving the concepts of elevation. However, seldom is fire protection hydraulics as simple as just applying the hydrostatic principles.

Kinetic Energy

In a situation where water is motionless, any energy existing was found to be potential energy. In a system where water is moving, some potential energy is still likely to exist. However, another component exists that is related to the velocity of the water—kinetic energy.

Work done is force applied over some distance [1]:

$$G = (F)(d)$$

And from Newton's Second Law [2]:

$$F = (m)(a)$$

P. Brock and V. Charter, *Fire Protection Hydraulics and Water Supply Analysis*, The Society of Fire Protection Engineers Series,
https://doi.org/10.1007/978-3-032-09877-1_3

Thus:

$$G = (m)(a)(d)$$

where:

G = work,

m = mass,

a = acceleration,

d = some distance.

Also, from physics [2], we know:

$$a = \frac{v - v_0}{t}$$

And:

$$d = \frac{(v + v_0)}{t}$$

where:

v = velocity at some time (t),

$\mathbf{v_o}$ = initial velocity at t = 0,

t = time,

d = distance traveled at time (t).

Substituting now gives us:

$$G = (m)\left(\frac{v - v_0}{t}\right)\left(\frac{v + v_0}{2}\right)(t)$$

$$G = \frac{m\left(v^2 - v_0^2\right)}{2} \tag{3.1}$$

And finally, where $v_o = 0$, (that is, the mass is initially at rest), the work may be expressed as:

$$G = \frac{(m)(v)^2}{2}$$

One half the product of the mass of a body and the square of its velocity is called the "*kinetic energy*" of the body [3]. These principles of physics apply to any body in motion and, of course, must also apply to water in motion. Therefore, an expression for the kinetic energy (**KE**) of some mass of water in motion is:

$$\mathrm{KE} = G = \frac{(m)(v)^2}{2}$$

where:

KE = kinetic energy,

m = some mass of water,

v = the velocity of the water.

As previously noted, where acceleration of a particle is due to gravity, mass can be written as:

$$m = \frac{W}{g}$$

where:

W = the weight of the particle,

g = is the acceleration due to gravity.

Finally, the kinetic energy of a particle may be expressed as:

$$\mathrm{KE} = \frac{(W)(v)^2}{(2)(g)} \tag{3.2}$$

Equation 3.2 is the expression used in the continuing discussion to represent the kinetic energy of water; that is, water particles.

The brief presentation of particle dynamics in this section was not intended to be exhaustive or comprehensive. Excellent presentations of these basic principles of physics are available in numerous basic physics texts, such as the one referenced in this section.

Conservation of Energy and Bernoulli's Equation

It has been established that there exists the possibility of two types of energy within any hydraulic system—potential energy and, if the water is moving, kinetic energy. Then, the total energy (**TE**) at any point in a system might be defined as the sum of the potential energy and kinetic energy at that point. Or:

$$TE = PE_t + KE$$

The Principle of Conservation of Energy states that the total energy within a system will remain constant. This necessarily implies that any change in the potential energy of a system must be matched by a corresponding change in kinetic energy, or

$$\Delta PE_t = \Delta KE_t$$

For example, if all of the water in the elevated tank in Fig. 3.1 were to fall from its elevation (**h**), all of its potential energy (**Wh**) would be converted to kinetic energy by the time all the water reached ground level.

Another important interpretation of the Principle of Conservation of Energy is the following: Because the total energy of a hydraulic system is constant, the total energy at all points within the system is also constant and equal. That is, for any two points within a system:

$$TE_1 = TE_2$$

And where total energy is represented by its various components:

$$PE_{h1} + PE_{p1} + KE_1 = PE_{h2} + PE_{p2} + KE_2$$

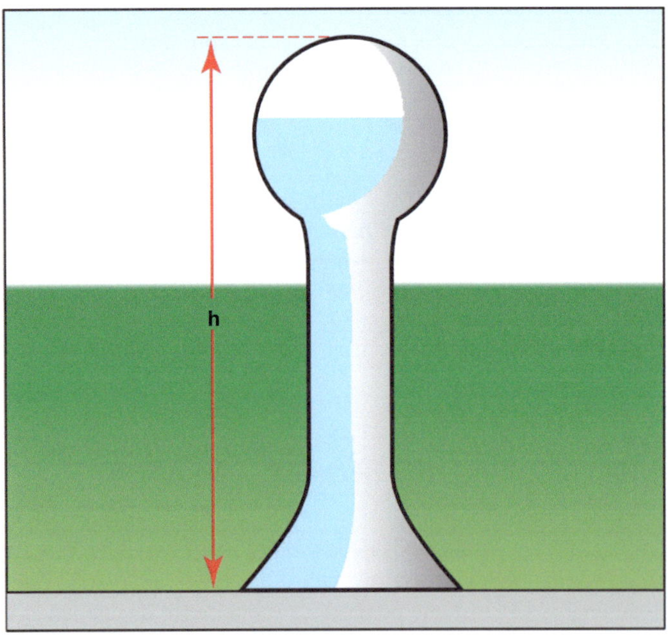

Fig. 3.1 Potential energy is converted to kinetic energy as water drains from the tank

Substituting for the general symbols gives:

$$(W)(h_1) + \frac{(W)(P_1)}{W} + \frac{(W)(v_1)^2}{(2)(g)} = (W)(h_2) + \frac{(W)(P_2)}{W} + \frac{(W)(v_2)^2}{(2)(g)}$$

This may be simplified further by dividing through by the weight of water (**W**) to give what is known as Bernoulli's equation, which is a fundamental concept to the fluid mechanics of incompressible fluids:

$$h_1 + \frac{P_1}{w} + \frac{v_1^2}{2g} = h_2 + \frac{P_2}{w} + \frac{v_2^2}{2g} \tag{3.3}$$

Note that in many textbook presentations, the components of potential energy attributed to elevation are represented by **Z** rather than **h** [4]. The interpretation is identical. Also, when dealing with water systems, the term **P/w** may be replaced by (2.31)(P) or (1019.7)(P) in S.I. units, because for water:

$$\frac{P}{w} = \frac{P(\text{lb./in}^2)}{62.4\,\text{lb./ft}^3}$$

$$\frac{P}{w} = \left(\frac{P(\text{ft}^3)}{62.4(\text{in}^2)}\right)\left(\frac{144(\text{in}^2)}{(\text{ft}^2)}\right)$$

$$\frac{P}{w} = \frac{144}{62.4}(P)(\text{ft})$$

$$\frac{P}{w} = (2.31)(P)(\text{ft})$$

where **P** is given in psi with the final units of (2.31)(P) being simply feet.

And in S.I. Units, the Solution is:

$$\frac{P}{w} = \frac{P(\text{bar})}{1.0\,\text{gm/cm}^3}$$

$$\frac{P}{W} = \left(\frac{P(\text{bar})}{1.0\,\text{gm/cm}^3)}\right)\left(\frac{1.0197\,\text{kg/cm}^2}{\text{bar}}\right)\left(\frac{1000\,\text{gm}}{\text{kg}}\right)$$

$$\frac{P}{W} = (1019.7)(P)(\text{cm})$$

where **P** is given in bar, the units of 1019.7 P will be centimeters.

It is important to understand that every term in Bernoulli's equation should have the same units. That is, all terms—h, (P/W, and $v^2/(2)(g)$—are in *units of feet of head (or centimeters).*

Figure 3.2 illustrates the basic principles defined by Bernoulli's equation—the total energy at Point 1 and Point 2 are the same. This is not to say that the potential energy or the kinetic energy at each point is the same, just the total energy.

Before looking at some examples of applications of Bernoulli's equation, note that Bourdon Gauge 1 reading pressure from a tap on a pipe is reading only potential energy components of the total energy existing in the system (Fig. 3.3). However, Gauge 2, attached to an internal pitot tube, is reflecting the total energy because it is also measuring the kinetic energy due to the water velocity. Gauge 3 shows only potential energy because the pitot tube opening is not oriented properly to measure the impact of velocity. Finally, Gauge 4 is reflecting total energy. At the point of measurement in an open stream, all of the energy is due to water velocity, and this is all kinetic.

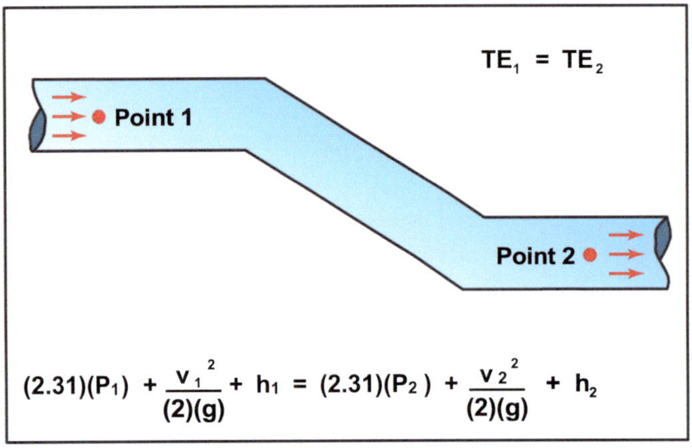

Fig. 3.2 Principle of conservation of energy as defined by Bernoulli's equation

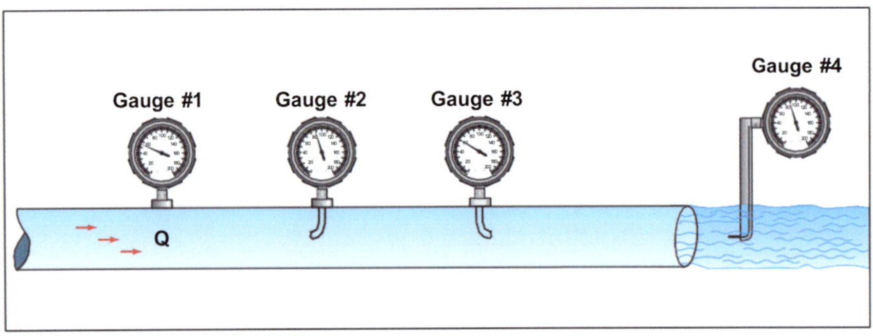

Fig. 3.3 Each gauge represents different energy components

Example 3.1

If the pressure gauge at P1 in Fig. 3.4 reads 80 psi (5.5 bar) and the gauge at P_2 reads 60 psi (4.1 bar), what is the water velocity at P_2 if v_1 is 10 ft per second (305 cm per second)? From Bernoulli's equation we know that:

$$(2.31)(P_1) + \frac{(v_1)^2}{(2)(g)} + h_1 + = (2.31)(P_2) + \frac{(v_2)^2}{(2)(g)} + h_2$$

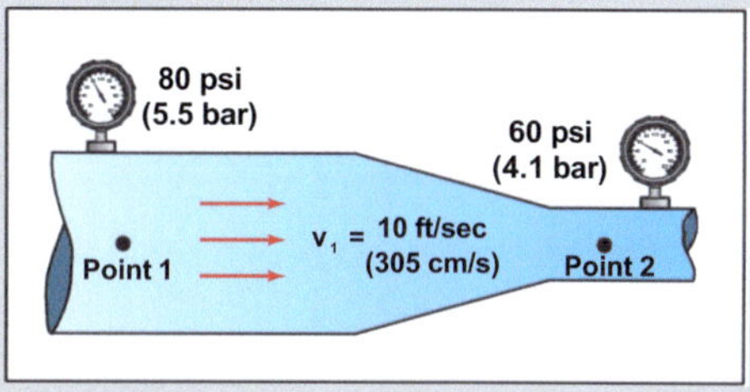

Fig. 3.4 Illustration for Example 3.1

And since:

$$h_1 = h_2$$

Rearranging gives:

$$(2.31)(P_1 - P_2) + \frac{(V_1)^2}{(2)(g)} = \frac{(V_2)^2}{(2)(g)}$$

Then:

$$(2.31)(2)(g)(P_1 - P_2) + (v_1)^2 = (v_2)^2$$

And finally:

$$v_2 = \sqrt{(2.31)(2)(g)(P_1 - P_2)(v_1)^2}$$
$$v_2 = \sqrt{(2.31\,\text{ft/psi})\left(64.4\,\text{ft/s}^2\right)(20\,\text{psi}) + 10^2\,\text{ft}^2/\text{s}^2}$$
$$v_2 = 55.46\,\text{ft/s}$$

In S.I. Units, the Solution is:

$$(1019.7)(P_1) + \frac{(v_1)^2}{(2)(g)} + h_1 = (1019.7)(P_2) + \frac{(v_2)^2}{(2)(g)} + h_2$$

And since:

$$h_1 = h_2$$

Rearranging gives:

$$(1019.7)(P_1 - P_2) + \frac{(V_1)^2}{(2)(g)} = \frac{(V_2)^2}{(2)(g)}$$

Then:

$$(1019.7)(2)(g)(P_1 - P_2) + (v_1)^2 = (v_2)^2$$

And finally:

$$v_2 = \sqrt{(1019.7)(2)(g)(P_1 - P_2) + (v_1)^2}$$
$$v_2 = \sqrt{(1019.7)(\text{cm/bar})(2)(9.8)\left(\text{m/s}^2\right)(100\ \text{cm/m})(5.5 - 4.1)(\text{bar}) + (305\ \text{cm/s})^2}$$
$$v_2 = 1700.3\,\text{cm/s}$$

Example 3.2

In the situation shown in Fig. 3.5, use Bernoulli's equation to solve for the pressure gauge reading at Point 1.

$$(2.31)(P_1) + \frac{(v_1)^2}{(2)(g)} + h_1 = (2.31)(P_2) + \frac{(v_2)^2}{(2)(g)} + h_2$$

$$(2.31)(P_1) = (2.31)(P_2) + \frac{(v_2)^2 - (v_1)^2}{(2)(g)} + h_2 - h_1$$

$$P_1 = P_2 + \frac{(v_2)^2 - (v_1)^2}{(2.31)(2)(g)} + \frac{h_2 - h_1}{2.31}$$

$$P_1 = 40\text{psi} + \frac{(7\,\text{ft/s})^2 - (6\,\text{ft/s})^2}{2.31\,\text{ft/psi}\left(64.4\,\text{ft/s}^2\right)} + \frac{0 - 25\,\text{ft}}{2.31\,\text{ft/psi}}$$

$$P = 29.3\,\text{psi}.$$

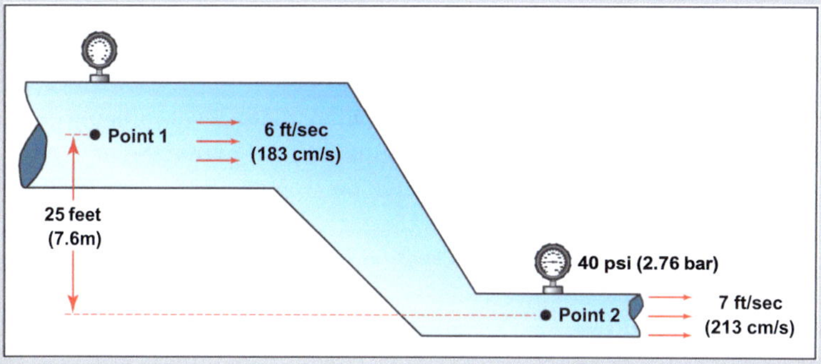

Fig. 3.5 Illustration for Example 3.2

In S.I. Units, the Solution is:

$$(1019.7)(P_1) + \frac{(v_1)^2}{(2)(g)} + h_1 = (1019.7)(P_2) + \frac{(v_2)^2}{(2)(g)} + h_2$$

$$(1019.7)(P_1) = (1019.7)(P_2) + \frac{(v_2)^2 - (v_1)^2}{(2)(g)} + h_2 - h_1$$

$$P_1 = P_2 + \frac{(v_2)^2 - (v_1)^2}{(1019.7\ \text{cm/bar})(2)(g)} + \frac{h_2 - h_1}{(1019.7\,\text{cm/bar})}$$

$$P_1 = (2.76\,\text{bar}) + \frac{(213\,\text{cm/s})^2 - (183\ \text{cm/s})^2}{(1019.7\,\text{cm/bar})(2)\left(9.8\ \text{m/s}^2\right)(100\,\text{cm/m})} + \frac{(0 - 7.6\,\text{m})(100\,\text{cm/m})}{1019.7\,\text{cm/bar}}$$

$$P = 2.02\,\text{bar}.$$

Conservation of Matter

The physical law of conservation of matter states that matter can neither be created nor destroyed. In practical application in hydraulics, this means that what goes in must come out. That is, the amount of water entering a system must leave a system. Carrying this concept further, if a fluid stream has a cross-sectional area (**A**) (Fig. 3.6), and water is passing through **A** at some velocity (**v**), then the flow rate (**Q**) through **A** may be represented by:

$$Q = (A)(v) \tag{3.4}$$

where **Q** is in some units of volume per unit of time, the exact units being determined by the units of **A** and **v**. The most common units for **Q** are *cubic feet per second (cfs)* and *gallons per minute (gpm), or in S.I. units, cubic meters per second (m^3/s) or liters per minute (L/min).*

Essentially, the law of matter conservation indicates that within a single system, **Q** is constant. Water will not be created nor will it disappear. If 1000 gpm (3785 L/min) enter a system, 1000 gpm (3785 L/min) must also leave a system and for any two points within a system:

$$Q_1 = Q_2$$

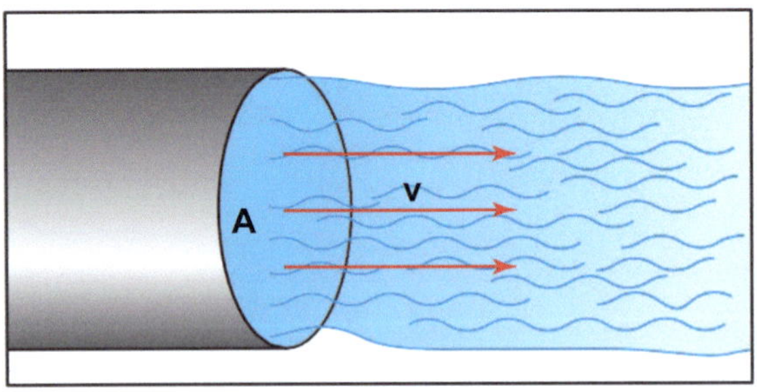

Fig. 3.6 The flow rate is the water velocity times the cross-sectional area of the stream

And so:

$$(A_1)(v_1) = (A_2)(v_2) \tag{3.5}$$

Equation 3.5 allows the development of several basic principles of water flow in piping systems. These principles are as follows:

Principle 1: If pipe size remains constant, water velocity within a system remains constant.

Principle 2: Within the same system, an increase in pipe diameter results in a reduction in water velocity.

Principle 3: Within the same system, a reduction in pipe size results in an increase in water velocity.

Principle 4: If pipe size within a system remains constant, water flowing uphill travels with the same velocity as water flowing downhill.

Understanding matter conservation and these four principles allows several useful calculations. The following examples demonstrate these mathematical applications.

Example 3.3

If the water velocity in the 6-in. (15 cm) pipe is 2 ft/s (0.61 m/s), what is the water velocity in the 2-in. (5 cm) pipe, and how many gpm (L/min) are flowing (Fig. 3.7)? We know that:

$$Q_1 = Q_2$$

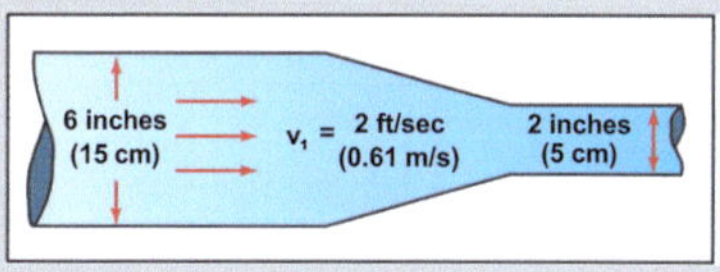

Fig. 3.7 Illustration for Example 3.3

And from Eq. 3.4 that:

$$(A_1)(v_1) = (A_2)(v_2)$$

Therefore:

$$v_2 = \frac{(A_1)(v_1)}{(A_2)}$$

where:

$$A_1 = (\pi)(r_1)^2$$

And:

$$A_2 = (\pi)(r_2)^2$$

Substituting gives the following:

$$v_2 = \frac{(\pi)(r_1)^2(v_1)}{(\pi)(r_2)^2}$$
$$v_2 = \left(\frac{(r_1)}{(r_2)}\right)^2 (v_1)$$
$$v_2 = \left(\frac{(3\text{ in.})}{(1\text{ in.})}\right)^2 (2\text{ ft/s}) = 18\text{ ft/s}$$

To solve for the flow rate, simply find and convert the units to gpm as follows:

$$Q = (A_1)(v_1) = (A_2)(v_2)$$
$$Q = (\pi)(r_1)^2(v_1)$$
$$Q = (\pi)(3\text{ in.})^2(2\text{ ft/s})$$
$$Q = (\pi)(9)\text{in}^2(2)\text{ft/s}$$
$$(9)(\pi)(\text{in}^2)\left(\frac{1\text{ ft}^2}{144\text{ in}^2}\right)(2)\left(\frac{\text{ft}}{\text{s}}\right)\left(\frac{60\text{ s}}{\text{min}}\right)\left(\frac{7.48\text{ gal}}{\text{ft}^3}\right)$$
$$Q = \left(\frac{(2)(9)(\pi)(60)(7.48)}{144}\right)\left(\frac{\text{gal}}{\text{min}}\right)$$
$$Q = 176.2\text{ gpm}$$

In S.I. Units, the Solution is:

$$v_2 = \left(\frac{r_1}{r_2}\right)^2 (v_1)$$

$$v_2 = \left(\frac{7.5\ \text{cm}}{2.5\ \text{cm}}\right)^2 (0.61\ \text{m/s}) = 5.49\,\text{m/s}$$

To solve for the flow rate, find and convert the units to liters/min as follows:

$$Q = A_1 v_1 = A_2 v_2$$
$$Q = \pi r_1^2 v_1$$
$$Q = (\pi)(7.5\ \text{cm})^2 (0.61\ \text{m/s})$$
$$Q = (176.7)(\text{cm})^2 (0.61\,\text{m/s})$$
$$Q = (176.7)(\text{cm})^2 (0.61\ \text{m/s})\left(\frac{100\ \text{cm}}{\text{m}}\right)\left(\frac{1.0\,\text{L}}{1000\,\text{cm}^3}\right)\left(\frac{60\,\text{s}}{\text{min}}\right)$$
$$Q = 646.7\,\text{L/min}$$

Example 3.4

In Fig. 3.8, if Q at Point 1 is 250 gpm (946 L/min), find v_1, v_2, and Q_2. From the conservation of matter:

$$Q_1 = Q_2 = 250\,\text{gpm}$$

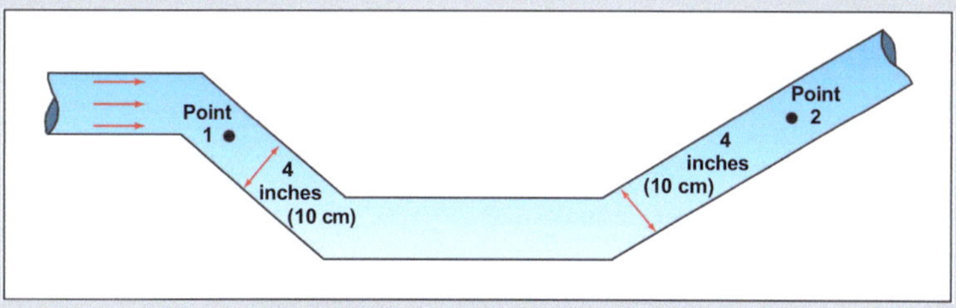

Fig. 3.8 Illustration for Example 3.4

To solve for the velocity at Point 1:

$$Q_1 = (A_1)(v_1)$$
$$v_1 = \frac{Q_1}{A_1}$$
$$v_1 = \frac{Q_1}{(\pi)(r_1)^2}$$
$$v_1 = \frac{250\,\text{gpm}}{(\pi)(2)^2\text{in.}^2}$$
$$v_1 = \frac{250}{(\pi)(4)}\left(\frac{\text{gal}}{(\text{min})(\text{in.})^2}\right)\left(\frac{1\,\text{ft}^3}{7.48\,\text{gal}}\right)\left(\frac{1\ \text{min}}{60\,\text{s}}\right)\left(\frac{144\,\text{in.}^2}{\text{ft}^2}\right)$$
$$v_1 = \frac{(250)(144)}{(\pi)(4)(7.48)(60)}\left(\frac{\text{ft}}{\text{s}}\right)$$
$$v_1 = 6.4\,\text{ft/s}$$

Also, since $A_1 = A_2$, v_1 must equal v_2.

In S.I. Units, the Solution is:

$$Q_1 = Q_2 = 66\ \text{L/min}$$

To solve for the velocity at Point 1:

$$Q_1 = (A_1)(v_1)$$
$$v_1 = \frac{Q_1}{A_1}$$
$$v_1 = \frac{Q_1}{(\pi)(r_1)^2}$$
$$v_1 = \frac{(66\ \text{L/min})}{(\pi)(5\ \text{cm})^2}$$
$$v_1 = \frac{946}{(\pi)(25)}\left(\frac{\text{L}}{(\text{min})(\text{cm})^2}\right)\left(\frac{1000\ \text{cm}^3}{\text{L}}\right)\left(\frac{1.0\,\text{m}}{100\,\text{cm}}\right)\left(\frac{1.0\ \text{min}}{60\,\text{s}}\right)$$
$$v_1 = 2.0\,\text{m/s}$$

Also, since $A_1 = A_2$, v_1 must equal v_2.

The importance of proper unit conversions has been illustrated in the previous examples. If the area is simply multiplied by the velocity, units of square inches feet per second (square centimeters meters per second) would result. Proper unit conversion is always necessary to get values that make sense.

Energy Loss Due to Friction

Bernoulli's equation is for an idealistic situation where friction does not exist. However, in the real world energy is lost to friction as water travels through a piping system. As shown in Chap. 5, Friction Loss in Piping Systems, the magnitude of the friction loss depends upon a number of factors such as pipe length, diameter, and roughness. Bernoulli's equation may be rewritten to reflect the existence of friction loss as follows:

$$\frac{P_1}{w} + \frac{v_1^2}{2g} + h_1 = \frac{P_2}{w} + \frac{v_2^2}{2g} + h_2 + h_f \tag{3.6}$$

where $\mathbf{h_f}$ is the energy lost to friction expressed in units of feet (meters) of water. Note the following examples.

Example 3.5

Suppose water is flowing through the piping system shown in Fig. 3.9. How much pressure is lost to friction if Gauge 1 reads 62 psi (4.3 bar) and Gauge 2 reads 53 psi (3.65 bar)?

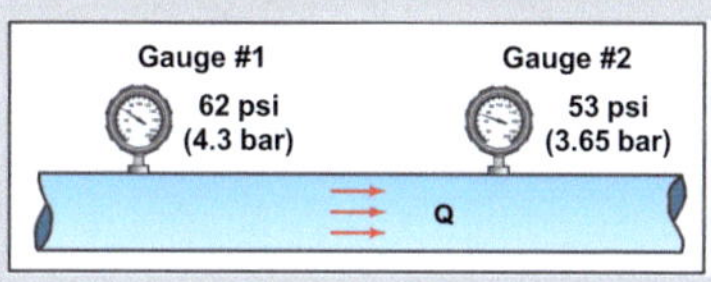

Fig. 3.9 Illustration for Example 3.5

Using Bernoulli's equation gives the following solution for the friction loss:

$$h_f = \frac{P_1 - P_2}{W} + \frac{(v_1)^2 - (v_2)^2}{(2)(g)} + h_1 - h_2$$

Since $Q_1 = Q_2$, and the pipe size does not change, v_1 must equal v_2. Also, since there is no change in elevation, $h_1 = h_2$. The equation then becomes:

$$h_f = \frac{P_1 - P_2}{W}$$

Or if the pressure lost is desired in units of psi,

$$h_f = P_1 - P_2$$
$$h_f = 62\,\text{psi} - 53\,\text{psi}$$

$$h_{\mathrm{f}} = 9\,\text{psi}$$

In S.I. Units, the Solution is:

$$h_{\mathrm{f}} = \frac{P_1 - P_2}{W}$$

If the pressure lost is desired in units of bar:

$$h_{\mathrm{f}} = P_1 - P_2$$
$$h_{\mathrm{f}} = 4.3\,\text{bar} - 3.65\,\text{bar}$$
$$h_{\mathrm{f}} = 0.65\,\text{bar}$$

Example 3.6

With water flowing in the water system shown in Fig. 3.10, p. 49, how much pressure is lost due to friction if Gauge 1 is reading 38 psi (2.6 bar) and Gauge 2 is reading 38 psi (2.6 bar)? Ordinarily, since Gauge 2 is lower in elevation, a higher reading would be expected at Gauge 2 than at Gauge 1. It is not true in this case because of friction loss, since pipe size and therefore velocity pressure does not change. Writing Bernoulli's equation and reflecting no change in velocities gives:

$$\frac{P_1}{W} + h_1 = \frac{P_2}{W} + h_2 + h_f$$

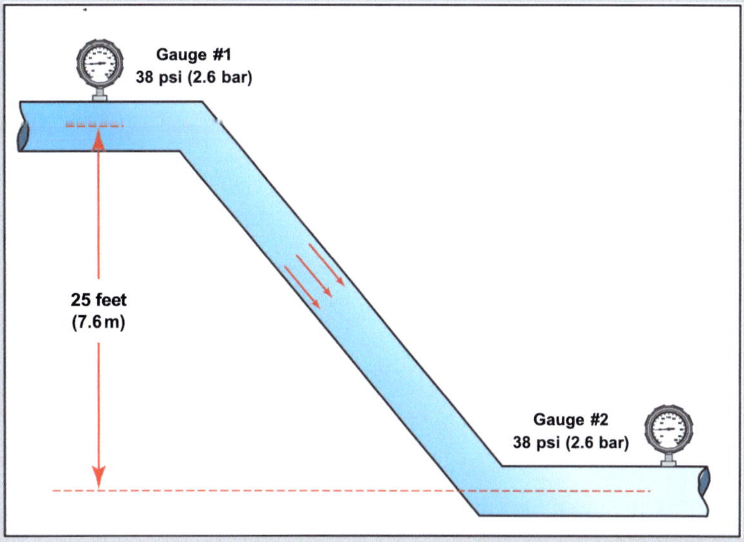

Fig. 3.10 Illustration for Example 3.6

And:

$$h_{\mathrm{f}} = \left(\frac{P_1 - P_2}{W}\right) + (h_1 - h_2)$$

Using Gauge 2 as the reference level and inserting the values of P_1 and P_2, gives:

$$h_{\mathrm{f}} = \left(\frac{38 - 38}{w}\right) + (25 - 0)$$

$$h_{\mathrm{f}} = 25\,\mathrm{ft}$$

If the pressure loss in psi is desired, then:

$$P_{\mathrm{f}} = (25\,\mathrm{ft})(0.433\,\mathrm{psi/ft})$$

$$P_{\mathrm{f}} = 10.8\,\mathrm{psi}$$

In S.I. Units, the Solution is:

$$h_\mathrm{f} = \left(\frac{P_1 - P_2}{W}\right) + (h_1 - h_2)$$

$$h_\mathrm{f} = \left(\frac{2.6 - 2.6}{w}\right) + (7.6 - 0)\,\mathrm{m}$$

$$h_\mathrm{f} = 7.6\,\mathrm{m}$$

If the pressure loss is desired in bar:

$$P_\mathrm{f} = (7.6\,\mathrm{m})(0.098\,\mathrm{bar/m})$$

$$P_\mathrm{f} = 0.74\,\mathrm{bar.}$$

Bernoulli's equation must also be adjusted when energy is added to the system from external sources such as pumps. Thus, in the comprehensive form, Bernoulli's equation may be written as follows:

$$\frac{P_1}{w} + \frac{v_1^2}{2g} + h_1 + h_p = \frac{P_2}{w} + \frac{v_2^2}{2g} + h_2 + h_f \tag{3.7}$$

The term $\mathbf{h_p}$ represents the energy added to the system by a pump and has units of feet (meters) of head.

Summary of Important Concepts

1. Hydrokinetics involves the study of water in motion.
2. The principles of hydrostatics will always be important, even in systems involving water in motion.
3. kinetic energy can be expressed as: $\frac{Wv^2}{2g}$
4. Bernoulli's equation illustrates the principle of conservation of energy and is expressed (without consideration of friction or external energy sources) as follows for water:

$$(2.31)(P_1) + \frac{(v_1)^2}{(2)(g)} + h_1 = (2.31)(P_2) + \frac{(v_2)^2}{(2)(g)} + h_2$$

5. Conservation of matter is expressed mathematically by: $Q_1 = Q_2$

And:

$$(A_1)(v_1) = (A_2)(v_2)$$

6. Proper unit conversion is essential.
7. In real-life application of Bernoulli's equation, consideration must be given to energy lost due to friction and energy added by pumps.

Exercises

3.1 What kinetic energy is generated by a cubic foot (28 L) of water traveling at 16.5 ft/s (5 m/s)? (**NOTE**: answer should be in units of foot-pounds or kilogram-meters.)

3.2 What kinetic energy is generated by 1 gallon (3.8 L) of gasoline traveling at 3.0 ft/s (0.9 m/s)?

3.3 What amount of work will be accomplished in transporting 500 gallons (1893 L) of water 1760 ft (536 m) at a rate of 500 gpm (1893 L/min)?

3.4 If oil with a specific gravity of 0.94 is flowing through a 12-in. (300 mm) pipe at a velocity of 3.5 ft/s (1.0 m/s), how many gpm (L/min) are flowing?

3.5 If oil with a specific gravity of 0.93 is flowing through an 8-in. (200 mm) pipe at a rate of 600 gpm (2270 L/min), what is the velocity of the oil in ft/s (m/s)?

3.6 Determine the pressure in psi (bar) at Point 1 in Fig. 3.11.

3.7 If the water velocity at Point 1 in Fig. 3.12 is 2 ft/s (0.6 m/s):

   a. What is Q_1 in gpm (L/min)?
   b. What is v_2 in ft/s (m/s)?
   c. What is Q_2 in gpm (L/min)?

3.8 In Fig. 3.13, find P_2.

$$\begin{aligned} \text{Given: } P_1 &= 83\,\text{psi}(5.7\,\text{bar}) \\ v_1 &= 3.9\,\text{ft/s}(1.2\ \text{m/s}) \\ v_2 &= 8.1\,\text{ft/s}(2.5\,\text{m/s}) \end{aligned}$$

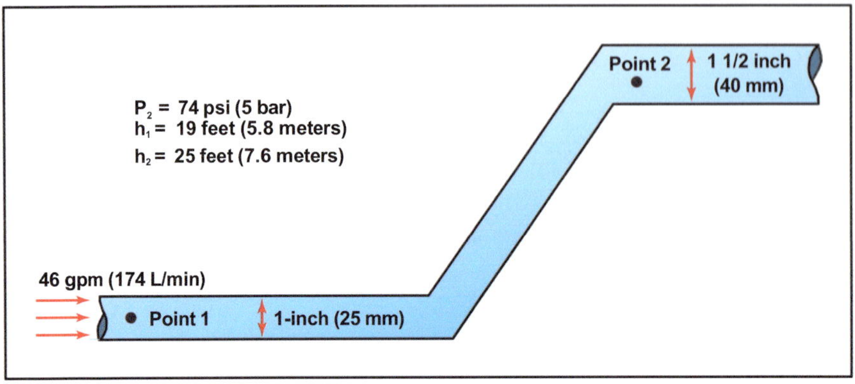

Fig. 3.11 Illustration for Exercise 3.6

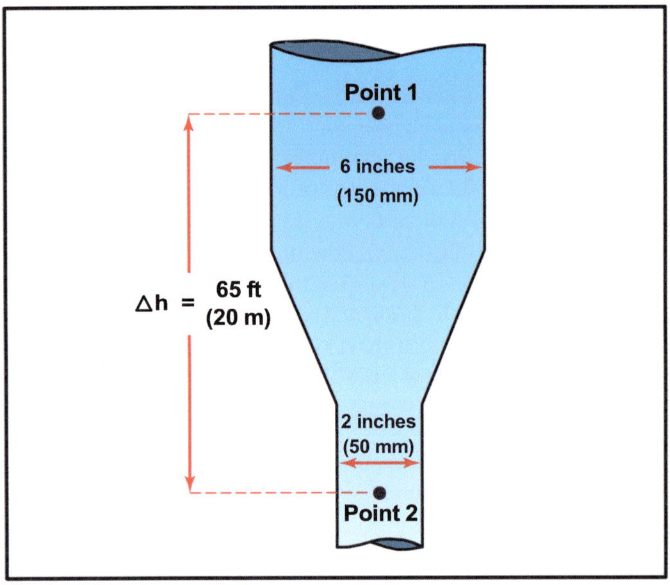

Fig. 3.12 Illustration for Exercise 3.7

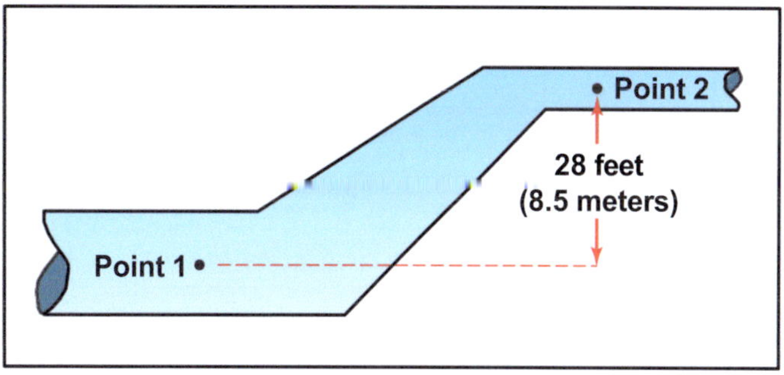

Fig. 3.13 Illustration for Exercise 3.8

3.9 If 645 gpm (2440 L/min) are flowing in Fig. 3.14, what is the friction loss if Gauge 1 reads 86 psi (5.9 bar) and Gauge 2 reads 55 psi (3.8 bar)?

3.10 In Fig. 3.15, if 230 ft (70 m) of pipe join Gauge 1,which reads 46 psi (3.2 bar), and Gauge 2, which reads 56 psi (3.86 bar), how much pressure is lost to friction if the pipe size is constant?

Fig. 3.14 Illustration for Exercise 3.9

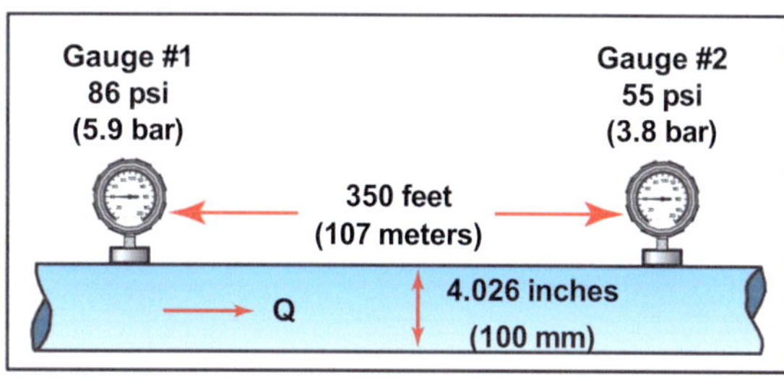

Fig. 3.15 Illustration for Exercise 3.10

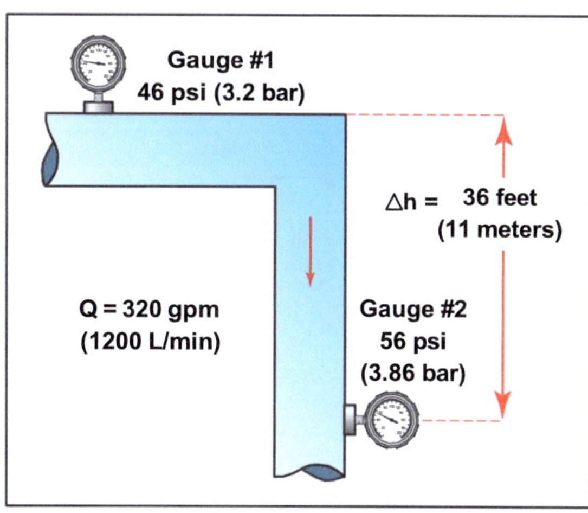

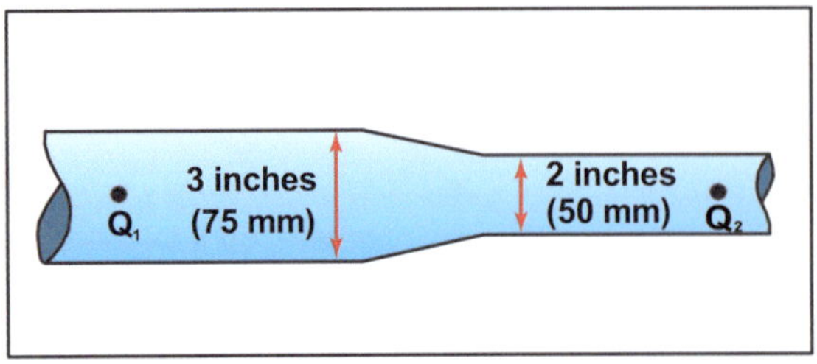

Fig. 3.16 Illustration for Exercise 3.11

3.11 In Fig. 3.16, if v_1 is 3 ft/s (0.9 m/s) and gasoline is flowing through the system, determine:

a. Q_1 in gpm (L/min)
b. Q_2 in gpm (L/min)
c. v_2 in ft/s (m/s)

3.12 In Fig. 3.17, if $v_1 = 8$ ft/s (2.4 m/s), what is v_2 and the flow in gpm (L/min)?

3.13 In Fig. 3.18, the water velocity in the 6-in. (150 mm) pipe is 2 ft/s (0.6 m/s), and the gauge on the 6-in. (150 mm) pipe reads 40 psi (2.75 bar). What will the gauge reading be on the 4-in. (100 mm) pipe? Assume 5 psi (0.3 bar) is lost to friction.

3.14 In Fig. 3.19, if Q_1 is 465 gpm (1760 L/min), find v_1, v_2, and Q_2.

3.15 In Fig. 3.20, find P_1, if P_2 is 36 psi (2.5 bar) and Q_1 is 805 gpm (3050 L/min).

3.16 In Fig. 3.21, if Q_1 is 90 gpm (340 L/min), d_1 is 1 in. (25 mm), and d_2 is 3 in. (75 mm):

a. What is Q_2?
b. What are v_1 and v_2?

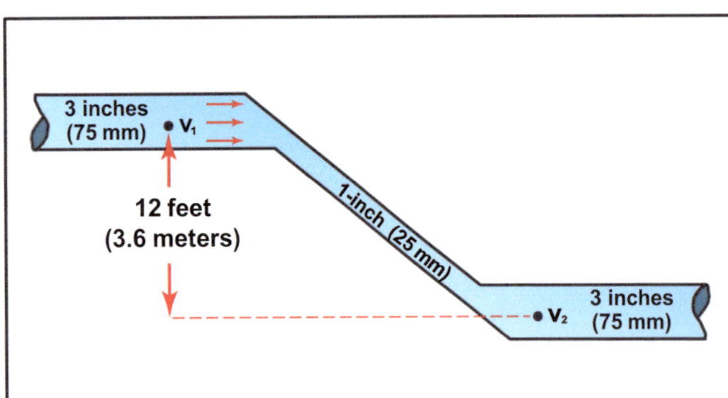

Fig. 3.17 Illustration for Exercise 3.12

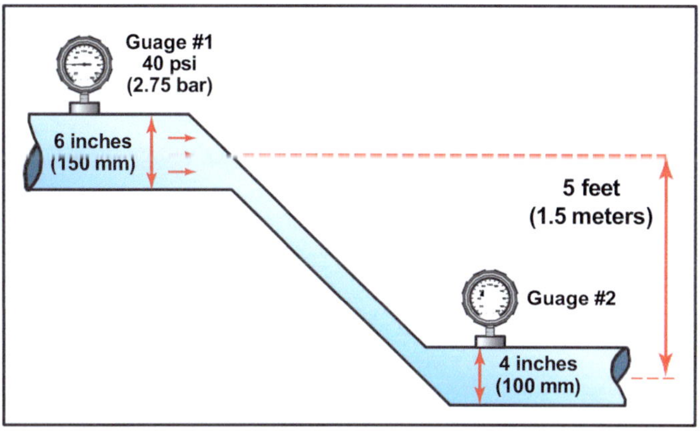

Fig. 3.18 Illustration for Exercise 3.13

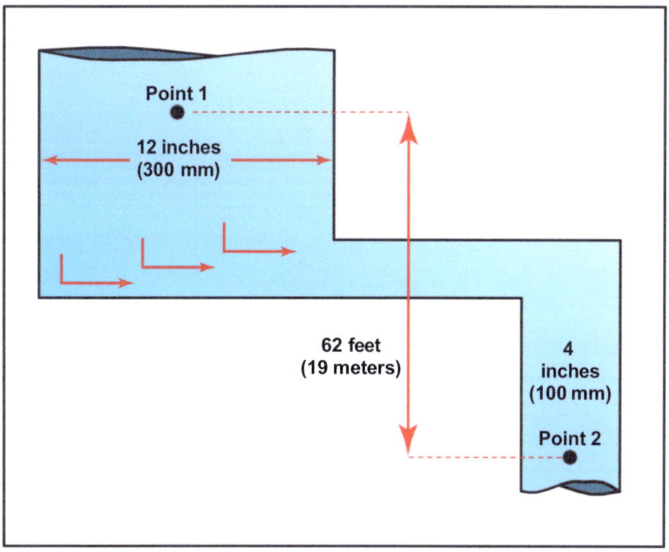

Fig. 3.19 Illustration for Exercise 3.14

3.17 If 100 gpm (378 L/min) are flowing in the piping system shown in Fig. 3.22, and Gauge 1 reads 120 psi (8.3 bar) and Gauge 2 reads 94 psi (6.5 bar), how much pressure in psi (bar) is lost to friction?

3.18 If Gauge 1 in Fig. 3.23 reads 24 psi (1.65 bar) and Gauge 2 reads 38 psi (2.6 bar), how much pressure is lost to friction?

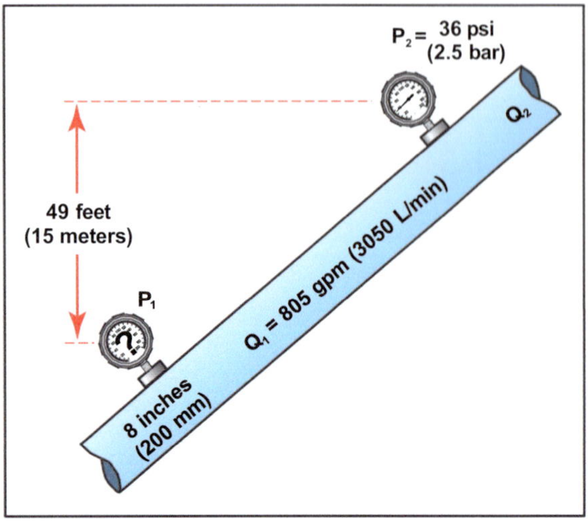

Fig. 3.20 Illustration for Exercise 3.15

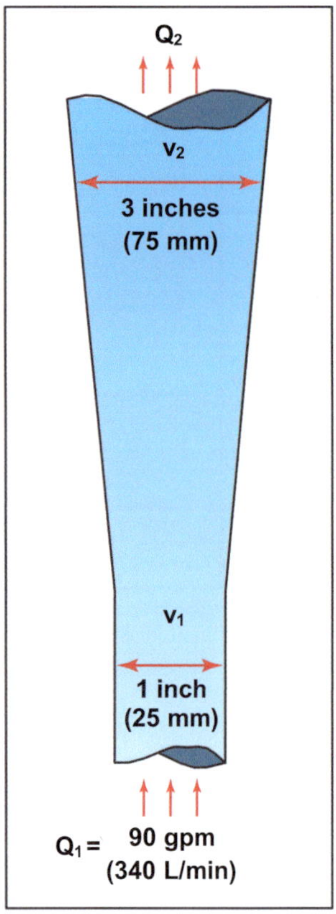

Fig. 3.21 Illustration for Exercise 3.16

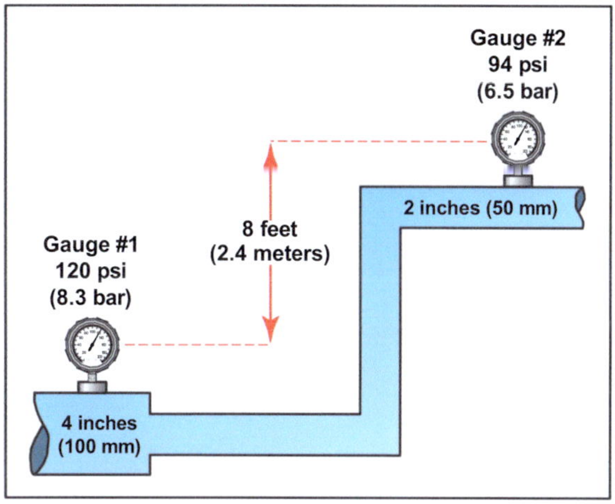

Fig. 3.22 Illustration for Exercise 3.17

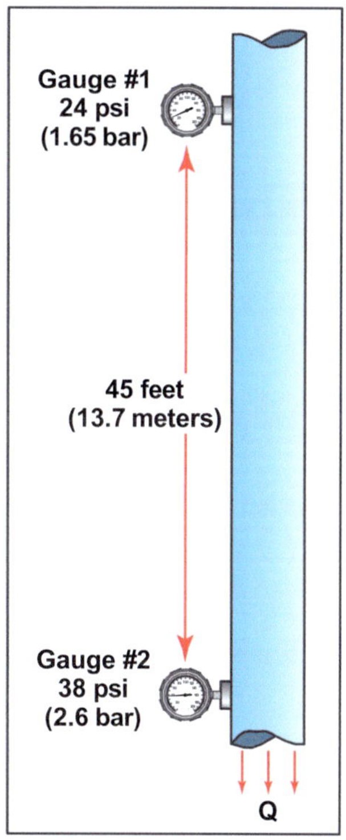

Fig. 3.23 Illustration for Exercise 3.18

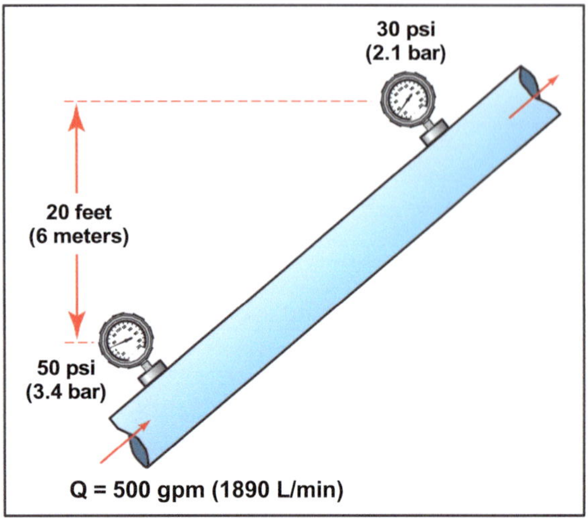

Fig. 3.24 Illustration for Exercise 3.19

3.19 Two gauges shown in Fig. 3.24 are placed 100 ft (30 m) apart on a 4-in. (100 mm) main. The main travels uphill such that the downstream gauge is 20 ft (6 m) higher than the upstream gauge. If the upstream gauge reads 50 psi (3.4 bar) and the downstream gauge reads 30 psi (2.1 bar), and 500 gpm (1890 L/min) is flowing, how much pressure is lost to friction in each foot (meter) of pipe between the two gauges?

3.20 If a 6-in. (150 mm) pipe is connected to an 8-in. (200 mm) pipe and the water velocity in the 6-in. (150 mm) pipe is 10 ft/s (3 m/s), what is the water velocity in the 8-in. (200 mm) pipe and what quantity in gpm (L/min) is flowing in each pipe?

Selected Answers Found in Appendix K.

References

1. Paul A. Tipler, *College Physics*. (New York: Worth Publishers Inc, 1987), p. 114.
2. Tipler 114.
3. Tipler 115.
4. Russell W. Henke, *Introduction to Fluid Mechanics* (Reading: Addison-Wesley Publishing Co., 1972), pp. 48-50.

Supplemental Readings

Hickey, Harry E. *Hydraulics for Fire Protection.* National Fire Protection Association: Quincy, MA, 1980.

National Fire Protection Association. *The Fire Protection Handbook.* National Fire Protection Association: Quincy, MA, 21st edition, Sec. 15, Ch.3.

Chapter 4
Flow Measuring Devices

Abstract This chapter introduces common flow measuring devices used in the fire protection industry. Hyrdostatic and hyrokinetic principles are used for the development of the devices as well as the hydraulic formulas to solve for flows mathematically. Bernoulli's equation is used for the application of flow measuring devices such as venturi meters and orifice plate meters.

A basic understanding of the principles of hydrostatics and hydrokinetics makes it easier to understand how various flow measuring instruments work. This chapter will give attention to some of the more common devices such as pitot tubes, venturi meters, and orifice plate meters. A more comprehensive study of the types contained in this chapter is presented by Hickey in his text [1].

Common Flow Measuring Devices

The Pitot Tube

The pitot tube is a common device used to measure velocity pressure and, thus, fluid velocity. The pitot tube consists of a small-diameter tube, usually about one-sixteenth of an inch (1.6 mm) in internal diameter. The tube is inserted into a fluid stream, parallel to the direction of flow, and aimed upstream against the flow. The tube may be connected to a piezometer tube (Fig. 4.1). The tube may also be connected to a Bourdon gauge, which is more common in fire protection applications (Fig. 4.2). In both Figs. 4.1 and 4.2, the pitot tube is used to measure "total energy." Therefore, when installed inside a closed conduit, the pitot tube alone cannot reveal the water velocity.

Suppose a Bourdon gauge is installed on the system in conjunction with a pitot tube and gauge (Fig. 4.3). If the two gauges are close enough together, friction loss is negligible. The pressure registered on the first gauge, P_1, reflects only components

P. Brock and V. Charter, *Fire Protection Hydraulics and Water Supply Analysis*, The Society of Fire Protection Engineers Series,
https://doi.org/10.1007/978-3-032-09877-1_4

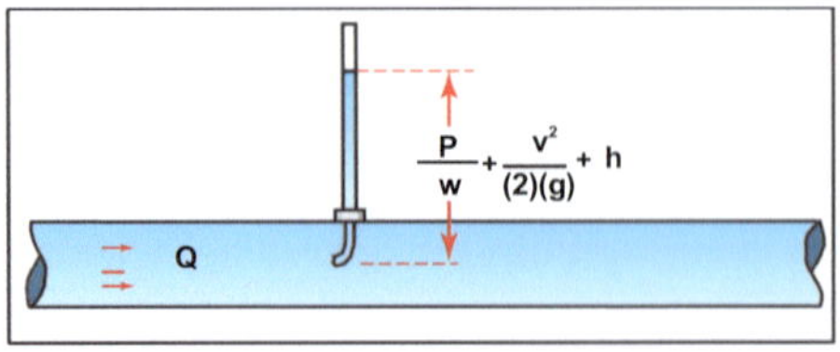

Fig. 4.1 Pitot tube attached to a piezometer tube

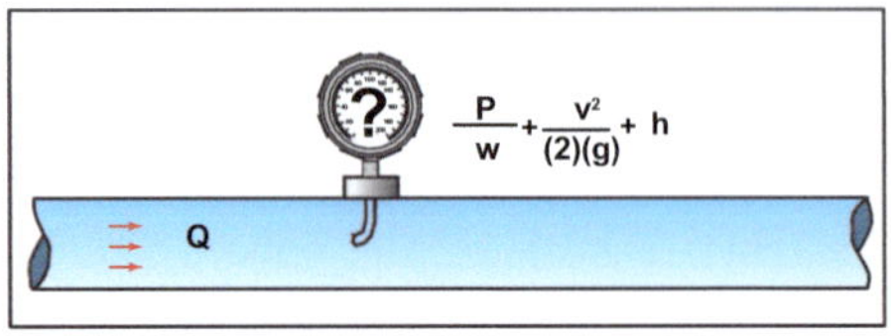

Fig. 4.2 Pitot tube with a Bourdon gauge

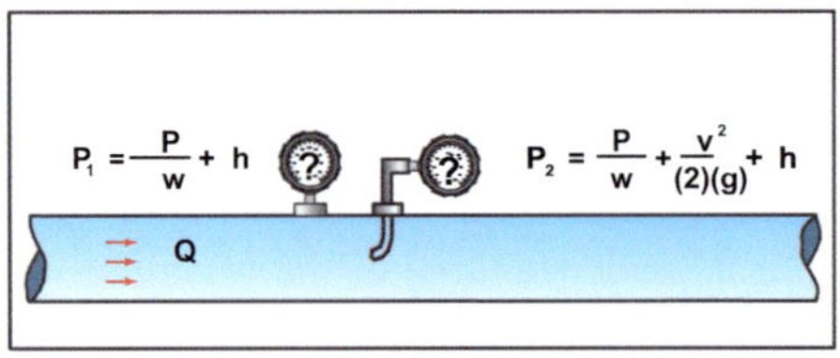

Fig. 4.3 One method for determining velocity pressure

of potential energy, or

$$P_1 = \frac{P}{w} + h$$

On the other hand, the pressure reflected on the second gauge, P_2, represents the total energy, or

$$p_2 = \frac{p}{w} + h + \frac{v^2}{(2)(g)}$$

Then:

$$p_2 - p_1 = \left(\frac{p}{w} + h + \frac{v^2}{(2)(g)}\right) - \left(\frac{p}{w} + h\right)$$

And, since the elevation of the pipe is constant:

$$p_2 - p_1 = \frac{v^2}{(2)(g)}$$

Therefore, the difference in readings on the two gauges represents the kinetic energy component, which is commonly referred to as the velocity pressure. If this velocity pressure is known, the velocity can be established as follows:

$$p = \frac{v^2}{(2)(g)}$$
$$v^2 = (2)(g)(P)$$
$$v = \sqrt{(2)(g)(P)}$$

If P has units of psi, some unit conversion will be required, so the velocity will have the appropriate units of fps (m/s). This is accomplished as follows:

$$v = \sqrt{(2)(g)\left(\frac{\text{ft}}{\text{s}^2}\right)(P)(\text{psi})}$$
$$v = \sqrt{(2)(g)(P)\left(\frac{\text{ft}}{\text{s}^2}\right)(\text{psi})\frac{2.31\ \text{ft}}{\text{psi}}}$$
$$v = \sqrt{(2)(32.17)(2.31)(P)\left(\frac{\text{ft}^2}{\text{s}^2}\right)}$$
$$v = (12.1912)\sqrt{P}(\text{ft/s}) \tag{4.1}$$

If P has units of bar, the unit conversion to yield a velocity in units of meters per second is as follows:

$$v = \sqrt{(2)(g)\left(\frac{\text{m}}{\text{s}^2}\right)(P)(\text{bar})}$$
$$v = \sqrt{(2)(9.8)\left(\frac{\text{m}}{\text{s}^2}\right)(P)(\text{bar})(10.197)\left(\frac{\text{m}}{\text{bar}}\right)}$$
$$v = \sqrt{(2)(9.8)(10.197)(P)\left(\frac{\text{m}^2}{\text{s}^2}\right)}$$
$$v = (14.14)\left(\sqrt{P}\right)\frac{\text{m}}{\text{s}} \tag{4.1a}$$

After the velocity is determined, the rate of flow in ft^3/s or gpm (m^3/s or L/min) can be calculated from:

$$Q = (A)(v)$$

If the pressure difference in Fig. 4.3 is expressed in units of feet (meters) of water, the expression for velocity becomes:

$$v = \sqrt{(2)(g)(h)} \tag{4.2}$$

This is the identical equation used in physics to calculate the velocity of a falling body that has fallen from some height (**h**) [2].

Perhaps the most common use of the pitot tube is to measure the velocity of water being discharged from an orifice. As pointed out previously, there are no potential energy components in an open stream. Therefore, the pitot tube inserted into an open stream of water, such as a stream being discharged from a fire hydrant, measures only velocity pressure.

After determining the water velocity, the flow rate should again be equal to **(A)(v)**, where **A** is now the cross-sectional area of the stream. However, a problem now arises. Observations and experiments reveal that the actual cross-sectional area of the stream being discharged from an orifice is not the same as the area of the orifice. The stream actually contracts after it is discharged from an orifice or nozzle. The magnitude of contraction depends on how the orifice or nozzle is constructed, but the point of maximum contraction occurs at a distance of one- half the orifice diameter away from the orifice. This point of maximum stream contraction is called the "vena contracta" [3].

A coefficient of contraction can be established to reflect the ratio between the orifice or nozzle cross-sectional area and the cross-sectional area of the stream. Thus:

$$C_c = \frac{A_s}{A_o}$$

where:

$\mathbf{C_c}$ = coefficient of contraction,

$\mathbf{A_s}$ = cross-sectional area of the stream at the vena contracta,

$\mathbf{A_o}$ = cross-sectional area of the orifice or nozzle.

So, now the equation for the flow rate (**Q**) becomes:

$$Q = (A_o)(C_c)(v)$$

And, substituting for **v**:

$$Q = (A_o)(C_c)\left(\sqrt{(2)(g)(h)}\right)$$

However, it has also been proven that the actual velocity of the stream is not equal to the theoretical $\sqrt{2gh}$, but is slightly less. Therefore, a velocity coefficient has also been defined as the ratio between the actual velocity and the theoretical velocity, such that:

$$C_v = \frac{V_a}{V_t}$$

where:

$\mathbf{C_v}$ = the velocity coefficient,

$\mathbf{v_a}$ = the actual stream velocity,

$\mathbf{v_t}$ = the theoretical stream velocity.

The equation for the flow rate now becomes:

$$Q = (C_c)(C_v)(A_o)\left(\sqrt{(2)(g)(h)}\right)$$

Now, if a single coefficient of discharge ($\mathbf{C_d}$) is defined as the product of $\mathbf{C_c}$ and $\mathbf{C_v}$, the equation for **Q** becomes:

$$Q = (C_d)(A_o)\left(\sqrt{(2)(g)(h)}\right) \tag{4.3}$$

The values of $\mathbf{C_d}$ for different orifices and nozzles vary. Henke points out that for a standard sharp-edged orifice, the approximate value of $\mathbf{C_d}$ is 0.60 (Fig. 4.4) [4]. Some other common nozzles and their corresponding coefficients of discharge are shown in (Fig. 4.5).

For a more practical application, Eq. 4.3 may be transformed through appropriate unit conversion into a form in which pressure readings in psi (bar) from the pitot tube can be used with the solution being directly in gpm (liters/min). This is accomplished as follows:

$$Q = (C_d)(A_o)\left(\sqrt{(2)(g)(h)}\right)$$

where:

$$A_o = (\pi)(r)^2(\text{in})^2$$

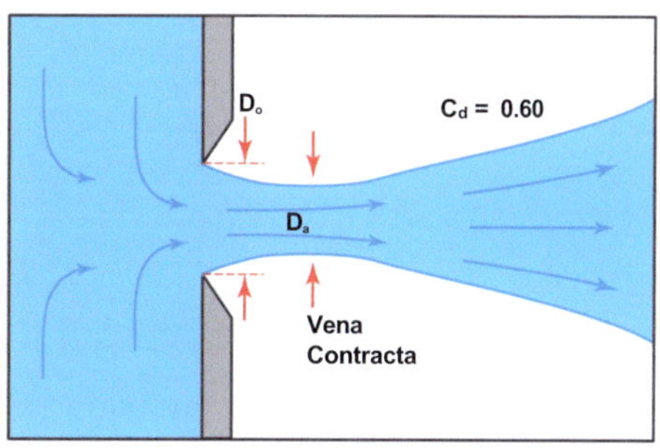

Fig. 4.4 A standard sharp-edged orifice

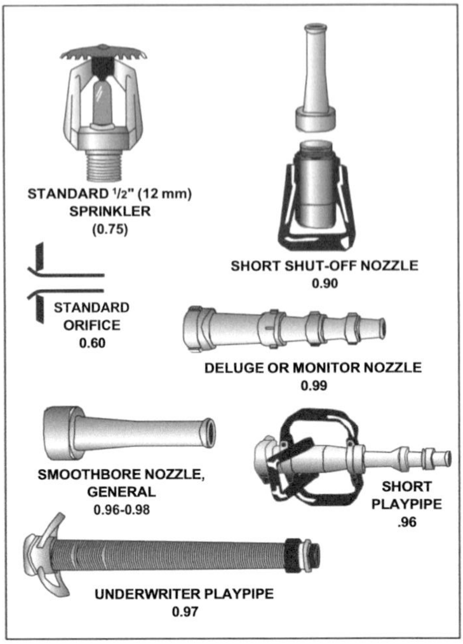

Fig. 4.5 Discharge coefficients of nozzles and common appliances

$$A_o = (\pi)\left(\frac{D^2}{4}\right)(\text{in})^2$$

With, using English units of measure:

D = internal orifice or nozzle diameter in inches.

Recalling also that $h = (2.31)\,(P)\,(\text{ft})$.

And if **P** is the velocity pressure measured with a pitot tube, then:

$$Q = \left(\frac{\pi}{4}\right)(C_c)(D)^2\left(\sqrt{(2)(2.31)(32.17)(P)(\text{ft})\left(\frac{\text{ft}}{\text{s}^2}\right)}\right)(\text{in})^2$$

$$Q = (0.785)(12.2323)(C_d)(D)^2\left(\sqrt{(P)}\right)\left(\frac{\text{ft}}{\text{s}}\right)\left(\frac{(1)(\text{ft})^2}{(144)(\text{in})^2}\right)(\text{in})^2$$

$$Q = (0.06645)(C_d)(D)^2\left(\sqrt{(P)}\right)\left(\text{ft}^3/\text{s}\right)$$

And finally, converting ft^3/s to gpm gives:

$$Q = (0.06645)(C_d)(D^2)\sqrt{(P)}(\text{ft}^3/\text{s})(60\,\text{s}/\text{min})(7.48\,\text{gal}/\text{ft}^3)$$

$$Q = (29.83)(C_d)(D)^2\sqrt{P} \tag{4.4}$$

In Standard International Units, if **D** is in millimeters, and recalling that:

$$h = 10.197(\text{m/bar})(P)(\text{bar})$$

Then:

$$Q = \left(\frac{\pi}{4}\right)(C_\text{d})(D)^2\left(\sqrt{(2)(9.8)(10.197)(P)(m)\left(\frac{\text{m}}{\text{s}^2}\right)}\right)(\text{mm})^2$$

$$Q = (0.785)(14.14)(C_\text{d})(D)^2\left(\sqrt{(P)}\right)\left(\frac{\text{m}}{\text{s}}\right)(\text{mm})^2\left(\frac{\text{m}^2}{(10^6)(\text{mm})^2}\right)$$

$$Q = (0.0000111)(C_\text{d})(D)^2\left(\sqrt{P}\right)\left(\frac{\text{m}^3}{\text{s}}\right)$$

And finally, converting $\left(\frac{\text{m}^3}{\text{s}}\right)$ to liters/minute gives:

$$Q = (0.0000111)(C_\text{d})(D)^2\left(\sqrt{P}\right)\left(\frac{\text{m}^3}{\text{s}}\right)\left(\frac{60\ \text{s}}{\text{min}}\right)\left(\frac{1000\ \text{L}}{\text{m}^3}\right)$$

$$Q = (0.666)(C_\text{d})(D)^2\left(\sqrt{P}\right) \tag{4.4a}$$

Equations 4.4 and 4.4a are important formulas to the fire protection specialist. These are the formulas used in converting pitot gauge readings into gpm or L/min during tests on hydrants, fire pumps, and fire department pumpers. Figure 4.6 shows several types of pitot tubes and gauges.

Example 4.1

If a pitot tube gauge reads 26 psi (1.8 bar) when the pitot tube is held in a stream from a 2½-in. (65 mm) hydrant outlet ($C_\text{d} = 0.9$), how many gpm (liters/min) are flowing?

Solution:

$$Q = (29.83)(C_\text{d})(D)^2\left(\sqrt{P}\right)$$

$$Q = (29.83)(0.9)(2.5)^2\left(\sqrt{26}\right)$$

$$Q = 855.6\ \text{gpm}$$

In S.I. Units the Solution is:

$$Q = (0.666)(C_\text{d})(D)^2(\sqrt{}P)$$

$$Q = (0.666)(0.9)(65)^2(\sqrt{}1.8)$$

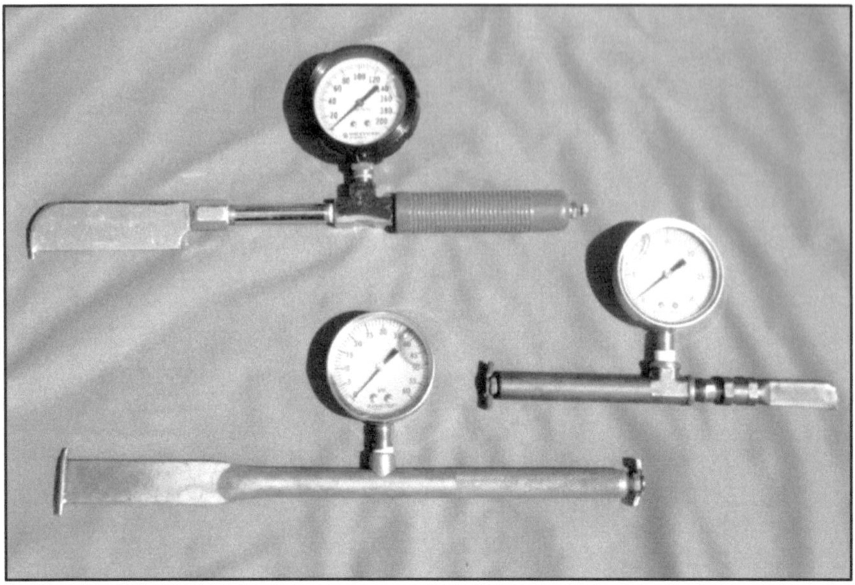

Fig. 4.6 Various styles of pitot tubes

$$Q = 3398\,\mathrm{L/min}$$

Example 4.2

What pitot tube gauge reading would represent a flow of 500 gpm (1893 L/min) from a short playpipe with a nozzle diameter of 1.75 in. (45 mm) and a coefficient of discharge of 0.96?

Solution:

In this case, it is required that the pressure be established instead of the flow. Therefore, Eq. 4.4 is first manipulated algebraically to solve for P:

$$Q = (29.83)(C_{\mathrm{d}})(D)^2\left(\sqrt{P}\right)$$

$$\sqrt{P} = \left(\frac{Q}{(29.83)(C_{\mathrm{d}})(D)^2}\right)$$

$$P = \left(\frac{Q}{(29.83)(C_{\mathrm{d}})(D)^2}\right)^2$$

Inserting the numerical values gives:

$$P = \left(\frac{500}{(29.83)(0.96)(1.75)^2} \right)^2$$
$$P = 32.5\,\text{psi}$$

In S.I. Units, the Solution is:

$$P = \left(\frac{Q}{(0.666)(C_d)(D)^2} \right)^2$$
$$P = \left(\frac{1893}{(0.666)(0.96)(45)^2} \right)^2$$
$$P = 2.14\,\text{bar}$$

Venturi Meter

The venturi meter, when coupled with a differential manometer, may be used to measure water velocity (Fig. 4.7). The device consists essentially of a piece of pipe in which the cross-sectional area has been constricted. The pipe is tapped at the points of both maximum and minimum diameter, and the manometer is attached at these points. The difference in the heights of the two mercury columns in the manometer represents the pressure difference existing inside the pipe at the two points. Because the manometer measures only potential energy components, the difference in mercury column heights, **Δh**, as shown in Fig. 4.7 will be:

$$\Delta h = \left(\frac{P_1}{w} + h_1 \right) - \left(\frac{P_2}{w} + h_2 \right)$$

And, if the venturi meter is installed horizontally, $h_1 = h_2$ and:

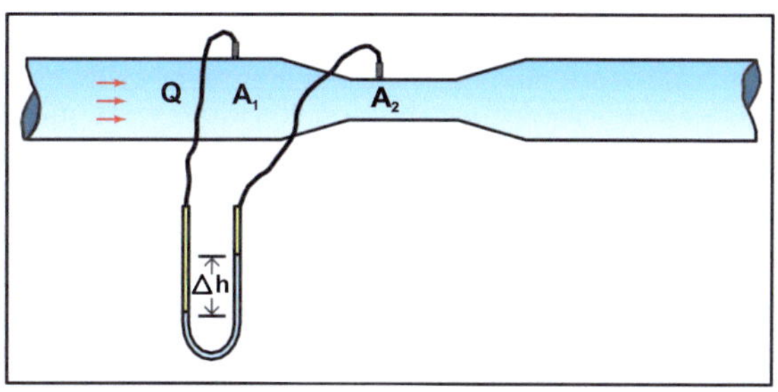

Fig. 4.7 Venturi meter with a differential manometer

$$\Delta h = \frac{P_1}{w} - \frac{P_2}{w}$$

Therefore, if the difference between the heights of the mercury columns is measured, and this measurement is converted to feet of water, $P_1/w–P_2/w$ will be known.

Since it is also known that $Q_1 = Q_2$ and thus:

$$(A_1)\,(v_1) = (A_2)\,(v_2)$$

Then:

$$v_2 = \frac{(A_1)(v_1)}{A_2}$$

Looking at Bernoulli's equation and substituting according to the relationship above gives

$$\frac{(P_1)}{w} + \frac{(v_1)^2}{(2)(g)} + h_1 + h_P = \frac{(P_2)}{w}\left(\frac{(v_1)^2}{(2)(g)}\right) + h_2 + h_f$$

where any effects of pumps are acting equally at both points, and where the friction loss is considered negligible due to the small distance between the taps, and **Δh** is used for $P_1/w–P_2/w$, the equation becomes:

$$\Delta h + \frac{(v_1)^2}{(2)(g)} = \left(\frac{A_1}{A_2}\right)^2\left(\frac{(v_1)^2}{(2)(g)}\right)$$

And solving for v_1 gives:

$$v_1 = \sqrt{\frac{(2)(g)(\Delta h)}{\left(\frac{A_1}{A_2}\right)^2 - 1}} \tag{4.5}$$

In the above equation, **Δh** is the difference in height between the mercury columns in the manometer after being converted to feet (meters) of water. $\mathbf{A_1}$ and $\mathbf{A_2}$ represent the cross-sectional areas of the meter at the two tap points, and **g** is the acceleration due to gravity. Once $\mathbf{v_1}$ has been calculated, the flow rate can be calculated from:

$$Q = (A_1)(v_1)$$

Labes noted that the actual velocity of water through the venturi meter is less than what would be theoretically predicted [5]. To produce more accurate results, a coefficient can be used similar to the coefficient of discharge (discussed in the section on pitot tubes). However, Eq. 4.5 may often be used without adjustment for an approximation of flow rate in a system.

Example 4.3

If a venturi meter placed in a 6-in. (150 mm) water line has a constriction with a 4-in. (100 mm) diameter, and the manometer shows a reading of 2 in. (51 mm) of mercury, how many gpm (L/min) are flowing?

Solution:

$$v_1 = \sqrt{\frac{(2)(g)(\Delta h)}{\left(\frac{A_1}{A_2}\right)^2 - 1}}$$

where:

$$\Delta h = (2 \text{ in of Hg})\left(\frac{1.13 \text{ ft of } H_2O}{\text{in of Hg}}\right) = 2.26 \text{ ft of } H_2O$$

And:

$$\frac{A_1}{A_2} = \frac{(\Pi)(r_1)^2}{(\Pi)(r_2)^2} = \frac{(r_1)^2}{(r_2)^2} = \frac{(3 \text{ in})^2}{(2 \text{ in})^2} = 2.25$$

Then:

$$v_1 = \sqrt{\frac{(2)(32.2)\left(\text{ft/s}^2\right)(2.26)(\text{ft})}{(2.25)^2 - 1}}$$

$$v_1 = 5.968 \text{ ft/s}$$

After calculating v_1, then Q_1 is:

$$Q_1 = (A_1)(v_1)$$

$$Q_1 = (\Pi)(r_1)^2(v_1)$$

$$Q_1 = (\Pi)(3 \text{ in})^2\left(\frac{1 \text{ ft}^2}{144 \text{ in}^2}\right)(5.986 \text{ ft/s})\left(\frac{449 \text{ gpm}}{\text{ft}^3/\text{s}}\right)$$

$$Q_1 = 527.7 \text{ gpm}$$

In S.I. units, the solution is:

$$v_1 = \sqrt{\frac{(2)(g)(\Delta h)}{\left(\frac{A_1}{A_2}\right)^2 - 1}}$$

where:

$$\Delta h = 51\,\text{mm of Hg}\left(\frac{13.6\,\text{mm of H}_2\text{O}}{1\,\text{mm of Hg}}\right)\left(\frac{1\,\text{m}}{1000\,\text{mm}}\right) = 0.694\,\text{m}$$

And:

$$\frac{A_1}{A_2} = \frac{(\Pi)(r_1)^2}{(\Pi)(r_2)^2} = \frac{(r_1)^2}{(r_2)^2} = \frac{(75\text{ mm})^2}{(50\text{ mm})^2} = 2.25$$

Then:

$$v_1 = \sqrt{\frac{(2)(9.8)\left(\text{m}/\,\text{s}^2\right)(0.694)(\text{m})}{(2.25)^2 - 1}}$$

$$v_1 = 1.83\,\text{m/s}$$

And finally:

$$Q_1 = (A_1)(V_1)$$

$$Q_1 = (\Pi)(r_1)^2(V_1)$$

$$Q_1 = (\Pi)(75\,\text{mm})^2\left(\frac{1\,\text{m}^2}{1 \times 10^6\text{mm}^2}\right)(1.83\,\text{m/s})\left(60\,\text{s}/\,\text{min}\right)\left(\frac{10^3\,\text{L}}{\text{m}^3}\right)$$

$$Q_1 = 1940\,\text{L}/\,\text{min}$$

Example 4.4

If a venturi meter using a mercury manometer is installed on a 6-in. (150 mm) test line from a fire pump, how many inches (mm) of mercury would represent a flow of 750 gpm (2840 L/min) if the venturi constriction is 5 in. (125 mm) in diameter?

Solution:

In this case the value of Δh must be established. First, solve for the velocity: $Q = A_1V_1$

$$Q = A_1 V_1$$

$$v_1 = \frac{Q}{A_1}$$

$$v_1 = \left(\frac{750\,\text{gpm}}{449\,\text{gpm/ft}^3/\text{s}}\right)\left(\frac{144\,\text{in}^2/\text{ft}^2}{\Pi(3\,\text{in})^2}\right)$$

$$v_1 = 8.5\,\text{fps}$$

Solving algebraically for Δh:

$$v_1 = \sqrt{\frac{(2)(g)(\Delta h)}{\left(\frac{A_1}{A_2}\right)^2 - 1}}$$

$$\left(v_1^2\right) = \frac{(2)(g)(\Delta h)}{\left(\frac{A_1}{A_2}\right)^2 - 1}$$

$$\left(v_1^2\right)\left[\left(\frac{A_1}{A_2}\right)^2 - 1\right] = (2)(g)(\Delta h)$$

$$\Delta h = \frac{\left(v_1^2\right)}{(2)(g)}\left[\left(\frac{A_1}{A_2}\right)^2 - 1\right]$$

Substituting the numerical values:

$$\Delta h = \frac{(8.5)^2}{(2)(32.2)}\left[\left(\frac{(\Pi)(3)^2}{(\Pi)(2.5)^2}\right)^2 - 1\right]$$

$$\Delta h = 1.2\,\text{ft of water}$$

Converting to inches of mercury:

$$\Delta h = \frac{(1.2\,\text{ft of water})}{(1.13\text{ ft of water per inch of Hg})}$$

$$\Delta h = 1.06\,\text{inches of Hg}$$

In S.I. Units, the solution is:

First, solving for the velocity:

$$Q = A_1 V_1$$

$$v_1 = \frac{Q}{A_1}$$

$$v_1 = (2840\,\text{L/min})\left(\frac{1\ \text{min}}{60\ \text{s}}\right)\left(\frac{10^6\text{mm}^2/\text{m}^2}{\Pi(75\ \text{mm})^2}\right)\left(\frac{1\ \text{m}^3}{10^3\ \text{L}}\right)$$

$$v_1 = 2.68\ \text{m/s}$$

Solving algebraically for Δh:

$$\Delta h = \frac{(v_1^2)}{(2)(g)}\left[\left(\frac{A_1}{A_2}\right)^2 - 1\right]$$

Substituting the numerical values:

$$\Delta h = \frac{(2.68)^2}{(2)(9.8)}\left[\left(\frac{(\Pi)(75)^2}{(\Pi)(62.5)^2}\right) - 1\right]$$

$$\Delta h = 0.393\ \text{meters of water}$$

Converting to millimeters of mercury:

$$\Delta h = (0.393\ \text{m of H}_2\text{O})\left(\frac{1.0\ \text{m of Hg}}{13.6\ \text{m of H}_2\text{O}}\right)\left(\frac{1000\ \text{mm}}{1.0\ \text{m}}\right)$$

$$\Delta h = 28.9\ \text{mm of mercury}$$

It is appropriate to note that the reduction in pressure created by the constriction in the venturi has additional practical applications. For example, if the difference in diameter is sufficient, and the water velocity is sufficient, the pressure at the point of constriction can actually become negative relative to atmospheric pressure and create a vacuum. This negative pressure can be used to suck foam concentrate from a container and induct it into a flowing hose stream. Figure 4.8 shows a common foam eductor, which works in accordance with the energy balance principles of Bernoulli's equation.

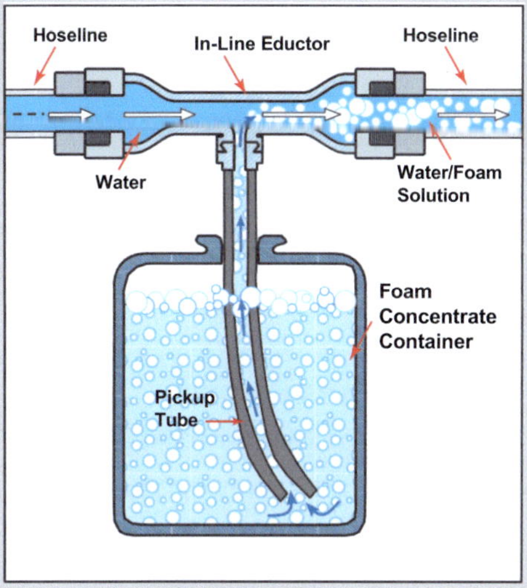

Fig. 4.8 Common inline foam eductor draws foam concentrate into the fire stream using the Venturi principle

Due to environmental and health problems associated with mercury, modern venturi flow meters more commonly convert the pressure differences directly into gpm. A flow meter of this type, which is commonly used on fire pump installations, is shown in Fig. 4.9.

Fig. 4.9 Venturi-type flowmeter used on fire pump installations. *Photo courtesy of Floyd Luinstra*

Orifice Plate Meter

An orifice plate meter is similar in principle to a venturi meter. However, the change of water velocity is accomplished by using a plate with an orifice that is smaller than the diameter of the pipe in which it is placed (Fig. 4.10). The theoretical expression

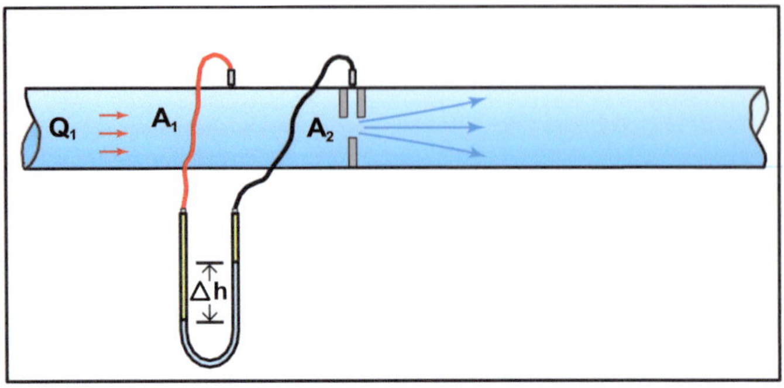

Fig. 4.10 Orifice plate meter with a differential manometer

for the velocity of water through the orifice in the plate, as derived from Bernoulli's equation, is the following:

$$v_2 = \sqrt{\frac{(2)(g)(\Delta h)}{1 - \left(\frac{A_2}{A_1}\right)^2}} \tag{4.6}$$

where

$\mathbf{\Delta h}$ = pressure difference as read from the manometer,

$\mathbf{A_2}$ = the cross-sectional area of the orifice,

$\mathbf{A_1}$ = the cross-sectional area of the pipe.

Again, the theoretical value of $\mathbf{v_2}$ does not equal the actual value. Also, the area of the stream is less than the area of the orifice. Therefore, a coefficient of discharge ($\mathbf{C_d}$) can be identified and Eq. 4.6 adjusted as follows to yield velocities closer to those experimentally determined.

$$v_2 = (C_d)\sqrt{\frac{(2)(g)(\Delta h)}{1 - (C_d)^2\left(\frac{A_2}{A_1}\right)^2}} \tag{4.7}$$

Then, the flow rate can be calculated from:

$$Q_2 = (A_2)\,(v_2)$$

Other Flow Meters

There are several other types of in-line flow meters available for fire protection applications. One type employs a small turbine which reflects the water velocity as it

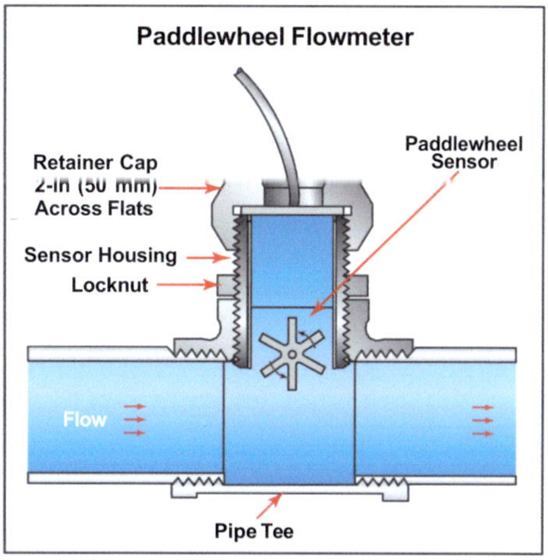

Fig. 4.11 This diagram shows the inner workings of a paddlewheel flowmeter

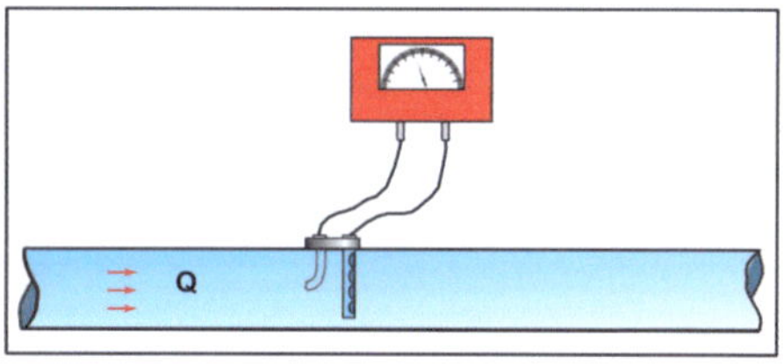

Fig. 4.12 This illustration shows the proper orientation of the probes on a flowmeter

turns. The rotational velocity of the turbine is translated into gpm and an electronic digital readout is provided. A similar type, as shown in Fig. 4.11, uses the rotation of a small paddle wheel to indicate water velocity.

Another type employs a pitot tube plus an additional pressure probe aimed downstream (Fig. 4.12). The probe aimed downstream measures only the potential energy while the pitot tube measures total energy. The difference between the two is the velocity pressure. This velocity pressure is translated directly into gpm and is indicated on a readout dial (Fig. 4.13).

Summary of Important Concepts

1. Hydrostatic and hydrokinetic principles have been used for the development of various flow measuring devices.
2. One of the more important hydraulics formulas for the fire protection specialist is $Q = (29.83)\,(C_d)\,(D)^2\,(\sqrt{P})$ and $Q = (0.666)\,(C_d)\,(D)^2\,(\sqrt{P})$ in S.I.units.
3. Bernoulli's equation is important in application of venturi meters and orifice plate meters as flow measuring devices.
4. Digital and other types of direct-reading flow measuring instruments are becoming commonplace.

Fig. 4.13 Flowmeter with dual probes and a direct gpm readout. *Photo courtesy of Floyd Luinstra*

Exercises

4.1. If a hydrant has a C_d of 0.8 and 2½-in. (65 mm) diameter outlets, what flow would be reflected by the following pitot readings?

a. 15 psi (1.0 bar)
b. 40 psi (2.75 bar)
c. 65 psi (4.5 bar)

4.2. If a hydrant has a C_d of 0.9 and 2 7/16-in. (62 mm) diameter outlets, what flow would be reflected by the following pitot readings?

a. 18 psi (1.24 bar)
b. 37 psi (2.55 bar)
c. 62 psi (4.27 bar)

4.3. What is the coefficient of discharge of a 2½-in. (65 mm) hydrant outlet that is delivering 816 gpm (3090 L/min) when the gauge on a pitot tube reads 40 psi (2.75 bar)?

4.4. What is the nozzle diameter of a straight stream nozzle that has a coefficient of discharge of 0.96 if the gauge on the pitot tube registers 33 psi (2.27 bar) when 257 gpm (970 L/min) are flowing?

4.5. If water is flowing from a 1¾-in. (45 mm) nozzle with a coefficient of discharge of 0.97 giving a pitot tube reading of 33 psi (2.27 bar).

   a. What is the kinetic energy of the open stream in units of feet (meters) of head?
   b. How many gpm (liters/min) are flowing?

4.6. In testing a fire pump, it is desired to flow 500 gpm (1890 L per min) from a 1¾-in. (45 mm) nozzle with a C_d of 0.97. What pitot pressure would indicate a flow of 500 gpm (1890 L/min)?

4.7. If a 575 gpm (2175 L/min) flow yields a pitot gauge reading of 78 psi (5.37 bar) from a nozzle with a coefficient of discharge of 0.97, what is the diameter of the nozzle?

4.8. A venturi meter with a 5-in. (125 mm) diameter restriction is located in a 6-in. (150 mm) pipe and gives a reading of 3.4 in. (8.6 cm) of mercury. How many gpm (liters per minute) are flowing?

4.9. What pitot tube pressure reading would represent a flow of 250 gpm (946 L/min) from a 1 1/8-in. (29 mm) nozzle with a coefficient of discharge of 0.98?

4.10. Solve for the coefficient of discharge of a device that delivers 45.5 gpm (172 L/min) from a 0.75-in. (20 mm) orifice when a pitot tube gives a reading of 15 psi (1.0 bar)?

4.11. If a venturi meter installed in a 1¼-in. (32 mm) pipe has a restriction of 0.65-in. (16.5 mm) diameter, what flow rate in gpm (liters/minute) is reflected by a 20-in. (50.8 cm) reading on the mercury manometer?

4.12. If a venturi meter installed in a 4-in. (100 mm) pipe has a 3-in. (80 mm) restriction and reads 2 in. (5 cm) of mercury, how many gpm (liters/minute) are flowing?

4.13. If a venturi meter installed in an 8-in. (200 mm) discharge line from a fire pump has a 6.75-in. (170 mm) restriction and gives a reading of 3.5 in. (8.9 cm) of mercury, how many gpm (liters/minute) are flowing?

4.14. What is the velocity of water in ft/s (m/s) through the restriction in a venturi meter yielding a manometer reading of 2.6 in. (6.6 cm) of mercury? The restriction is 1¾ in. (45 mm) in diameter, and the pipe in which the venturi meter is installed is 2½ in. (65 mm) in diameter.

4.15. What reading in inches (cm) of mercury would represent a fire pump flow of 2250 gpm (8500 L/min) on the venturi meter described in Exercise 4.13?

4.16. What reading in inches (cm) of mercury would represent a flow of 750 gpm (2840 L/min) on the venturi meter described in Exercise 4.12?

4.17. What is the velocity of water in ft/s (m/s) through the restriction in a venturi meter yielding a manometer reading of 4.0 in. (10.2 cm) of mercury if the restriction has a diameter of 3 in. (80 mm) and the pipe in which it is installed has a diameter of 4 in. (100 mm)?

4.18. A fire pump discharges water into an 8-in. (200 mm) pipe in which a venturi meter is installed. If the venturi restriction has a diameter of 6.5 in. (162 mm), how many gpm (L/min) would be reflected by a reading of 4 in. (10.2 cm) of mercury on the differential manometer?

4.19. What flow would be represented by an orifice plate meter with a coefficient of 0.72 if the manometer reads 3 in. (7.6 cm) of mercury, the orifice is 2.5 in. (65 mm) in diameter, and it is installed in a 4-in. (100 mm) pipe?

4.20. What flow would be represented by an orifice plate meter with a coefficient of 0.68 if the manometer reads 4.5 in. (11.4 cm) of mercury, the orifice is 4.0 in. (100 mm) in diameter, and it is installed in a 6-in. (150 mm) pipe?

Selected answers found in Appendix K.

References

1. Harry E. Hickey, *Hydraulics for Fire Protection* (Quincy: National Fire Protection Association, 1980), pp. 191–202.
2. Paul A. Tipler, *College Physics*. (New York: Worth Publishers Inc., 1987), p. 27.
3. Russell W. Henke, *Introduction to Fluid Mechanics* (Reading: Addison-Wesley Publishing Co. 1972), p. 83.
4. Henke 83.
5. Willis G. Labes, *Hydraulic Systems for the Fire Protection Engineer* (Chicago: Chicago Chapter of the Society of Fire Protection Engineers, 1959).

Chapter 5
Friction Loss in Piping Systems

Abstract This Chapter introduces two different methods to calculate friction loss through piping systems. The Darcy–Weisbach formula is introduced and is one of the most widely used methods of calculating friction loss in piping systems. The Hazen Williams formula introduced is the fire protection industry's standard for calculating friction loss.

The concept of energy loss due to friction was introduced in Chap. 3, Hydrokinetics. A term was included in the Bernoulli's energy balance equation to compensate for these losses, resulting in Eq. 3.6. Measuring the magnitude of this energy loss between two points in a piping system is a relatively simple task. Piezometer tubes, differential manometers, or Bourdon gauges may be used to measure the head or pressure differences between points in a system. Any pressure difference between two points is a result of friction alone if pipe size is constant, the elevation does not change, and no pumps are located between the points considered.

Determining Friction Loss

While measuring pressure losses due to friction is rather simple, predicting the magnitude of friction loss using formulas and calculations is not a trivial matter. The magnitude of friction loss is affected by such variables as specific weight and viscosity of the fluid, pipe diameter, pipe length, pipe roughness, and fluid velocity. Because fluids react differently to friction at different velocities, no single formula has been established that can accurately predict friction loss over all velocity ranges. In fact, all friction loss formulas are empirical in nature; the formulas were developed by trying to match calculations to test results. For this reason, friction loss formulas should be considered to yield approximate results at best.

P. Brock and V. Charter, *Fire Protection Hydraulics and Water Supply Analysis*, The Society of Fire Protection Engineers Series,
https://doi.org/10.1007/978-3-032-09877-1_5

Reynolds Number

Water travels through a piping system in a *laminar* state, a *turbulent* state, or some transition state in between (Fig. 5.1). At low velocities, in which the fluid flows in parallel layers, the flow is called laminar. The flow is turbulent at higher velocities, where there is no definite pattern to the direction of the water particles.

The state of flow of a fluid has traditionally been established mathematically through the calculation of a factor called the Reynolds number ($\mathbf{R_E}$). The Reynolds number is calculated as follows:

$$R_E = \frac{(V)(D)}{v} \tag{5.1}$$

where

$\mathbf{R_E}$ = the Reynolds number.

$\boldsymbol{V}$ = mean fluid velocity in fps.

$\boldsymbol{D}$ = pipe diameter in feet.

$\mathbf{v}$ = kinematic viscosity of the fluid in ft^2/sec.

The kinematic viscosity of a fluid is the ratio of its absolute viscosity (lb. sec/ft^2) to its mass density (lb. sec/ft^4). Tables 5.1 and 5.2 contain values of the kinematic viscosities for certain liquids. Note that the kinematic viscosity varies with temperature.

If the flow rate ($\boldsymbol{Q}$) is given in gallons per minute and the pipe diameter ($\boldsymbol{D}$) is in inches, the Reynolds number may also be calculated from:

$$R_E = \frac{(0.034)(Q)}{(v)(D)} \tag{5.1a}$$

In S.I. units the Reynolds number equation is the same, but the diameter is in meters, and the kinematic viscosity is in m^2/sec.

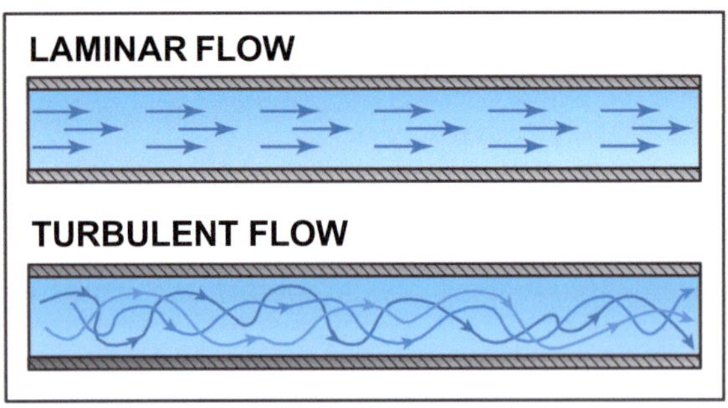

Fig. 5.1 Laminar and turbulent flow

Table 5.1 Kinematic viscosity of certain liquids in ft^2/sec

Temp °F	Water	Medium fuel oil	Gasoline	Propylene glycol solution (40% by weight)
40	1.664	6.55	0.81	10.23
50	1.41	5.55	0.765	7.53
60	1.217	4.75	0.73	5.38
70	1.059	4.12	0.69	4.41
80	0.93	3.65	0.66	3.77
90	0.826	3.19	0.63	3.01
100	0.739	2.78	0.60	2.69

Kinematic viscosity = tabular value × 10^{-5}

Table 5.2 Kinematic viscosity of certain liquids in m^2/sec

Temp °C	Water	Medium fuel oil	Gasoline	Propylene glycol solution (40% by weight)
4.4	0.155	0.609	0.075	0.95
10	0.131	0.516	0.071	0.7
15.6	0.113	0.441	0.068	0.5
21.1	0.098	0.383	0.064	0.41
26.7	0.086	0.339	0.061	0.35
32.2	0.077	0.296	0.059	0.28
37.8	0.069	0.258	0.056	0.25

Kinematic viscosity = tabular value × 10^{-5}

However, if the flow rate (***Q***) is expressed in liters/minute and the pipe diameter (***D***) is in millimeters, the Reynolds number is expressed as follows:

$$\mathrm{R_E} = \frac{(0.0209)(Q)}{(v)(D)} \tag{5.1b}$$

Different texts use different values of the Reynolds number to define laminar and turbulent flow. Jeppson considers a fluid to be in the state of laminar flow if its Reynolds number is 2100 or less [1]. For Reynolds numbers in excess of 2100, the fluid is in the turbulent state. It has been established that fluids exhibit different friction loss characteristics in different flow states. Thus, for the most accurate friction loss calculations, different formulas for the different flow states must be used.

The Darcy–Weisbach Method

The Darcy–Weisbach formula is widely used and yields good results. Historically, it has received little acceptance in fire protection circles with respect to evaluating or designing water systems.

However, it has become the required method for calculating losses in antifreeze sprinkler systems and for losses involving the flow of foam concentrates in foam fire extinguishing systems.

The Darcy–Weisbach equation for evaluating the loss of head in a piping system is as follows:

$$h_{\mathrm{f}} = \frac{(f)(V)^2(L)}{(2)(g)(D)} \tag{5.2}$$

where

$\boldsymbol{h_{\mathrm{f}}}$ = head loss due to friction in feet.

$\boldsymbol{f}$ = a dimensionless friction factor.

$\boldsymbol{L}$ = pipe length in feet.

$\boldsymbol{V}$ = fluid velocity in fps.

$\boldsymbol{D}$ = pipe diameter in feet.

g = acceleration due to gravity (32.2 ft/sec^2).

In S.I. Units an expression of the Darcy–Weisbach equation would be:

$$h_{\mathrm{f}} = \frac{(1000)(f)(V)^2(L)}{(2)(g)(D)} \tag{5.2a}$$

where:

$\boldsymbol{h_{\mathrm{f}}}$ = head loss due to friction in meters.

$\boldsymbol{f}$ = a dimensionless friction factor.

$\boldsymbol{L}$ = pipe length in meters.

$\boldsymbol{V}$ = fluid velocity in meters per second.

$\boldsymbol{D}$ = pipe diameter in mm.

$\boldsymbol{g}$ = acceleration due to gravity (9.8 m/sec^2).

For laminar flow ($R_E < 2100$), the friction factor is:

$$f = \frac{64}{R_E} \tag{5.3}$$

For turbulent flow, Nikuradse in 1933 established that f depends both on the magnitude of the Reynolds number and on the relative roughness of the interior surface of the pipe [2]. Several different formulas have been proposed for application in situations involving different Reynolds numbers and pipe surface conditions. However, the Colebrook equation is considered a practically reliable formula for calculating the friction factor for turbulent flow conditions and all pipe surface conditions [3]. The Colebrook equation is as follows:

$$\frac{1}{(f)^{0.5}} = (-2)\left[\log\left(\frac{e}{(3.7)(D)}\right) + \left(\frac{2.51}{(\mathrm{R_E})(f)^{0.5}}\right)\right] \tag{5.4}$$

where

f = the dimensionless friction factor.

$\boldsymbol{D}$ = actual inside pipe diameter in feet or millimeters.

$\boldsymbol{e}$ = diameter of pipe surface imperfections in feet or millimeters.

Solving for f using Eq. 5.4 is difficult and can best be accomplished through computer assistance. Jeppson provides a simple Fortran algorithm in his text [4]. However, charts are also available that will yield values of f when $\mathbf{R_E}$, $\boldsymbol{e}$, and $\boldsymbol{D}$ are known. The charts are known as Moody diagrams (Fig. 5.2).

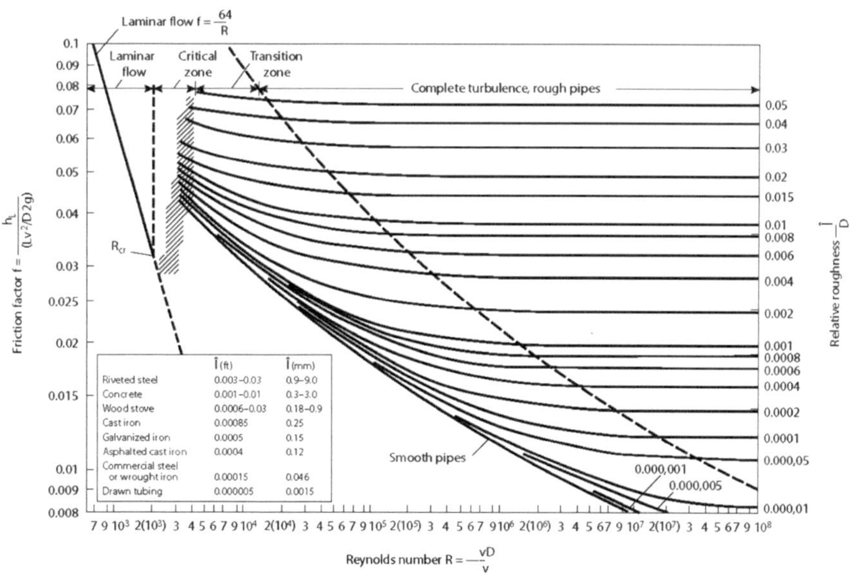

Fig. 5.2 A Moody diagram. *Courtesy of Springer Nature.*

Example 5.1

If 5 gpm (19 L/min) of water at 60 °F (15.6 °C) is flowing in an 8 in. (200 mm) pipe, what is the Reynolds number? Is the flow laminar or turbulent?

Solution:

$$
\begin{aligned}
\mathrm{R_E} &= \frac{(V)(D)}{\nu} \\
Q &= (A)(V) \\
V &= \frac{Q}{(\pi)(r)^2} \\
V &= \left(\frac{5gpm}{(\pi)(4\,\text{in.})^2}\right)\left(\frac{1\,\text{ft}^3/\text{sec}}{449\,\text{gpm}}\right)\left(\frac{144\,\text{in.}^2}{\text{ft}^2}\right) \\
V &= 0.032\,\text{ft/sec}
\end{aligned}
$$

Then:

$$
\begin{aligned}
\mathrm{R_E} &= \frac{(V)(D)}{\nu} \\
\mathrm{R_E} &= \frac{(0.032\,\text{ft/sec})(8/12\,\text{ft})}{\left(1.217\times 10^{-5}\right)\text{ft}^2/\text{sec}} \\
\mathrm{R_E} &= 1753
\end{aligned}
$$

Since 1753 is less than 2100, the flow is laminar.

In S.I. Units the solution is:

$$
\begin{aligned}
\mathrm{R_E} &= \frac{(0.0209)(Q)}{(\nu)(D)} \\
\mathrm{R_E} &= \frac{(0.0209)(19\,\text{L/min})}{\left(0.113\times 10^{-5}\right)(200\,\text{mm})} \\
\mathrm{R_E} &= 1757
\end{aligned}
$$

Flow is laminar.

Example 5.2

If water at 70 °F (21.1 °C) is flowing at a rate of 10 ft/sec (3.05 m/sec) in a 6 in. (150 mm) pipe, what is the Reynolds number? Is this laminar or turbulent flow?

Solution:

$$R_E = \frac{(V)(D)}{v}$$
$$R_E = \frac{(10\,\text{ft/sec})(6/12\,\text{ft})}{\left(1.059 \times 10^{-5}\right)\text{ft}^2/\text{sec}}$$
$$R_E = 472{,}144$$

Since R_E exceeds 2100, the flow is turbulent.

In S.I. Units the solution is:

$$150\,\text{mm} = 0.15\,\text{m}$$
$$R_E = \frac{VD}{v}$$
$$R_E = \frac{(3.05\,\text{m/sec})(0.15\,\text{m})}{0.098 \times 10^{-5}}$$
$$R_E = 466{,}837$$

Flow is turbulent.

Example 5.3

If water at 60 °F (15.6 °C) is flowing at a rate of 5 gpm (19 L/min) through 2500 ft (770 m) of 8 in. (200 mm) pipe, how much energy, expressed in psi (bar), will be lost due to friction?

Solution:

From *Example* 5.1 it is known that this is laminar flow and that $R_E = 1753$. Thus:

$$h_f = \frac{(f)(V)^2(L)}{(2)(g)(D)}$$

And where, for laminar flow:

$$f = \frac{64}{R_E}$$
$$f = \frac{64}{1753}$$
$$f = 0.0365$$

Then:

$$h_\mathrm{f} = \frac{(0.0365)(0.032\,\mathrm{ft/sec})^2(2500\,\mathrm{ft})}{(2)(32.2\,\mathrm{ft/sec^2})(8/12\,\mathrm{ft})}$$

$$h_\mathrm{f} = \frac{(0.0365)(0.001024)(2500)(\mathrm{ft})}{(2)(32.2)(0.6667)}$$

$$h_\mathrm{f} = 0.0022\,\mathrm{ft}$$

And, in psi:

$$P_\mathrm{t} = (0.0022\,\mathrm{ft})(0.433\,\mathrm{psi/ft})$$

$$P_\mathrm{t} = 0.00095$$

In S.I. Units the solution is:
Since the flow is laminar with $\mathrm{R_E} = 1757$

$$f = \frac{64}{\mathrm{R_E}}$$

$$f = \frac{64}{1757}$$

$$f = 0.0364$$

Then:

$$h_\mathrm{f} = \frac{(1000)(f)(V)^2(L)}{(2)(g)(D)}$$

$$h_\mathrm{f} = \frac{(1000)(0.036)(0.00975\,\mathrm{m/sec})^2(770\,\mathrm{m})}{(2)(9.8\,\mathrm{m/sec^2})(200\,\mathrm{mm})}$$

$$h_\mathrm{f} = 0.0007\,\mathrm{m}$$

And, in bar:

$$P_\mathrm{f} = (0.0007\,\mathrm{m})(0.098\,\mathrm{bar/m})$$

$$P_\mathrm{f} = 0.00007\,\mathrm{bar}$$

Example 5.4

What is the pressure lost, due to friction in psi (bar), when water at 70 °F (21.1 °C) flows through 300 ft (91 m) of 6 in. (150 mm) cast-iron pipe at a rate of 10 ft/sec (3.05 m/sec)?

Solution:

From *Example* 5.2 it is known that this is turbulent flow and $R_E = 472{,}143$. From the chart in the lower left-hand corner of the Moody diagram, the design value for ***e*** is found to be 0.00085 ft (see Figure 5.2). Next, ***e*/*D*** is calculated to be the following:

$$\frac{e}{D} = \frac{0.00085\,\text{ft}}{(6/12)\,\text{ft}}$$

$$\frac{e}{D} = 0.0017$$

The nearly horizontal curves on the Moody diagram represent ***e*/*D***. The curve most closely representing the calculated value of 0.0016 must be located. Interpolation between curves is appropriate. When the proper curve is located, it is followed to its point of intersection with a vertical line drawn above the value of the Reynolds number, in this case about 470,000. Note that this will be between the 4 and 5 to the right of 10^5. Once this point of intersection is established, the value of ***f*** is read from the left-hand or right-hand margin of the diagram horizontally from the intersection point. In this example, ***f*** is about 0.023. Then,

$$h_f = \frac{(f)(V)^2(L)}{(2)(g)(D)}$$

$$h_f = \frac{(0.023)\left(10\,\text{ft/sec}\right)^2(300\,\text{ft})}{(2)\left(32.2\,\text{ft/sec}^2\right)(6/12\,\text{ft})}$$

$$h_f = \frac{(0.023)(100)(300)(\text{ft})}{(2)(32.2)(0.5)}$$

$$h_f = 21.4\,\text{ft}$$

And, in psi:

$$P_f = (21.4\,\text{ft})(0.433\,\text{psi/ft})$$

$$P_f = 9.3\,\text{psi}$$

In S.I. Units the solution is:

$$\frac{e}{D} = \frac{(0.00085\,\text{ft})(304.8\,\text{mm/ft})}{150\,\text{mm}}$$

$$\frac{e}{D} = 0.0017$$

And from the Moody Diagram (Fig. 5.2) at the intersection of *e*/*D* and a Reynolds Number of about 470,000:

$$f = 0.023$$

Then:

$$h_f = \frac{(1000)(f)(V)^2(L)}{(2)(g)(D)}$$

$$h_f = \frac{(1000)(0.023)(3.05\,\text{m/sec})^2(91\,\text{m})}{(2)\left(9.8\,\text{m/sec}^2\right)(150\,\text{mm})}$$

$$h_f = 6.62\,\text{m}$$

And, in bar:

$$P_f = (6.62\,\text{m})(0.098\,\text{bar/m})$$

$$P_f = 0.65\,\text{bar}$$

Example 5.5

What is the pressure lost to friction in psi (bar) when a medium fuel oil at 40 °F (4.4 °C) flows through 125 ft (38 m) of 1 1/4-in. (32 mm) commercial steel pipe at a rate of 60 gpm (227 L/min)?

Solution:

The flow rate must first be converted into feet per second and then the Reynolds number calculated:

$$Q = AV$$

$$V = \frac{Q}{A}$$

$$V = \left(\frac{60\,\text{gpm}}{(\pi)(0.625\,\text{in.})^2}\right)\left(\frac{1\,\text{ft}^3/\text{sec}}{449\,\text{gpm}}\right)\left(\frac{144\,\text{in.}^2}{\text{ft}^2}\right)$$

$$V = 15.68\,\text{ft/sec}$$

Then:

$$R_E = \frac{(V)(D)}{v}$$

$$R_E = \frac{(15.68\,\text{ft/sec})(1.25\,\text{in.})(1\,\text{ft}/12\text{in.})}{6.55 \times 10^{-5}\,\text{ft}^2/\text{sec}}$$

$$R_E = 24{,}936$$

Since the flow is turbulent, the Moody diagram must be used to establish the friction factor:

$$\frac{e}{D} = \frac{0.00015}{(1.25\,\text{in.})(1\,\text{ft}/12\,\text{in.})} = 0.00144$$

Using R_E and *e*/*D*, the Moody diagram yields a value of 0.029 for f. The friction loss is then calculated:

$$h_f = \frac{(f)(V)^2(L)}{(2)(g)(D)}$$

$$h_f = \frac{(0.029)(15.68\,\text{ft/sec})^2(125\,\text{ft})}{(2)\left(32.2\,\text{ft/sec}^2\right)(1.25\,\text{in.})(1\,\text{ft}/12\,\text{in.})}$$

$$h_f = 133\,\text{ft}$$

And, in psi:

$$P_f = (133\,\text{ft})(0.433\,\text{psi/ft}) = 57.6\,\text{psi}$$

In S.I. Units the solution is:

$$R_E = \frac{(0.0209)(\text{Q})}{(v)(\text{D})}$$

$$R_E = \frac{(0.0209)(227\text{L/min})}{\left(0.609 \times 10^{-5}\right)(32\text{mm})}$$

$$R_E = 24,345$$

Since the flow is turbulent, Fig. 5.2 must be used to establish f:

$$\frac{e}{D} = \frac{(0.00015\,\text{ft})(304.8\,\text{mm/ft})}{32\,\text{mm}} = 0.00143$$

Using R_E and *e*/*D*, the Moody diagram yields a value of 0.03 for f. Then:

$$h_{\mathrm{f}} = \frac{(1000)(f)(V)^2 L}{(2)(g)(D)}$$

$$h_{\mathrm{f}} = \frac{(1000)(0.029)(4.78\,\mathrm{m/sec})^2(38\,\mathrm{m})}{(2)\left(9.8\,\mathrm{m/sec^2}\right)(32\,\mathrm{mm})}$$

$$h_{\mathrm{f}} = 40.14\,\mathrm{m}$$

And, in bar:

$$P_{\mathrm{f}} = (40.14\,\mathrm{m})(0.098\,\mathrm{bar/m}) = 3.93\,\mathrm{bar}$$

Although used in other disciplines, the Darcy–Weisbach equation is rarely used in fire protection applications. However, NFPA 11, *Standard for Low, Medium and High-Expansion Foam*, is one of the standards that requires the Darcy–Weisbach technique to be used in calculating the pressure loss to friction as a foam concentrate is delivered from the storage tank to the proportioning equipment. Then, after the concentrate is mixed with water, the formula which is most familiar to fire system designers is required to be used. This formula is called the Hazen–Williams formula.

The Hazen–Williams Formula

The Hazen–Williams formula is considered less accurate than the Darcy–Weisbach formula, but is simpler to use and is the fire protection industry standard. Therefore, to comply with most nationally recognized standards, the Hazen–Williams formula must be used.

In the first years of the twentieth century, Gardner Williams and Allen Hazen began to examine recorded friction loss measurements produced by several dozen experimenters [5]. Their studies resulted in the development of a completely empirical formula that we now know as the Hazen–Williams friction loss formula. This formula, along with friction loss tables computed from the formula, was published in 1903. The original form of the equation is as follows:

$$v = (1.318)(C)(R)^{0.63}(S)^{0.54} \tag{5.5}$$

where:

$\boldsymbol{v}$ = water velocity in ft/sec.

$\boldsymbol{R}$ = the hydraulic radius in inches.

$\boldsymbol{S}$ = the slope of the hydraulic gradient in feet of head loss per foot of pipe.

$\boldsymbol{C}$ = Hazen–Williams coefficient of roughness.

In S.I. Units the original formula is as follows:

$$v = (0.8492)(C)(R)^{0.63}(S)^{0.54}$$

where:

$\boldsymbol{v}$ = water velocity in m/sec.

$\boldsymbol{C}$ = Hazen–Williams coefficient of roughness.

$\boldsymbol{R}$ = the hydraulic radius in meters.

$\boldsymbol{S}$ = the slope of the hydraulic gradient in meters of head loss per meter of pipe.

Direct application of this formula will allow calculation of the water velocity giving a specified head loss. In the early 1900s (and even in the early 1970s) using such an exponential equation was tedious because it required the use of logarithms and slide rules. In fact, in 1905 Hazen and Williams began to market a slide rule designed to simplify the calculation. Countless other slide rules have been developed since that time.

Calculators available today make application of the formula routine. However, the formula in its original form is seldom of practical value to the fire protection specialist. It is more useful to have an equation which will yield the friction loss if the pipe size, pipe roughness, and flow in gallons per minute (L/min) are known. Proper manipulation of the original formula, including changing water velocity in feet per second (m/sec) to gallons per minute (L/min), and feet (meters) of head to psi (bar) yields the forms of the equation commonly used in fire protection circles [6].

That is

$$P_{\mathrm{f}} = \frac{(4.52)(Q)^{1.85}}{(C)^{1.85}(D)^{4.87}} \tag{5.6}$$

where

$\boldsymbol{P_{\mathrm{f}}}$ = the pressure lost to friction in psi/ft of pipe.

$\boldsymbol{Q}$ = the flow rate in gpm.

$\boldsymbol{C}$ = the Hazen–Williams coefficient of roughness.

$\boldsymbol{D}$ = the internal pipe diameter in inches.

In S.I. Units the form of the equation is as follows:

$$P_{\mathrm{f}} = \frac{(6.05)(Q)^{1.85}(10)^{5}}{(C)^{1.35}(D)^{4.87}} \tag{5.6a}$$

where:

$\boldsymbol{P_f}$ = the pressure lost to friction in bar per meter of pipe.

$\boldsymbol{Q}$ = the flow rate in liters per minute.

$\boldsymbol{C}$ = the Hazen–Williams coefficient of roughness.

$\boldsymbol{D}$ = the internal pipe diameter in millimeters.

Before the widespread availability of electronic calculators, this form of the equation was no easier to use than the original form. Therefore, tables, graphs, and slide rules were developed. Tables of friction loss factors generated from the equation are contained in Appendices B through F of this text. However, the direct use of the formula with calculators or computers is encouraged for greater versatility and to avoid interpolation.

From examining this formula, much can be learned about friction loss. For example, $\boldsymbol{Q}$ is raised to the 1.85 power in the equation. That is almost $\boldsymbol{Q}^2$. When the flow rate is doubled, if all other things remain constant, the friction loss will increase by about four times. If the flow triples, friction loss will be almost nine times greater. Note that the pipe diameter is in the denominator of the right-hand side of the equation. Any increase in pipe size will reduce friction loss if all other factors remain the same. If the diameter is doubled, the friction loss will be reduced by a factor near 1/32, since $\boldsymbol{D}$ is raised almost to the fifth power. If the diameter is tripled, the friction loss will be reduced to about 1/243 of its original value. Pipe diameter is the single variable with the greatest impact on friction loss.

The American piping industry designates pipe size in units of inches. In the international community, pipe sizes are most often designated in units of millimeters (mm). In both cases, the size indicated will be the nominal inside diameter of the pipe. Pipe classes and schedules are also used to identify types of piping. For example, 4-in. (100 mm) Schedule 40 steel pipe will have an actual internal diameter of 4.026 in. (102.3 mm); however, Schedule 10 pipe will have a diameter of 4.26 in. (108.2 mm), while they both will have the same external diameter. Use of the actual interior diameter with the Hazen–Williams formula will yield different results than if the nominal diameter is used. Appendix A contains nominal and actual internal diameters for various pipe sizes and materials.

The Hazen–Williams coefficient of roughness ($\boldsymbol{C}$) is a reflection of the interior condition of the pipe. Being in the denominator, large values of $\boldsymbol{C}$ indicate smooth pipe (reduced friction loss), and smaller values of $\boldsymbol{C}$ reflect rough or deteriorating pipe (greater friction loss). Old, unlined cast-iron pipe may have a $\boldsymbol{C}$ as low as 40, while new plastic pipe will have a $\boldsymbol{C}$ as high as 150.

See Tables 5.3, 5.4, and 5.5 for sample values of $\boldsymbol{C}$ for various ages and types of pipe.

The Hazen–Williams formula, which is empirical, yields approximate results. However, the fire protection industry believes that the formula gives results close enough to reality for practical applications. Indeed, NFPA 13, *Standard for the Installation of Sprinkler Systems*, requires use of the Hazen–Williams formula in calculating friction loss in hydraulic calculations, [6] and NFPA 24, *Standard for the*

Table 5.3 Guide for estimating the Hazen–Williams coefficient of roughness, *C*, for situations involving moderately corrosive water

Pipe material	C-factor
Cast iron, unlined:	
10 years old	90
15 years old	75
20 years old	65
30 years old	55
50 years old	50
Cast iron, unlined, new	120
Cast iron, cement lined	140
Cast iron, bitumastic enamel lined	140
Average steel, new	110
Asbestos-cement	140
Reinforced concrete	140
Plastic	150

Table 5.4 Design C-factors for above ground fire protection system piping

Pipe material	C-factor
Black steel or galvanized steel used on wet systems including deluge	120
Black steel or galvanized steel used on dry or preaction systems	100
Black steel or galvanized steel used on dry or preaction systems using nitrogen, vacuum pressure, or vapor corrosion inhibitor	120
All plastic, copper tube, brass, or stainless steel	150

Table 5.5 Design C-factors for underground piping

Pipe material	C-factor
Unlined cast or unlined ductile iron	100
All plastic piping	150
Cement-Lined cast or ductile iron	140
Asbestos-cement	140

Installation of Private Fire Service Mains and Their Appurtenances, also requires its use [7].

Example 5.6

How much pressure is lost to friction as 850 gpm (3200 L/min) travel through 400 ft (120 m) of 6-in. (150 mm) Schedule 40 steel pipe having a C of 100?

Solution:

$$P_{\mathrm{f}} = \frac{(4.52)(Q)^{1.85}}{(C)^{1.85}(D)^{4.87}}$$
$$P_{\mathrm{f}} = \frac{(4.52)(850)^{1.85}}{(100)^{1.85}(6.065)^{4.87}}$$
$$P_{\mathrm{f}} = 0.036\,\mathrm{psi/ft}$$

And, for the total loss over the 400 ft:

$$P_{\mathrm{T}} = (0.036\,\mathrm{psi/ft})(400\,\mathrm{ft})$$
$$P_{\mathrm{T}} = 14.4\,\mathrm{psi}$$

In S.I. Units the solution is:

$$P_{\mathrm{f}} = \frac{(6.05)(Q)^{1.85}\left(10^{5}\right)}{(C)^{1.85}(D)^{4.87}}$$
$$P_{\mathrm{f}} = \frac{(6.05)(3200)^{1.85}\left(10^{5}\right)}{(100)^{1.85}(154.1)^{4.87}}$$
$$P_{\mathrm{f}} = 0.008\,\mathrm{bar/m}$$

And, for the total loss over the 120 m:

$$P_{\mathrm{f}} = (0.008\,\mathrm{bar/m})(120\,\mathrm{m})$$
$$P_{\mathrm{f}} = 0.96\,\mathrm{bar}$$

Example 5.7

What size of Schedule 10 steel pipe should be specified to limit friction loss to a maximum of 0.02 psi/ft (0.0014 bar/m) in delivering 1000 gpm (3785 L/min). Assume $C = 140$.

Algebraically manipulating the Hazen–Williams formula permits the calculation of the required diameter as follows:

$$P_{\mathrm{f}} = \frac{(4.52)(Q)^{1.85}}{(C)^{1.85}(D)^{4.87}}$$
$$D = \left(\frac{(4.52)(Q)^{1.85}}{(C)^{1.85}(P_{\mathrm{f}})}\right)^{1/4.87}$$

$$D = \left(\frac{(4.52)(1000)^{1.85}}{(140)^{1.85}(0.02)} \right)^{1/4.87}$$

$$D = 6.42 \text{ in.}$$

The 6.42 in. exceeds the internal diameter of 6-in. Schedule 10 piping. Therefore, 8-in. pipe must be specified since there is no standard pipe size between 6 and 8 in.

In S.I. Units the equation is:

$$P_\mathrm{f} = \frac{(6.05)(Q)^{1.85}(10)^5}{(C)^{1.85}(D)^{4.87}}$$

$$D = \left(\frac{(6.05)(Q)^{1.85}(10)^5}{(P_\mathrm{f})(C)^{1.85}} \right)^{1/4.87}$$

$$D = \left(\frac{(6.05)(3785)^{1.85}(10)^5}{(0.0014)(140)^{1.85}} \right)^{1/4.87}$$

$$D = 207.6 \text{ mm}$$

Since Schedule 10, 200 mm pipe has an internal diameter of 209.5 mm, 200 mm pipe will do the job.

Application and Manipulation of the Hazen–Williams Formula

The Hazen–Williams formula can be manipulated to develop relationships and equations which are useful in analysis of water systems. The Hazen–Williams formula in the form shown in Eqs. 5.6 and 5.6a gives friction loss in psi/ft and bar/m respectively. By adding the term (L) to the numerator to represent pipe length, the loss in any length of pipe can be represented, and the equation becomes

$$P_\mathrm{T} = \frac{(4.52)(Q)^{1.85}(L)}{(C)^{1.85}(D)^{4.87}} \tag{5.7}$$

In S.I. Units the equation is:

$$P_\mathrm{T} = \frac{(6.05)(Q)^{1.85}(10)^5(L)}{(C)^{1.85}(D)^{4.87}} \tag{5.7a}$$

Assuming that two different situations are to be represented, the friction loss for each would be as follows:

$$P_{T1} = \frac{(4.52)(Q_1)^{1.85}(L_1)}{(C_1)^{1.85}(D_1)^{4.87}} \tag{5.8}$$

$$P_{T2} = \frac{(4.52)(Q_2)^{1.85}(L_2)}{(C_2)^{1.85}(D_2)^{4.87}} \tag{5.9}$$

If, for example,
$L_1 = L_2$, $C_1 = C_2$, and $C_1 = C_2$, and if Eq. 5.8 is divided by Eq. 5.9, the following relationship results:

$$\frac{P_{T1}}{P_{T2}} = \left(\frac{Q_1}{Q_2}\right)^{1.85}$$

or

$$\mathrm{P_{T1}} = (\mathrm{P_{T2}})\left(\frac{\mathrm{Q_1}}{\mathrm{Q_2}}\right)^{1.85} \tag{5.10}$$

The results are identical using the equation for international units, so Eq. 5.10 applies to any system of units.

The significance of Eq. 5.10 is the following: In a given piping situation, if the friction loss for one flow is known, the friction loss for any other flow can be quickly computed.

Example 5.8

As 2100 gpm (7950 L/min) flows through a section of pipe, 58.8 psi (4.05 bar) are lost to friction. What would the friction loss be for this same pipe at a flow of 1500 gpm (5675 L/min)?

$$P_{T1} = (P_{T2})\left(\frac{Q_1}{Q_2}\right)^{1.85}$$

$$P_{T1} = (58.8\,\text{psi})\left(\frac{1500\,\text{gpm}}{2100\,\text{gpm}}\right)^{1.85}$$

$$P_{T1} = 31.6\,\text{psi}$$

In S.I. Units the solution is:

$$P_{T1} = (P_{T2})\left(\frac{Q_1}{Q_2}\right)^{1.85}$$

$$P_{T1} = (4.05)\left(\frac{5675\,\text{L/min}}{7950\,\text{L/min}}\right)^{1.85}$$

$$P_{T1} = 2.17\,\text{bar}$$

The same kind of manipulation, holding three of the five variables constant, yields the following additional relationships, all of which are applicable to any system of units:

$$Q_1 = (Q_2)\left(\frac{P_{T1}}{P_{T2}}\right)^{0.54} \tag{5.11}$$

$$L_1 = (L_2)\left(\frac{D_1}{D_2}\right)^{4.87} \tag{5.12}$$

Equation 5.11 is valuable in evaluating the results of hydrant flow tests, which will be discussed in Chap. 8, Testing and Analysis of Water Supply Systems.

Example 5.9

If a hydrant flow test revealed that pressure dropped 38 psi (2.62 bar) from the beginning or static pressure while 960 gpm (3630 L/min) were flowing, how many gpm (L/min) would be flowing if the pressure dropped 52 psi (3.58 bar)?

$$Q_1 = (Q_2)\left(\frac{P_{T1}}{P_{T2}}\right)^{0.54}$$

$$Q_1 = (960\,\text{gpm})\left(\frac{52\,\text{psi}}{38\,\text{psi}}\right)^{0.54}$$

$$Q_1 = 1137\,\text{gpm}$$

In S.I. Units the solution is:

$$Q_1 = (Q_2)\left(\frac{P_{T1}}{P_{T2}}\right)^{0.54}$$

$$Q_1 = (3630\,\text{L/min})\left(\frac{3.58\,\text{bar}}{2.62\,\text{bar}}\right)^{0.54}$$

$$Q_1 = 4297\,\text{L/min}$$

Equation 5.12 is useful in developing equivalent pipe lengths. For example, it might be desired to know what length of 8-in. (200 mm) pipe would give the same friction loss as some existing length of 6-in. (150 mm) pipe. Or in different terms, what is the 8-in. (200 mm) equivalent length of some length of 6-in. (150 mm) pipe? *Example* 5.10 illustrates this concept.

Example 5.10

What length of 8-in. (200 mm) Schedule 40 pipe would give the same friction loss as 1100 ft (335 m) of 6-in. (150 mm) Schedule 40 pipe if both have a ***C*** of 100?

Solution:

Using Eq. 5.12

$$L_2 = (L_1)\left(\frac{D_2}{D_1}\right)^{4.87}$$

$$L_2 = (1100\,\text{ft})\left(\frac{7.981\,\text{in.}}{6.065\,\text{in.}}\right)^{4.87}$$

$$L_2 = 4188\,\text{ft}$$

Therefore, 4188 ft of 8-in. pipe with $C = 100$ will give the same friction loss as 1100 ft of 6-in. pipe with $C = 100$, and for any flow rate.

In S.I. Units the solution is:

$$L_2 = (L_1)\left(\frac{D_2}{D_1}\right)^{4.87}$$

$$L_2 = (1100\,\text{ft})\left(\frac{7.981\,\text{in.}}{6.065\,\text{in.}}\right)^{4.87}$$

$$L_2 = 4188\,\text{ft}$$

Therefore, 1273 m of 200 mm pipe with $C = 100$ will give the same friction loss as 335 m of 150 mm pipe with $C = 100$ for any flow rate.

The concept of equivalent pipe lengths is commonly used in hydraulic calculations. One of its most common applications is in establishing losses through pipe fittings such as elbows, tees, and valves. Tables and charts give the equivalent lengths for the various valves and fittings that might be found on a hydraulic system. The concept of using an equivalent pipe length to represent a pipe fitting when hydraulically designing an automatic sprinkler system is discussed in detail in Chap. 13.

Summary of Important Concepts

1. Accurate calculation of friction loss through piping systems is difficult. Empirical formulas are generally used which yield acceptable approximations.

2. The Darcy–Weisbach formula is one of the most widely used methods of calculating friction loss in piping systems.
3. The Hazen–Williams formula is the fire protection industry's standard for calculating friction loss.
4. The Hazen–Williams formula can be manipulated to yield relationships useful in analyzing hydraulic systems.

Exercises

5.1 Water is moving at a rate of 1 ft^3/sec (28 L/sec) through a 24-in. (600 mm) main. What is the Reynolds number? Is the flow laminar or turbulent? Liquid temperature is 70 °F (21.1 °C).

5.2 What is the Reynolds number of gasoline flowing through a 1-in. (22 mm) rubber hose at a rate of 6 gpm (22.7 L/min)? Is the flow laminar or turbulent. Liquid temperature is 70 °F (21.1 °C).

5.3 What is the Reynolds number of a medium fuel oil being transported through an 8-in. (200 mm) pipe at a rate of 650 gpm (2460 L/min)? Is the flow laminar or turbulent? Liquid temperature is 70 °F (21.1 °C).

5.4 What is the Reynolds number of a medium lubricating oil flowing through a 2-in. (50 mm) pipe at a rate of 22 gpm (83 L/min) if the oil temperature is 50 °F (10 °C)?

5.5 A typical automatic sprinkler system might be designed to deliver 22.5 gpm (85 L/min) through a 1-in. (25 mm) pipe. If the water temperature is 70 °F (21.1 °C), is the flow laminar or turbulent?

5.6 In a typical Class I standpipe system, the required flow for one standpipe is 500 gpm (1890 L/min). Such standpipes are usually 6-in. (150 mm) in diameter. For water temperature of 80 °F (26.7 °C), is the flow laminar or turbulent?

5.7 Use nominal pipe diameters and the Darcy–Weisbach equation to establish the friction loss in psi (bar) in the following situations:

   a. Water at 60 °F (15.6 °C) flowing at 400 gpm (1500 L/min) through 1000 ft (305 m) of 6-in. (150 mm) unlined cast-iron pipe.
   b. Water at 70 °F (21.1 °C) flowing at 1500 gpm (5680 L/min) through 600 ft (180 m) of 8-in. (200 mm) unlined cast-iron pipe.
   c. Water at 70 °F (21.1° C) flowing at 1725 gpm (6530 L/min) through one-half mile (805 m) of 12-in. (300 mm) asphaltic lined cast-iron pipe.
   d. Water at 70 °F (21.1 °C) flowing at 300 gpm (1135 L/min) through 150 ft (46 m) of 4-in. (100 mm) commercial steel pipe.
   e. Water at 70 °F (21.1 °C) flowing at 100 gpm (380 L/min) through 15 ft (4.6 m) of 2-in. (50 mm) copper tubing.

5.8 Use nominal pipe diameters and the Darcy–Weisbach equation to establish the friction loss in psi (bar) in the following situations:

a. Water at 60 °F (15.6 °C) flowing at 40 gpm(150 L/min) through 12 ft (3.7 m) of 1-in. (25 mm) commercial steel pipe.
b. Water at 70 °F (21.1 °C) flowing at a rate of 1125 gpm (4260 L/min) through150ft (46m) of 6-in. (150mm)commercial steel pipe.
c. Water at 60 °F (15.6 °C) flowing at 650 gpm (2460 L/min) through 380 ft (116 m) of 4-in. (100 mm) unlined cast-iron pipe.
d. Water at 70 °F (21.1 °C) flowing at a rate of 26 gpm (98 L/min) through 70 ft (21 m) of 3/4-in. (20 mm) copper tubing in a residential sprinkler system.
e. Water at 50 °F (10 °C) flowing through 10,000 ft (3050 m) of 12-in. (300 mm) asphalt lined castir on pipe at a rate of 2000 gpm (7570 L/min).

5.9 If gasoline at 80 °F (26.7 °C) is flowing through 800 ft (245 m) of 8-in. (200 mm) galvanized iron pipe, how much pressure in psi (bar) is lost to friction at 1,000 gpm (3785 L/min)? (Use nominal diameter)

5.10 If a medium fuel oil is heated to 100 °F (37.8 °C), how much pressure in psi (bar) will be lost to friction if the oil is transported at a rate of 300 gpm (1135 L/min) through 250 ft (76 m) of 3-in. (80 mm) commercial steel pipe? (Use nominal diameter)

5.11 What minimum diameter of unlined cast-iron pipe could be used to transport a medium lubricating oil at 100 °F (37.8 °C) 750 ft (230 m) at a rate of 500 gpm (1890 L/min) with a resultant friction loss not exceeding 20 psi (1.4 bar)? (Hint: you must use trial and error)

5.12 What minimum size of asphalt-lined cast-iron pipe could be specified to transport a medium fuel oil at 100 °F (37.8 °C) at a rate of 1000 gpm (3785 L/min) a distance of 1000 ft (305 m) and not lose more than 20 psi (1.4 bar) to friction? (Hint: you must use trial and error)

5.13 Use nominal pipe diameters and the Hazen–Williams formula with the following C-factors and recalculate the friction loss for the situations described in Exercise 5.7:

a. $C = 100$
b. $C = 100$
c. $C = 140$
d. $C = 120$
e. $C = 150$

5.14 Use nominal pipe diameters and the Hazen–Williams formula with the following C-factors and recalculate the friction loss for the situations described in Exercise 5.8:

a. $C = 120$
b. $C = 120$
c. $C = 100$
d. $C = 150$
e. $C = 140$

5.15 From Exercises 5.7 and 5.13, what general conclusions can be drawn relative to friction loss as calculated by the two methods promoted in this chapter?

5.16 From Exercises 5.8 and 5.14, what general conclusions can be drawn relative to friction loss as calculated by the two methods promoted in this chapter?

5.17 If 0.8 psi (0.06 bar) is lost to friction as 65 gpm (246 L/min) of water flows through a 1 1/4-in. (32 mm) elbow, what would the equivalent 1 1/4-in. (32 mm) length of pipe be for this elbow? (Assume $C = 120$ and nominal pipe diameter)

5.18 If a beginning pressure of 80 psi (5.51 bar) is available at Point A when 1000 gpm (3785 L/min) are flowing, what minimum size of enamel-lined cast-iron pipe must be used to deliver water at 1000 gpm (3785 L/min) to Point B which is 1000 ft (305 m) away from Point A and have a pressure of 55 psi (3.79 bar) remaining at Point B?

5.19 If 6 psi (0.41 bar) is lost to friction in a section of pipe 800 ft (240 m) long while 700 gpm (2650 L/min) of water are flowing, what is the C-factor of the pipe if the pipe diameter is 8.23 in. (209 mm)?

5.20 It is desired to have 100 psi (6.89 bar) available at the top of a standpipe 40 ft (12 m) high while 500 gpm (1895 L/min) of water are flowing. If the pressure available at ground level is 140 psi (9.65 bar) with 500 gpm (1895 liters/min) flowing, what minimum size of Schedule 40 steel pipe with a C of 120 will be needed?

Selected answers found in Appendix K.

References

1. Roland W. Jeppson, *Analysis of Flow in Pipe Networks* (Ann Arbor: Ann Arbor Science, 1977), p. 26.
2. Jeppson 25–26.
3. R. Giles, J.B. Evett and C Liu, *Schaum's Outline of Theory in Problems of Fluid Mechanics and Hydraulics* (New York: McGraw-Hill Book Co., 1994), p. 141.
4. Jeppson 39.
5. Hunter Rouse, *Hydraulics in the United States* (Ames: Institute of Hydraulics Research, University of Iowa, 1976), pp. 74-76.
6. NFPA 13, *Standard for the Installation of Sprinkler Systems* (Quincy: National Fire Protection Association, 2025).
7. NFPA 24, *Standard for the Installation of Private Fire Service Mains and Their Appurtenances* (Quincy: National Fire Protection Association, 2025).

Supplemental Readings

Hickey, Harry E. *Hydraulics For Fire Protection.* National Fire Protection Association: Quincy, MA, 1980, pp. 81–102.

Jeppson, Roland W. *Analysis of Flow in Pipe Networks*. Ann Arbor Science: Ann Arbor, MI, 1977, pp. 25–46.

National Fire Protection Association. *The Fire Protection Handbook*. National Fire Protection Association: Quincy, MA, 21st edition, Sec. 15, Ch. 3.

Chapter 6
Analysis of Looped and Gridded Piping Systems

Abstract This Chapter introduces the concepts of looped and gridded piping systems and how to analyze them. Simple loops and how to calculate the flow split between the two lines is described. The use of equivalent lengths formula is also introduced. The Hardy Cross Method is introduced on how to calculate flows and pressure losses within a gridded system.

This chapter is a continuation of Chap. 5, Friction Loss in Piping Systems, and deals with the more complicated issues of determining how much pressure is lost as water travels through looped or gridded piping systems. In either piping arrangement, determining pressure loss becomes an energy balancing exercise. Bernoulli's equation and the concept of conservation of energy are valuable in many applications, such as solving friction losses in loops and grids.

Determining Friction Loss in Loops and Grids

The traditional approach to solving for pressures lost to friction in loops and grids is to first determine how the water divides into each leg or pipe of the loop or grid. How the water gets from one point to another is not important, only that it gets there. However, we must know how the water divides to compute the friction loss. The techniques for determining how water divides through simple loops will be discussed first.

Simple Loops

A simple loop is defined by the following:

- Exactly one inflow point and one outflow point exists.

P. Brock and V. Charter, *Fire Protection Hydraulics and Water Supply Analysis*, The Society of Fire Protection Engineers Series,
https://doi.org/10.1007/978-3-032-09877-1_6

- Exactly two paths exist between the inflow and outflow points.

Figure 6.1 illustrates several simple loops. Note that there is never more than one inflow or more than one outflow point. Notice that there are only two possible ways to get from the inflow point to the outflow point.

Any deviation from the definition will give erroneous results when using the techniques presented in this chapter. For example, if there are three paths or two outflow points, the methods for calculating friction loss presented in this first section will not work.

Additional concepts must be understood to successfully deal with simple loops. These are as follows:

- The total flow coming into the loop must equal the flow leaving the loop. This is the Principle of Conservation of Matter. Water must not be created nor disappear somewhere within the loop.
- The total flow must be equal to the sum of the flow in Path 1 plus the flow in Path 2. Or,

$$Q_T = Q_1 + Q_2$$

- The friction loss must be equal in both paths. This is the concept of conservation of energy. For any beginning pressure at the inflow point, if different magnitudes of friction loss existed in each path, two different pressures would exist at the outflow point. It is simply not possible to have two different pressures existing at the same point at the same time. Thus, the friction loss in each leg or path must be the same.

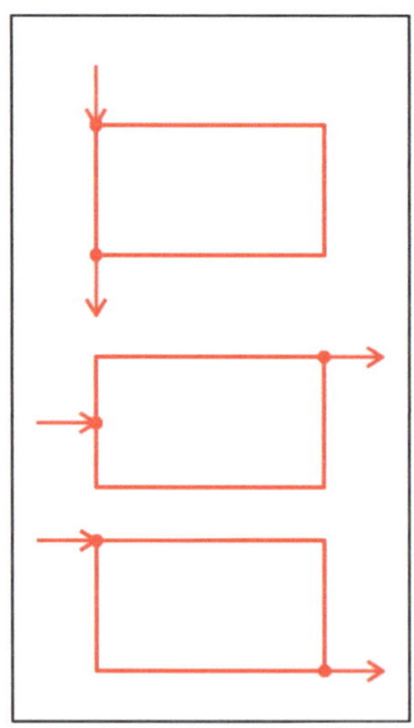

Fig. 6.1 Examples of simple loops

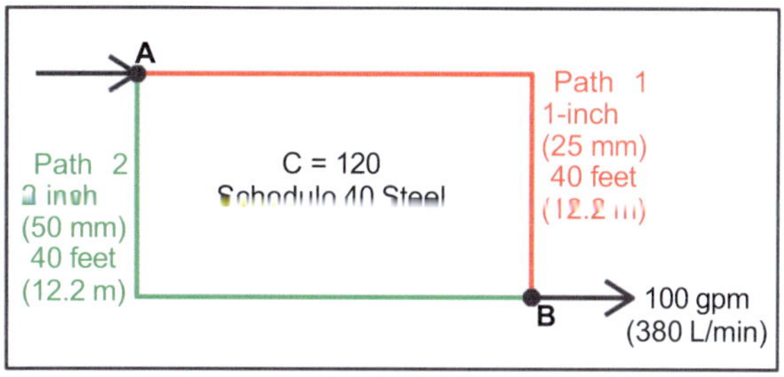

Fig. 6.2 A simple loop with two different pipe diameters

- The total friction loss across a simple loop is the friction loss in either path, not the sum of the friction losses in the two paths. Or,

$$P_{\mathrm{ft}} = P_{\mathrm{f1}} = P_{\mathrm{f2}}$$

With the definition of a simple loop and an understanding of these four principles, techniques can be promoted to solve for the friction loss across a simple loop. For example, consider the situation illustrated in Fig. 6.2. In this case, it is desired to know how much pressure is lost to friction as 100 gpm (380 L/min) flows from Point A to Point B. Several methods are available to solve the problem. Two methods will be introduced in this text.

Method I

The first is a trial and error method that is very simple. A guess is made as to how the water flow might divide into the two paths. New guesses are made until the two flows yield equal friction loss.

Using the Hazen–Williams formula, what would be the friction loss calculations if the first guess is 35 gpm (132 L/min) through Path 1 and 65 gpm (248 L/ min) through Path 2?

Path 1:

$$P_{\mathrm{f1}} = \frac{(4.52)(35\,\mathrm{gpm})^{1.85}(40\,\mathrm{ft})}{(120)^{1.85}(1.049)^{4.87}} = 14.8\,\mathrm{psi}$$

Path 1, in bar:

$$P_{\mathrm{f1}} = \frac{(6.05)(132\,\mathrm{L/min})^{1.85}(12.2\,\mathrm{m})(10)^{5}}{(120)^{1.85}(26.6\,\mathrm{mm})^{4.87}} = 1.01\,\mathrm{bar}$$

Path 2:

$$P_{f2} = \frac{(4.52)(65\,\text{gpm})^{1.85}(40\,\text{ft})}{(120)^{1.85}(2.067)^{4.87}} = 1.7\,\text{psi}$$

Path 2, in bar:

$$P_{f2} = \frac{(6.05)(248\,\text{L/min})^{1.85}(12.2\,\text{m})(10)^5}{(120)^{1.85}(52.5\,\text{mm})^{4.87}} = 0.12\,\text{bar}$$

It is obvious that 1.7 psi (0.12 bar) is not equal to 14.8 psi (1.01 bar), so a solution has not been reached. Another guess must be made, and another until the friction loss in each leg is equal. From a practical standpoint, it is usually considered acceptable if balance is achieved to within 0.5 psi (0.03 bar) [1]. There is little real value in pursuing answers more accurate than that. The final solution to the example problem reveals that approximately 14 gpm (53 L/min) will flow through Path 1 and 86 gpm (327 L/min) through Path 2 yielding friction loss of 2.7 psi (0.19 bar) and 2.9 psi (0.2 bar), respectively. Although 2.7 (0.19) does not equal 2.9 (0.2), the two are considered close enough, and the higher of the two would be used. Therefore, the total pressure lost to friction as 100 gpm (380 L/min) flows from Point A to Point B would be 2.9 psi (0.2 bar) (not 2.9 + 2.7).

Using Method I is slow and tedious at best. It was presented here to reinforce the principle that the flow will split in such proportions that the resulting pressure loss will be equal in each path. Method I is rarely used since faster and more direct methods exist. Let's turn our attention to one of these.

Method II

The solution to the previous problem can be reached directly through application of the following equation, which is valid for any system of units:

$$Q_1 = \frac{Q_T}{1 + \left[\frac{L_1}{L_2}\right]^{0.54}} \tag{6.1}$$

where

$\boldsymbol{Q_1}$ = The flow through Path 1.

$\boldsymbol{Q_T}$ = the total flow across the loop.

$\boldsymbol{L_1}$ = the length of Path 1.

$\boldsymbol{L_2}$ = the length of Path 2.

An important problem exists with this formula. Note that no consideration is given to pipe diameters or pipe roughness. Therefore, Eq. 6.1 is only good for loops where all pipe is of the same diameter and has the same C-factor. If the two paths or legs have different diameters or C-factors, one of the pipes must be converted mathematically into some equivalent length of pipe with an identical diameter and C-factor. This may be accomplished by using the following equation, which is simply an expanded version of Eq. 5.12.

$$L_e = L_1 \left(\frac{D_e}{D_1}\right)^{4.87} \left(\frac{C_e}{C_1}\right)^{1.85} \tag{6.2}$$

where

$\boldsymbol{L_e}$ = the new equivalent length.

$\boldsymbol{L_1}$ = the original length.

$\boldsymbol{D_1}$ = the original diameter.

D = the diameter of the new equivalent length.

$\boldsymbol{C_1}$ = the original C-factor.

$\boldsymbol{C_e}$ = the C-factor of the new equivalent length.

In the example illustrated in Fig. 6.2, one of the pipes must be changed into some equivalent length having the same diameter and C-factor as the other. For example, we can change the 2-in. pipe into some equivalent length of 1-in. pipe. Since both have the same C-factor, the calculation would be as follows:

$$L_e = (40\text{ ft}) \left(\frac{1.049\text{ in}}{2.067\text{ in}}\right)^{4.87} \left(\frac{120}{120}\right)^{1.85} = 1.47\text{ ft}$$

This calculation tells us that 1.47 ft of 1-in. pipe gives the same friction loss as 40 ft of 2-in. pipe. With this information, Eq. 6.1 can now be used to calculate the flow split across the loop. If the original 1-in. leg is considered to be Path 1, then the flow through Path 1 is

$$Q_1 = \frac{100\text{ gpm}}{1 + \left(\frac{40\text{ ft}}{1.47\text{ ft}}\right)^{0.54}} = 14.4\text{ gpm}$$

The flow through Path 2 is most easily calculated from the following:

$$Q_2 = Q_T - Q_1 = 100 - 14.4 = 85.6\text{ gpm}$$

The pressure lost to friction can then be calculated using either flow, since the friction loss in either path must be the same. Also, by calculating the friction loss in both legs and comparing the results, a check can be made for calculation errors. In this case, the pressure lost to friction through Path 2 is:

$$P_{f2} = \frac{(4.52)(85.6\text{ gpm})^{1.85}(40\text{ ft})}{(120)^{1.85}(2.067\text{ in})^{4.87}} = 2.8\text{ psi}$$

Note that the original diameter and length were used in the calculation. The same answer would have resulted if the equivalent length of 1.47 ft would have been used with the 1.049 in. diameter. However, a common mistake is mixing equivalent lengths with original diameters or using the flow for Path 1 with the diameter and length of Path 2.

The solution is accomplished in the same manner using S.I. Units as follows:

Changing the 50 mm pipe into some equivalent length of 25 mm pipe gives:

$$L_e = (12.2\,\text{m})\left(\frac{26.6\,\text{mm}}{52.5\,\text{mm}}\right)^{4.87}\left(\frac{120}{120}\right)^{1.85} = 0.44\,\text{m}$$

Then:

$$Q_1 = \frac{(380\,\text{L/min})}{1 + \left(\frac{12.2\,\text{m}}{0.44\,\text{m}}\right)^{0.54}} = 54\,\text{L/min}$$

$$Q_2 = 380\,\text{L/ min} - 54\,\text{L/ min} = 326\,\text{L/ min}$$

And:

$$P_{f2} = \frac{(6.05)(326\,\text{L/min})^{1.85}(12.2\,\text{m})(10)^5}{(120)^{1.85}(52.5\,\text{mm})^{4.87}} = 0.2\,\text{bar}$$

If the calculation result from Method II above is compared to the result from Method I, you will see that the results are essentially the same. This should provide some confidence that both methods yield good results.

Example 6.1

Compute the pressure lost to friction between Points A and B (Fig. 6.3). Since all pipe is 4-in. (100 mm) and $C = 100$,

$$Q_1 = \frac{780}{1 + \left(\frac{100}{300}\right)^{0.54}} = 502\,\text{gpm}$$

$$Q_2 = 780 - 502 = 278\,\text{gpm}$$

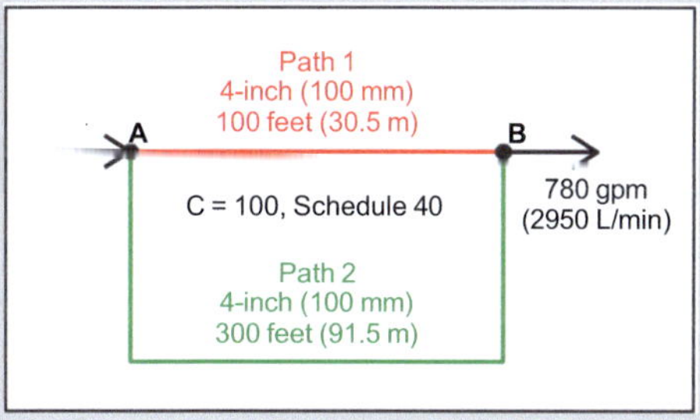

Fig. 6.3 Illustration for Example 6.1

Friction loss through Path 1 is then:

$$P_{\mathrm{f}} = \frac{(4.52)(502)^{1.85}(100)}{(100)^{1.85}(4.026)^{4.87}} = 10.1\,\mathrm{psi}$$

To check the results for correctness, friction loss through the other path may also be computed as follows:

$$P_f = \frac{(4.52)(278)^{1.85}(300)}{(100)^{1.85}(4.026)^{4.87}} = 10.2\,\mathrm{psi}$$

Since the difference in friction loss in the two paths is less than 0.5 psi, the answers are considered acceptable.

In S.I. Units the solution is:

Since all pipe is 100 mm and $C = 100$:

$$Q_1 = \frac{2950}{1 + \left(\frac{30.5}{91.5}\right)^{0.54}} = 1900\,\mathrm{L/min}$$

$$Q_2 = 2950 - 1900 = 1050\,\mathrm{L/min}$$

And:

$$P_{\mathrm{f1}} = \frac{(6.05)(1900)^{1.85}(30.5)(10)^5}{(100)^{1.85}(102.3)^{4.87}} = 0.7\,\mathrm{bar}$$

$$P_{\mathrm{f2}} = \frac{(6.05)(1050)^{1.85}(91.5)(10)^5}{(100)^{1.85}(102.3)^{4.87}} = 0.7\,\mathrm{bar}$$

Equal pressure loss in both paths indicates a correct solution.

Example 6.2

Determine the friction loss between Points A and B (Fig. 6.4). Since the two pipes have different diameters and C-factors, one must be changed into an equivalent length having the same diameter and C-factor as the other. Changing the 3-in. into 2-in. pipe gives

$$L_e = (80)\left(\frac{2.067}{3.068}\right)^{4.87}\left(\frac{120}{100}\right)^{1.85} = 16.4\,\text{ft}$$

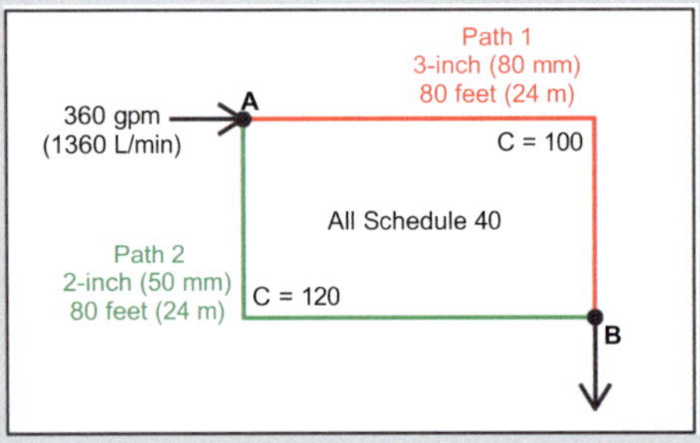

Fig. 6.4 Illustration for Example 6.2

Therefore, 16.4 ft of 2-in. pipe with a C of 120 gives the same friction loss as 80 ft of 3-in. pipe with a C of 100. Now, solving for the flow in Path 1 gives

$$Q_1 = \frac{360}{1 + \left(\frac{16.4}{80}\right)^{0.54}} = 253\,\text{gpm}$$

And:

$$Q_2 = 360 - 253 = 107\,\text{gpm}$$

Finally

$$P_f = \frac{(4.52)(253)^{1.85}(16.4)}{(120)^{1.85}(2.067)^{4.87}} = 8.6\,\text{psi}$$

In S.I. Units the solution is as follows:

First, changing the 80 mm pipe into an equivalent length of 50 mm pipe gives:

$$L_e = (24)\left(\frac{52.5}{77.9}\right)^{4.87}\left(\frac{120}{100}\right)^{1.85} = 4.92\,\text{m}$$

Therefore, 4.92 m of 50 mm pipe with a C of 120 gives the same friction loss as 24 m of 80 mm pipe with a C of 100. Then, solving for the flow in Path 1 gives:

$$Q_1 = \frac{1360}{1 + \left(\frac{4.92}{24}\right)^{0.54}} = 954\,\text{L/min}$$

So:

$$Q_2 = 1360 - 054 = 406\,\text{L/min}$$

And finally:

$$P_f = \frac{(6.05)(954)^{1.85}(4.92)(10)^5}{(120)^{1.85}(52.5)^{4.87}} = 0.58\,\text{bar}$$

Complex Loops

Attention will now be turned to the more complicated issue of complex loops. Piping arrangements involving multiple loops are often referred to as *grids*. Complex loops typically have a more practical application since city water supply systems are all of gridded design. Also, many industrial underground fire systems or building sprinkler systems are also of gridded design.

Piping Systems

A *complex loop* (which will also be referred to as a *grid* in this text) is a piping system characterized by one or more of the following:

- More than one inflow point exists.
- More than one outflow point exists.
- More than two paths exist between inflow and outflow points.

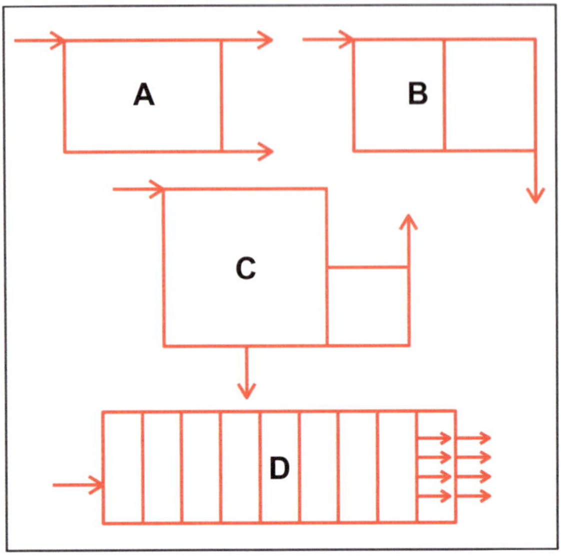

Fig. 6.5 Examples of complex loops

Notice that the existence of any one of the above is sufficient to characterize the system as complex. All three of the situations do not have to exist. Therefore, all of the systems illustrated in Fig. 6.5 qualify as complex.

Hardy Cross Method

The Hardy Cross Method is a technique used for solution of these gridded systems. Professor Hardy Cross developed this method in the era between the World Wars when computers were nonexistent. A manual and practical technique was needed to solve the complicated problems involving gridded underground water supply systems.

Professor Hardy Cross developed an iterative energy balance approach. Essentially this means that you keep on trying, getting closer and closer, until finally you reach an acceptable stopping point. This method has been the most commonly used approach until the development of more efficient mathematical techniques assisted by computers. Today some modern computer programs still employ the Hardy Cross approach.

The following discussion is not intended to provide an exhaustive coverage of the method. Many fluid mechanics texts provide excellent coverage of the methodology. However, the fundamentals are presented here because some automatic sprinkler design computer programs use the method, and a knowledge of the technique is currently required for completion of portions of the professional engineering examination for the fire protection discipline and for certification as a NICET Level IV sprinkler system designer.

The following summarizes the calculation steps that appeared in the paper titled, "A Rapid Method of Analyzing Flow in Water Distribution Systems Based on the Theory Developed by Professor Hardy Cross." This paper was written by T. Francis O'Conner who was a sanitary engineer for the U.S. War Department. This method turns out to be anything but rapid. It is a tedious and time-consuming process full of opportunities to make mistakes. Any mistake, of course, makes the process more tedious and time-consuming.

The use of the Hardy Cross Method requires the following 10 steps to complete the process. Two example problems will be solved in Examples 6.3 and 6.4.

Step 1: Always start by making a simple line diagram showing all inflow points and outflow points, all pipe diameters, pipe lengths, and pipe C-factors.

Step 2: Number each line or pipe with an Arabic numeral. Each pipe of different flow, diameter, or C-factor must be numbered. A pipe common to two loops will have only a single number.

Step 3: For systems with multiple loops, number each loop with a Roman numeral or with an alphabetic designation. For any complex system, except those defined by the simple existence of multiple inflows or outflows (for example, type A from Fig. 6.5), one or more imaginary loops must be created. The number of imaginary loops should equal the number of inflow points. The purpose of the imaginary loops is to make sure that every pipe appears in two loops. Imaginary loops will be on the outside of the system and will connect the inflow points.

Step 4: Next, estimates of flows in every pipe must be made. One of the most critical requirements is that conservation of matter must exist for the system and at each junction or intersection point. That is, the flow coming into any point must equal the flow going out. Further, the flow coming into the system must equal the flow going out. On the drawing, show these first estimates and indicate their direction with an arrow.

Step 5: All data must next be copied onto the Hardy Cross work sheet. Where multiple loops exist, the data for the individual loops are grouped together with at least two spaces maintained between the loops for separation. It is not necessary to input data for each pipe twice, but every pipe must be shown to exist in two loops.

Step 6: An essential aspect of the process is that flows must be designated as positive or negative. Flows in a clockwise direction around a loop are considered positive and in a counterclockwise direction negative.

Remember that in multiple loop systems, each pipe appears in two loops. In one loop, the flow will be positive and in the adjacent loop, it will be negative. The flow in an individual pipe will never have the same sign in different loops. This is why you should use arrows on your drawing to indicate flow direction.

Step 7: With all data copied onto the work sheet, the friction loss for each pipe is calculated from the Hazen–Williams formula. Nominal pipe diameters may be used. A value for 1.85 $\mathbf{P_f/Q}$ must also be calculated. The value of $\mathbf{P_f}$ is the friction loss for the pipe, and **Q** is the flow through the pipe.

The Hazen-Williams formula, of course, yields friction loss in psi or bar. It is important to give a positive or negative sign to the friction loss to correspond with the sign of the flow. However, the value of 1.85 $\mathbf{P_f/Q}$ is always positive.

Step 8: The values for the friction in the pipes are added algebraically to get a separate total for each loop. The values for 1.85 $\mathbf{P_f/Q}$ are also summed for each individual loop.

Step 9: This step applies only for a system without multiple loops (for example, type A from Fig. 6.5). The sum of the friction loss in the loop is divided by the sum of the values for 1.85 $\mathbf{P_f/Q}$.

The result of this division is the correction factor. The correction factor should be rounded off to the nearest 10 gpm or 10 L/min, the *sign must be changed*, and then this correction is added algebraically to the previous flow estimates for each pipe. Each pipe now has a new corrected flow and Steps 7, 8, and 9 are repeated. This repetitive process is continued until the absolute value of the sum of the friction loss is less than or equal to 1.0 psi or 0.1 bar. When this is accomplished, a good estimate of the flow split and friction loss has been established. Because the process is so tedious, for manual calculations balance to 1.0 psi and 0.1 bar rather than 0.5 psi and 0.03 bar is considered an acceptable stopping point.

Step 10: The process for the multiple loop systems is more complicated because a double correction is actually required for each pipe. A correction for each loop is created in an identical manner described in Step 9. However, for each pipe, a second correction is taken by finding the second loop in which the pipe appears and bringing the correction for that loop, *without changing the sign*, up into the first loop in which the pipe appears. Thus, each pipe will have a double correction. Note that the total correction for different pipes in the same loop need not be the same. After corrections are made, it is important to check again to see that the flows coming into each point equal the flows going out. This process is repeated and continued for *all* loops until again the sum of the absolute value of the friction loss in *all* loops is less than or equal to 1.0 psi or 0.1 bar. It is important to note that corrections must continue to be made in all loops as long as corrections are needed in *any* loop.

Examples are usually necessary to understand most mathematical procedures. Three examples will be presented. First, let's look at the less complicated case of a complex system of the simplest type involving only a single loop, but more than one outflow point.

Example 6.3

Determine the pressure lost to friction between Points A and B as 500 gpm (1890 L/min) is delivered to both Points B and C (Fig. 6.6).

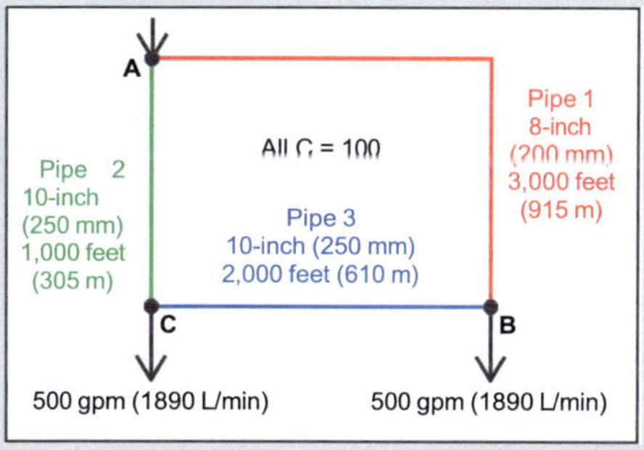

Fig. 6.6 Illustration for Example 6.3

After drawing the line diagram and labeling the pipes (Steps 1–3), the flow estimate for each pipe must be established (Step 4). As long as conservation of matter exists, the values of these first estimates are not important, although intelligent guesses will allow arriving at a solution in fewer iterations.

In this example, 400 gpm (1510 L/min) is estimated to flow from the inflow point into Pipe 1 toward Point B and being clockwise it is given a positive sign. This leaves 600 gpm (2270 L/min) to flow from A to C (1000–400 gpm) or (3780–1510 L/min) in Pipe 2 and will bear a negative sign. If 500 gpm (1890 L/min) goes out at Point C, this leaves 100 gpm (380 L/min) to flow to Point B through Pipe 3. This 100 gpm (380 L/min) will also be negative and will combine with the 400 gpm (1510 L/min) in Pipe 1 to give the total outflow of 500 gpm (1890 L/min) at Point B. Notice the conservation of matter at all junction points.

All values are then copied onto the worksheet (Fig. 6.7) and friction loss and 1.85 $\mathbf{P_f/Q}$ are established for each pipe, and the corrections made. The solution is achieved in this case within one correction. The friction loss between A and B may be computed over either path and is found to be approximately 3.4 psi (0.22 bar)

Loop No.	Line No.	Size	C	Length	Q	P_f	$\frac{1.85P_f}{Q}$	Correction	Q	P_f	$\frac{1.85P_f}{Q}$	Correction	Q	P_f	$\frac{1.85P_f}{Q}$	Correction
	1	8	100	3000	+400	+7.0	.032	-130	+270	+3.4						
	2	10	100	1000	-600	-1.7	.005	-130	-730	-2.4						
	3	10	100	2000	-100	-0.1	.002	-130	-230	-0.6						
						+5.2 /	.039 =	+133		+0.4	←	Stop				
	In	S.I.	Units:													
	1	200	100	915	+1510	+0.52	.0006	-560	+950	+0.22						
	2	250	100	305	+2270	-0.12	.0001	-560	-2830	-0.18						
	3	250	100	610	-380	-0.01	.0000	-560	-940	-0.05						
						+0.39 /	.0007 =	+557		-0.01	←	Stop				

Fig. 6.7 Worksheet for Example 6.3

Example 6.4

Determine the pressure lost to friction between Points A and B in the piping system shown in (Fig. 6.8).

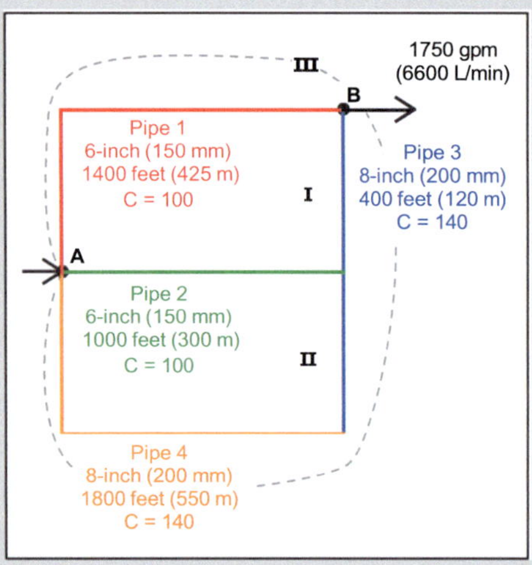

Fig. 6.8 Illustration for Example 6.4

The first step is drawing a diagram and numbering all pipes and all loops. Since only a single point of supply exists, only one imaginary loop is designated. That is Loop III, which starts and ends at Point A and includes all pipes around the outside of the system (Pipes 1, 3, and 4). Notice that with this imaginary loop, each pipe now appears in two loops. The initial flow estimates are then made and given the appropriate sign notation, keeping in mind conservation of matter at each junction point. The flow leaving each junction point must equal the amount entering. The sign notation can be difficult to visualize in imaginary loops. However, if the sign is applied to the flow in the real loops, it will be just the opposite for the same pipe in an imaginary loop. All data is entered on the worksheet. Notice that redundant information need not be included. The friction loss, $\mathbf{P_f}$, and 1.85 $\mathbf{P_f/Q}$ are calculated for each leg in every loop before any corrections are made. The worksheets for this problem are shown in Fig. 6.9a and b. Notice that a double correction is made for **each** pipe. It is difficult to remember (but you must) that when bringing a correction up into the loop where it originates, the sign is changed; however when bringing the second correction up from a lower loop, the sign is not changed. It is important to understand that a solution has not been reached until the absolute value of the sum of the friction loss in **all** loops is 1.0 psi (0.1 bar) or less. You cannot stop calculating any loop until all loops are at the desired stopping point.

Loop No.	Line No.	Size	C	Length	Q	P_f	$\frac{1.85P_f}{Q}$	Correction	Q	P_f	$\frac{1.85P_f}{Q}$	Correction	Q	P_f	$\frac{1.85P_f}{Q}$	Correction
	1	6	100	1400	+460	+17.3	0.07	-30_{-30}	+400	+13.3	0.062	$+20_{+20}$	+440	+15.9	0.067	-10_{-10}
I	2	6	100	1000	-390	-9.1	0.043	-30_{-20}	-440	-11.4	0.048	$+20_{+10}$	-410	-10.0	0.045	-10_{0}
	3	8	140	400	-1290	-4.4	0.006	-30_{-30}	-1350	-4.8	0.007	$+20_{+20}$	-1310	-4.5	0.006	-10_{-10}
						+3.8 /	.119=	+32		-2.9 /	.117=	-24.7		+1.4 /	.118=	+12
II	2					+9.1	0.043			+11.4	0.048			+10.0	0.045	
	4	8	140	1800	-900	-10.2	0.021	$+20_{-30}$	-910	-10.4	0.021	-10_{+20}	-900	-10.2	0.021	0_{-10}
						-1.1 /	.064 =	-17		+1.0 /	.069 =	+14.49		-0.2 /	.066 =	-3
	1					-17.3	.07			-13.3	0.062			-15.9	0.067	
III	3					+4.4	.006			+4.8	0.007			+4.5	0.006	
	4					+10.2	.021			+10.4	0.021			+10.2	0.021	
						-2.7 /	.097 =	-28		+1.9/	.09 =	+21		-1.2 /	.094 =	-13
(continuing)																
	1				+420	+14.6										
	2				-420	-10.4										
	3				-1330	-4.6										
						-.4	Stop									
	2					+10.4										
	4				-910	-10.4										
						0	Stop									
	1					-14.6										
	3					+4.6										
	4					+10.4										
						+.4	Stop									

Fig. 6.9 **a** Worksheet for Example 6.4. **b** Worksheet for Example 6.4 (in S.I. units)

Loop No.	Line No.	Size	C	Length	Q	P_f	$\frac{1.85P}{Q}$	Correction	Q	P_f	$\frac{1.85P}{Q}$	Correction	Q	P_f	$\frac{1.85P}{Q}$	Correction
	1	150	100	425	+1750	+1.29	.0014	-120 -120	+1510	+.99	.0012	+70 +90	+1670	+1.19	.0013	-60 -60
I	2	150	100	300	-1500	-.69	.0009	-120 0	-1620	-.79	.0009	+70 0	-1550	-.73	.0009	-60 -20
	3	200	140	120	-4850	-.32	.0001	-120 -120	-5090	-.35	.0001	+70 +90	-4930	-.33	.0001	-60 -60
						+.28 /	.0024 =	+117		-.15 /	.0022 =	-68		+.13 /	.0023 =	+57
II	2					+.69	.0009			+.79	.0009			+.73	.0009	
	4	200	140	550	-3350	-.74	.0004	-0 -120	-3470	-.79	.0004	0 +90	-3380	-.75	.0004	+20 -60
						-.05 /	.013 =	-4		0 /	.0013 =	0		-.02 /	.0013 =	-15.4
	1					-1.29	.0014			-.99	.0012			-1.19	.0013	
III	3					+.32	.0001			+.35	.0001			+.33	.0001	
	4					+.74	.0004			+.79	.0004			+.75	.0004	
						-.23 /	.0019 =	-121		+.15 /	.0017 =	+88		-.11 /	.0018 =	-61
(cont.)																
	1				+1550	+1.03	.0012	+50 +50	+1650	+1.16						
	2				-1630	-.80	.0009	+50 +20	-1560	-.74						
	3				-5050	-.34	.0001	+50 +50	-4950	-.33						
						-.11 /	.0022 =	-50		+.09	Stop					
	2					+.80	.0009			+.74						
	4				-3420	-.77	.0004	-20 +50	-3390	-.75						
						+.03 /	.0013 =	+23		-.01	Stop					
	1					-1.03	.0012			-1.16						
	3					+.34	.0001			+.33						
	4					+.77	.0004			+.75						
						+.08 /	.0017 =	+47		-.08	Stop					

Fig. 6.9 (continued)

After the solution has been reached, the friction loss between A and B may be computed by adding the losses across any combination of pipes joining the two points (e.g., Pipe 1, Pipes 2 and 3, or Pipes 4 and 3). Mistakes can be made in this step if simple addition is used rather than considering how the water is actually flowing. Remember that each flow is both positive and negative depending upon which loop is being considered. By adding the losses in Pipes 2 and 3 from the final friction loss column on the worksheet the loss is found to be approximately 15 psi (1.05 bar).

Example 6.5

Determine the pressure lost to friction between Points A and B in the piping system shown in Fig. 6.10. There are two sources of supply, A and C, delivering 3000 gpm (11,350 L/min) to Point B.

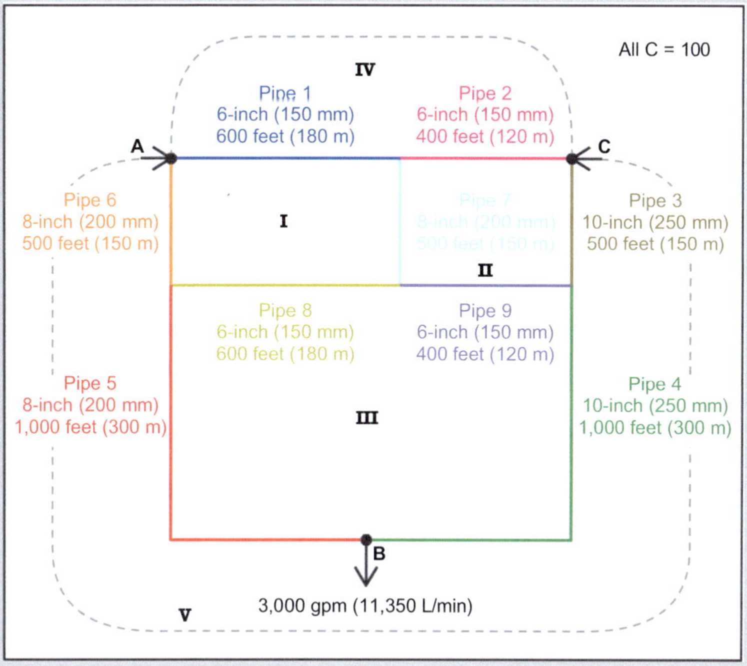

Fig. 6.10 Illustration for Example 6.5

First steps again involve making a line drawing similar to Fig. 6.10. Note that all pipes have been labeled as well as the loops. However, in this multiple-looped system, two imaginary loops have been created because of the two inflow points. Imaginary Loop IV is composed of Pipes 1and 2. Imaginary Loop V is composed of Pipes 3, 4, 5, and 6. Notice that with these two imaginary loops, every pipe will now appear in two loops. Next, flow estimates are made keeping in mind conservation of matter and proper sign notation.

Sign notation in imaginary loops is often difficult to visualize. If this is the case, simply go to the real loops to define the sign. The sign notation for any pipe is always the opposite when it appears in its second loop. With the flow estimates in hand, the friction loss and 1.85 $\mathbf{P_f/Q}$ are established for each leg in every loop before any correction is made. The work sheet and computations for this problem are shown in Fig. 6.11a and b. Notice that a double correction is made for each pipe. Notice that the solution has not been arrived at until all loops have achieved an absolute value of the sum of the friction loss of 1.0 psi (0.1 bar) or less. After the solution has been reached, the friction loss between A and B may be computed by adding the friction loss across any path joining the two points. Again, mistakes can be made in this step if simple algebraic addition is used rather than considering how the water is actually flowing.

Remember, each flow has both a positive and negative sign, depending upon which loop is being considered.

Loop No.	Line No.	Size	C	Length	Q	P_f	$\frac{1.85P_f}{Q}$	Correction	Q	P_f	$\frac{1.85P_f}{Q}$	Correction	Q	P_f	$\frac{1.85P_f}{Q}$	Correction
	1	6	100	600	+200	+1.6	.015	+10 −10	+200				+220	+1.9		
I	6	8	100	500	-860	-4.8	.010	+10 −30	-880				-890	-5.2		
	7	8	100	500	+480	+1.6	.006	+10 −10	+480			➤	+510	+1.8		
	8	6	100	600	+180	+1.3	.013	+10 +20	+210				+180	+1.3		
						-.3 /	.044 =	-7						-.2	Stop	
	2	6	100	400	-280	-2.0	.013	+10 −10	-280				-290	-2.1		
II	3	10	100	500	+1660	+5.5	.006	+10 −30	+1640				+1600	+5.1		
	7					-1.6	.006					➤		-1.8		
	9	6	100	400	-300	-2.2	.014	+10 +20	-270				-230	-1.4		
						-.3 /	.039 =	-8						-.2	Stop	
	4	10	100	1000	+1960	+15.0	.014	-20 −30	+1910		After		+1930	+14.6		
III	5	8	100	1000	-1040	-13.8	.025	-20 −30	-1090		several		-1070	-14.5		
	8					-1.3	.013				more			-1.3		
	9					+1.1	.010				corrections			+1.4		
						+1.0 /	.062 =	+16						+.2	Stop	
IV	1					-1.6	.015							-1.9		
	2					+2.0	.013					➤		+2.1		
						-.4 /	.028 =	-14						+.2	Stop	
	3					-5.5	.006							-5.1		
V	4					-15.0	.014							-14.6		
	5					+13.8	.025					➤		+14.5		
	6					+4.8	.010							+5.2		
						-1.9 /	.055 =	-34						0	Stop	

Fig. 6.11 **a** Worksheet for Example 6.5 showing the first correction and the final solution. **b** Worksheet for Example 6.5 (in S.I. units)

Loop No.	Line No.	Size	C	Length	Q	P_f	$\frac{1.85P_f}{Q}$	Correction	Q	P_f	$\frac{1.85P_f}{Q}$	Correction	Q	P_f	$\frac{1.85P_f}{Q}$	Correction
	1	150	100	180	+750	+.11	.0003	$+30_{+70}$	+850	+.14	.0003	-30_{-40}	+780	+.12		
I	6	200	100	150	-3250	-.35	.0002	$+30_{-130}$	-3350	-.37	.0002	-30_{+70}	-3310	-.37		
	7	200	100	150	+1820	+.12	.0001	$+30_{-50}$	+1800	+.12	.0001	-30_{+50}	+1820	+.12		
	8	150	100	180	+680	+.10	.0002	$+30_{+120}$	+830	+.14	.0003	-30_{-90}	+710	+.10		
						-.02 /	.0008 =	-25		+.03	.0009 =	+33		-.03	Stop	
	2	150	100	120	-1070	-.15	.0003	$+50_{+70}$	-950	-.12	.0002	-50_{-40}	-1040	-.14		
II	3	250	100	150	+6280	+.40	.0001	$+50_{-130}$	+6200	+.39	.0001	-50_{+70}	+6220	+.40		
	7					-.12	.0001			-.12	.0001			-.12		
	9	150	100	120	-1140	-.17	.0003	$+50_{+120}$	-970	-.12	.0002	-50_{-90}	-1110	-.16		
						-.04 /	.0008 =	-50		+.03	.0006 =	+50		-.02	Stop	
	4	250	100	300	+7420	+1.10	.0003	-120_{-130}	+7170	+1.03	.0003	$+90_{+70}$	+7330	+1.07		
III	5	200	100	300	-3930	-1.01	.0005	-120_{-130}	-4180	-1.13	.0005	$+90_{+70}$	-4020	-1.05		
	8					-.1	.0002			-.14	.0003			-.10		
	9					-.17	.0003			+.12	.0002			+.16		
						+.16 /	.0013 =	+123		-.12	.0013 =	-92		+.08	Stop	
IV	1					-.11	.0003			-.14	.0003			-.12		
	2					+.15	.0003			+.12	.0002			+.14		
						+.04 /	.0006 =	+67		-.02	.0005			+.02	Stop	
	3					-.40	.0001			-.39	.0001			-.40		
V	4					-1.10	.0003			-1.03	.0003			-1.07		
	5					+1.01	.0005			+1.13	.0005			+1.05		
	6					+.35	.0002			+.37	.0002			+.37		
						-.14 /	.0011 =	-127		+.08	.0011 =	+73		-.05	Stop	

Fig. 6.11 (continued)

By adding the losses in Legs 6 and 5 from the final friction loss column on the worksheet, the friction loss is found to be approximately 19.7 psi (1.42 bar).

After a few exercises using the Hardy Cross Method, students are typically at the point of changing majors or lynching professors. Most of the profanity I know I learned from students trying to work the Hardy Cross Method. However, computer programs are now available to simplify the process. Understanding of the manual method will assist in understanding the input required for many programs and will contribute toward a greater appreciation for the computer and computer programmers.

Other iterative techniques, such as the Newton-Rhapson Method, are also utilized with the computer. In most cases, these techniques provide more efficient use of the computer so that the number of iterations required to converge toward a solution are minimized. Jeppson provides a good summary as well as several programming approaches in his text [2].

Blank flow analysis worksheets may be found in Appendix J at the end of this text.

Summary of Important Concepts

1. The flow through one leg of a simple loop can be directly computed if all pipe diameters and C-factors are the same from:

$$Q_1 = \frac{Q_T}{1 + \left(\frac{L_1}{L_2}\right)^{0.54}}$$

2. A simple equation for changing a pipe of one diameter and C-factor into some equivalent length of another diameter and C-factor is:

$$L_e = L_1 \left(\frac{D_e}{D_1}\right)^{4.87} \left(\frac{C_e}{C_1}\right)^{1.85}$$

3. Gridded systems have traditionally been solved by a manual iterative approach known as the Hardy Cross Method.
4. Computer programs have replaced the tedious and time- consuming manual method of solving for friction loss in gridded systems.

Exercises

6.1. Solve for the flow split and pressure lost to friction in the simple loops shown in Fig. 6.12a through f:

6.2. Solve for the flow split and pressure lost to friction in the simple loops shown in Fig. 6.13a through f.

6.3. What length of 8-in. (200 mm) cement-lined cast-iron pipe with $C = 140$ will give the same friction loss as 420 ft (130 m) of 6-in. (150 mm) unlined cast-iron pipe with $C = 100$? For what flows will this equivalent length give the same friction loss?

6.4. What length of 4-in. (100 mm) unlined cast-iron pipe with $C = 100$ will give the same friction loss as 2500 ft (760 m) of 12-in. (300 mm) unlined cast-iron pipe with $C = 100$? Verify your answer at flows of 500, 1000 and 1500 gpm (1890, 3780 and 5670 L/min).

6.5. Use the Hardy Cross method to determine the pressure lost to friction between Points A and B in the systems shown in Fig. 6.14a through c.

6.6. Use the Hardy Cross method to determine the pressure lost to friction between Points A and B in the systems shown in Fig. 6.15a through c.

Selected answers found on page Appendix K.

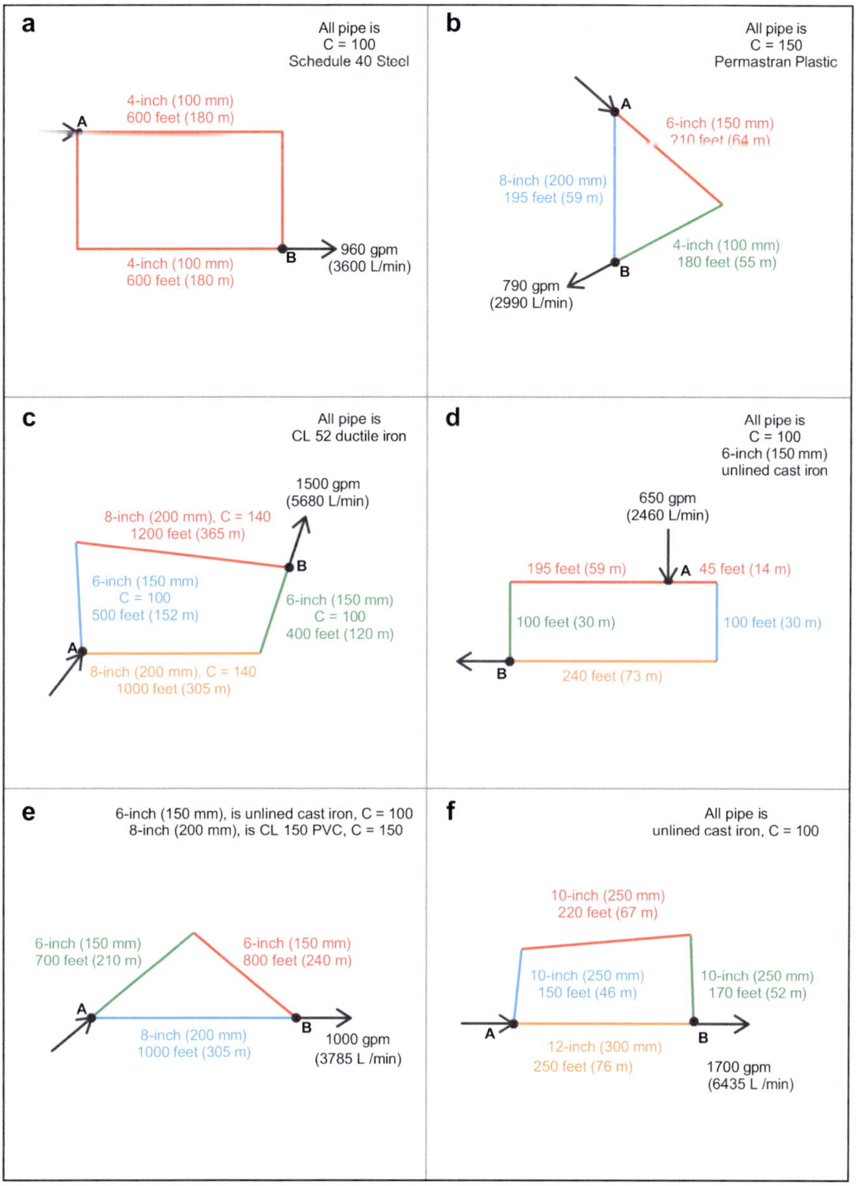

Fig. 6.12 **a–f** Illustration for Exercise 6.1

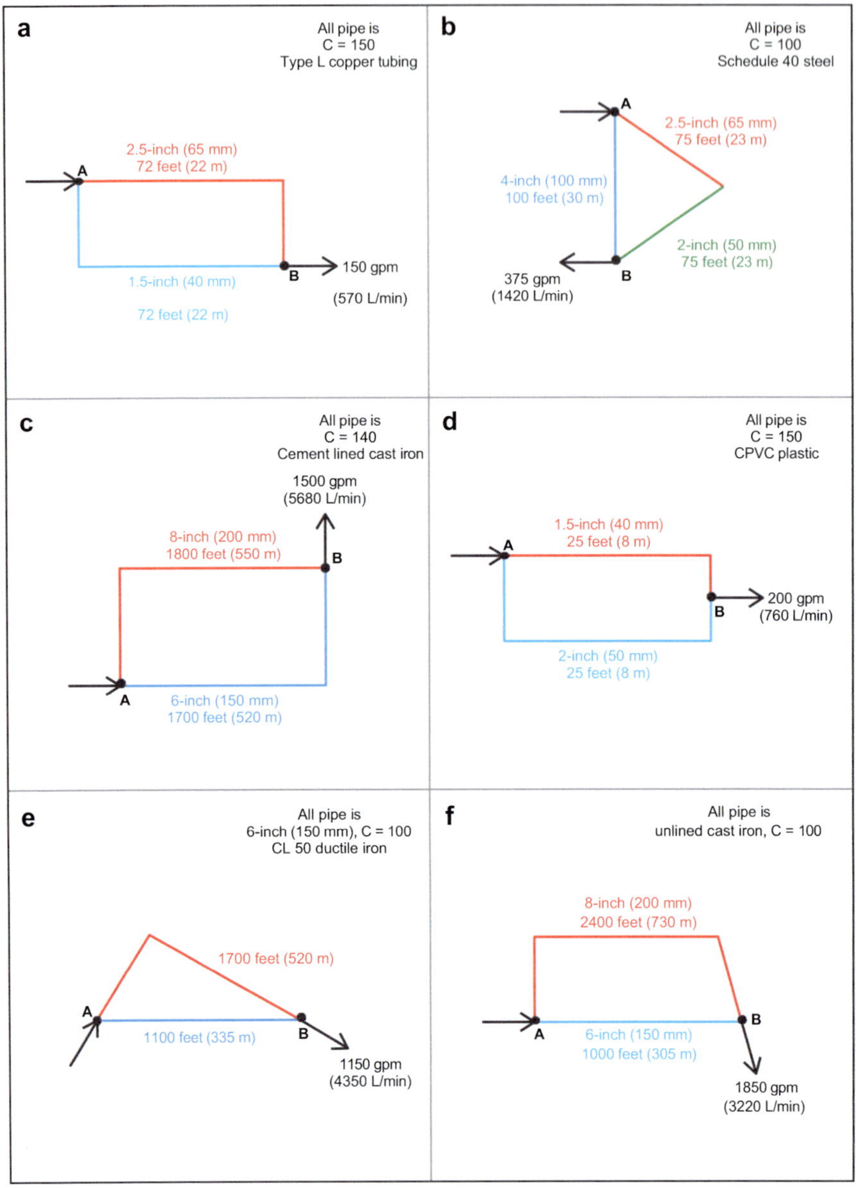

Fig. 6.13 a–f Illustration for Exercise 6.2

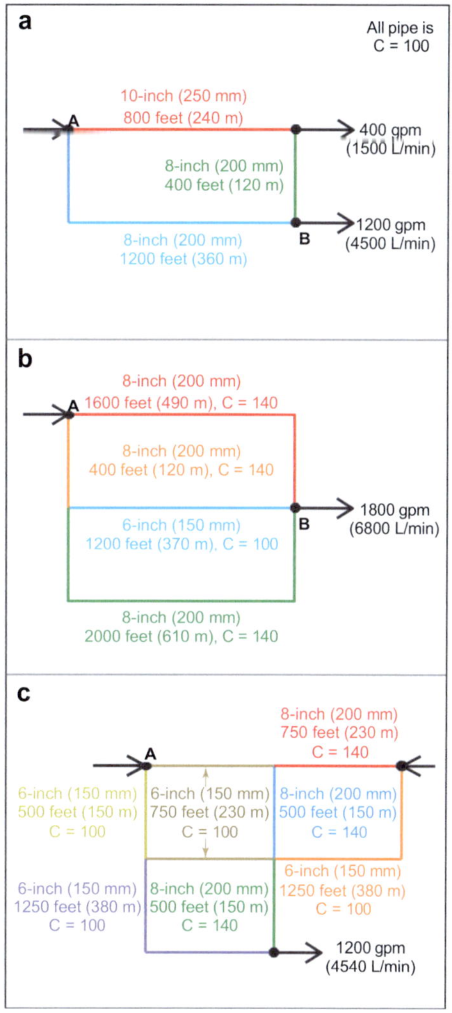

Fig. 6.14 a–f Illustration for Exercise 6.5

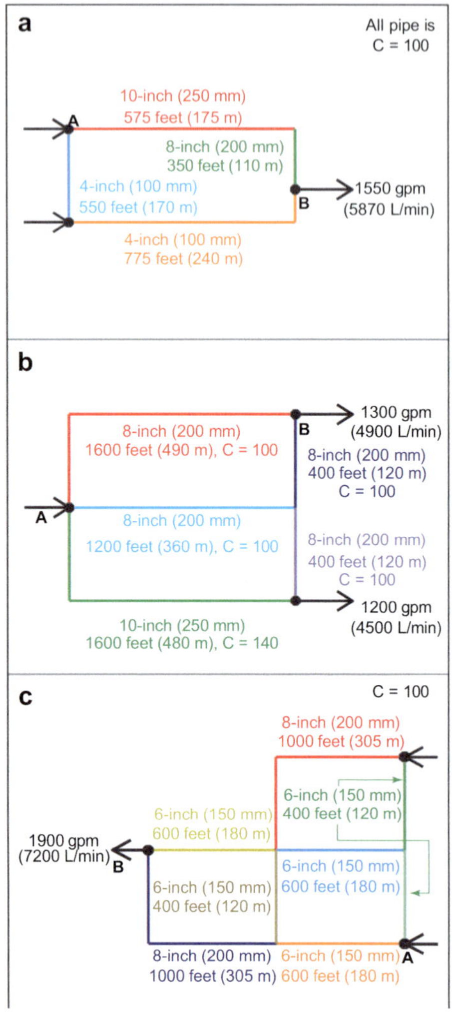

Fig. 6.15 a–f Illustration for Exercise 6.6

References

1. NFPA 13, *Standard for the Installation of Sprinkler Systems* (Quincy: National Fire Protection Association, 2025).
2. Roland W. Jeppson, *Analysis of Flow in Pipe Networks* (Ann Arbor, Michigan: Ann Arbor Science, 1977), p. 37–41.

Chapter 7
Components of Water Supply Systems

Abstract This chapter introduces the components of water supply systems, from water treatment facilities, water storage systems, pipe sizing, and valving. Fire hydrants are introduced including common color codes used in communities. Reliable water supply systems must be able to provide adequate flow and pressure to meet domestic and industrial consumption requirements as well as fire protection demands.

It has become apparent to over many years of teaching that some people do not realize that the drinking water in a community and the fire hydrants in a community are most often supplied by the same water system. Of course, at some very large industrial facilities there may exist a raw, untreated water system for fire protection in addition to treated water for industrial and human consumption. In our cities and towns, however, our water systems are usually dual purpose. For this reason, the fire protection specialist should have at least a basic understanding of water treatment and distribution systems.

The best water supply systems with respect to fire protection will always possess several specific characteristics. They will always be able to deliver adequate quantities of water with adequate pressure. They will also be reliable so that adequate water supplies are available at all times during the year. Finally, the best water supplies will be accessible for use. That is, they will possess an adequate number of hydrants well distributed throughout the system. This chapter introduces the basic components of water supply systems for the purpose of assisting the fire protection specialist in evaluating the adequacy, reliability, and accessibility of fire protection water supply systems.

P. Brock and V. Charter, *Fire Protection Hydraulics and Water Supply Analysis*, The Society of Fire Protection Engineers Series,
https://doi.org/10.1007/978-3-032-09877-1_7

Water Treatment Facilities

There are some communities that get their water supply from wells or springs where the water is of such good quality that very little in the way of treatment is necessary. However, in most communities, especially where the water comes from rivers and lakes, water treatment is necessary to make the water safe to drink and pleasing to the eye. Fire protection specialists are not expected to be experts in water purification and for that reason the discussion in this section is superficial at best. The Supplemental Readings listed at the end of this chapter provide a much more in-depth study. There are, however, two issues of importance relative to fire protection associated with water treatment facilities. They are:

- Does the treatment plant have adequate capacity to deliver the maximum daily domestic and industrial consumption plus the maximum fire flow?
- Is the treatment facility sufficiently reliable?

The following discussion is intended to address these issues.

In the typical treatment plant, the raw water is delivered to the plant by either gravity feed or from pumps through one or more pipelines (Fig. 7.1). The water is delivered to settling basins where the velocity of flow across the basin is slow enough to allow gravity to settle out much of the suspended matter in the water. To accelerate this process, the water may be delivered to flocculation and coagulation basins, where aluminum sulfate or similar chemicals are added to the water. The aluminum sulfate acts as a coagulant by attracting the very small, suspended particles into larger masses, which will fall out of the water more quickly. The water is then directed to sand filters, which may include activated charcoal. Sand filtration removes most remaining suspended particles, clarifies the water to make it acceptable in appearance, and removes much of the bacteria that may exist. At one or more points along the way, the water will be chlorinated for added disinfection. The water will then typically be delivered to storage tanks from which it will be pumped into the distribution system. A simple flow diagram of the treatment process is shown in Fig. 7.2.

The amount of water that a treatment facility needs to deliver is dependent uponthe maximum domestic, commercial, and industrial demand of the community plus the maximum fire protection demand. It is also affected by the amount of storage existing on the distribution system. The more elevated and ground-level storage that exists on the system, the smaller the treatment capacity may be. Factors that affect how much water can be treated include:

- The amount of water that can be delivered to the plant, which depends upon the number and size of pumps and pipelines
- The size and capacity of the sedimentation and coagulation basins
- The number, size, and filtration rate of sand filters. Filters are typically rated in gallons per minute per square foot (liters per minute per square meter) of filter

Fig. 7.1 A typical water treatment and distribution plant

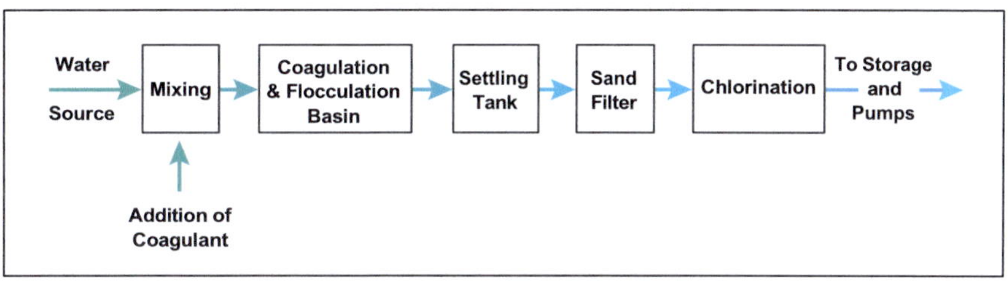

Fig. 7.2 A flow diagram of water through a typical treatment plant

area and may have rates ranging from 0.13 to 0.5 gpm per square foot (0.49–1.99 L/min per square meter) for slow sand filters or 1.0–2.0 gpm per square foot (3.78–7.57 L/min per square meter) for rapid sand filters [1].

- The number and size of pumps delivering water to the distribution system.

It might be noted that water plant capacity is often given in units of millions of gallons per day (MGD) or millions of liters per day (MLD). Because of this, fire protection demands must typically be converted from gpm (L/min) for the purpose of comparison. Note the following example.

Example 7.1

What water plant capacity must be available to meet a fire protection demand of 5000 gpm (18,900 L/min)?

Solution:

Converting to units of MGD yields the following:

$$Q = (5000\,\text{gpm})(60\,\text{min/h})(24\,\text{h/day}) = 7.2\,\text{MGD}$$

In S.I. Units the solution is:

Converting to units of MLD gives:

$$Q = (18{,}900\,\text{L/ min})(60\,\text{min/h})(24\,\text{h/day}) = 27.2\,\text{MLD}$$

Example 7.2

If a community of 30,000 people has a maximum domestic and industrial demand of 200 gallons (760 L) per person per day and a maximum calculated fire protection demand of 3500 gpm (13,250 L/min), what capacity is needed from the water treatment plant?

Solution:

$$Q = (3500\,\text{gpm})(60\,\text{min/h})(24\,\text{h/day}) + (30{,}000\,\text{people})(200\,\text{gal/person/day})$$
$$Q = 11.04\,\text{MGD}$$

In S.I. Units the solution is:

$$Q = (13{,}250\,\text{L/min})(60\,\text{min/h})(24\,\text{h/day}) + (30{,}000\,\text{people})(760\,\text{L/day})$$
$$Q = 41.9\,\text{MLD}$$

The reliability of the treatment facility is affected by the following features:

- The number of pumps and pipes delivering raw water to the plant. The worst case is the water plant with a single incoming raw water line.
- The number of sedimentation and coagulation basins. Redundancy is always necessary. If a basin is out of service for cleaning or repair, the remaining basins must be adequate to meet system needs.
- The number of sand filters.

- The number of pumps and types of power to the pumps. Having some pumps not dependent on electricity is a good idea.
- The number of discharge pipes leading to the water distribution system. The worst case is a facility with a single discharge line coming out of the plant.
- The susceptibility of the treatment facility to flooding or other natural disasters, including the location of pumps below ground level making them subject to flooding by broken pipes.

In evaluating the reliability of treatment plants, it is useful to assess the impact of losing the largest single basin, filter, pump, or pipe. If the largest single component can be lost and the plant can continue to deliver the maximum needs of the community, this is an indication of a dependable facility.

Water Distribution System

Having a large capacity and reliable treatment facility is of little value if the water cannot be delivered in adequate amounts and with adequate pressure to the point of use. Whether or not this is possible is dependent in part upon the capacity and pressure rating of the pumps at the treatment plant and pumping stations that may be distributed throughout a community, as well as the extent of elevated storage. However, the feature having the greatest impact is the piping distribution system. The performance of the distribution system is affected by several variables including the following:

- Piping material
- Pipe diameter
- Piping arrangement.

Piping Material

Let's look first at piping materials. In the earliest days, systems of wooden piping were installed (Fig. 7.3). There is very little of this piping left in the ground. The primary disadvantage of wooden pipe was the limit on interior diameters. Of course it was from the wooden piping that the expression "fireplug" evolved. To access water for firefighting, holes were drilled in the wood piping and, to stop the flow, plugs were hammered in the holes.

Most communities of any age will have some unlined cast iron piping. This piping corroded and tuberculated easily and was often of small diameter, all resulting in poor flow characteristics. Unlined cast iron is rarely installed today. To improve the performance of the cast iron pipe, various linings have been used such as cement or asphalt. These linings retard corrosion and improve the hydraulic characteristics of the piping.

Fig. 7.3 A cross-sectional view of a wooden water main

A very popular piping material for many years was asbestos cement (Fig. 7.4). Millions of miles of this piping were installed all over the world. This piping would not corrode or tuberculate and maintained excellent hydraulic characteristics for years. However, the problem with the release of asbestos fibers into the air when the pipe was cut or drilled has resulted in its being rarely used in new installations.

Modern piping systems are likely to be of polyvinylchloride plastic. However, ductile iron and welded steel piping are also commonly used. More important than piping material, however, is pipe diameter.

Fig. 7.4 A section of asbestos cement water pipe

Pipe Diameter

Chapters 12 and 13 will explain that gpm demands for fire protection application will rarely be less than 500 gpm (1890 L/min). Even if fighting fires in residential areas, fire departments like to have at least 1000 gpm (3780 L/min) available. Application of the Hazen–Williams formula from Chap. 5 indicates that at a rate of 1000 gpm (3780 L/min), 20–40 psi (1.4–2.8 bar) can be lost in each 100 ft (30.5 m) of 4-in. (100 mm) pipe. For this reason, piping 4-in. (100 mm) and smaller should never be used in water distribution systems supplying fire hydrants. Piping as small as 6-in. (150 mm) should not be used unless the piping is looped or gridded. In general, water distribution piping should be at least 8-in. (200 mm) in diameter, especially if the piping is dead end. Unfortunately, the water systems in many communities were constructed with little attention given to fire protection demands. It is, therefore, common to see 4-in. (100 mm) pipe supplying fire hydrants in the older sections of many cities and towns.

Piping Arrangement

The looping and gridding of piping was mentioned in the previous paragraph. Piping arrangement is a third feature that affects water distribution system performance. Chapter 6 introduced the impact of loops and grids. The reduction in pressure loss to friction can be dramatic when dead-end mains are tied together to create a gridded distribution system. For example, if the 8-in. (200 mm) dead- end main is converted to a gridded system as illustrated in Fig. 7.5, the pressure loss in delivering 2000 gpm (7570 L/min) a distance of 1000 ft (305 m) is reduced from 46 psi (3.2 bar) to about 7 psi (0.5 bar). A well-gridded system using pipe sizes 8-in. (200 mm) and larger are identifying features of a strong distribution system.

The typical water distribution system will also contain pumping stations to boost the pressure to outlying parts or higher parts of a community. These pumping stations often include ground-level storage tanks from which the pumps take suction. Elevated storage tanks are also common components. Many people mistakenly believe that the pressure within a water system comes from these elevated tanks. They believe this without thinking about how the tanks are filled. The pressure on most water systems comes from the pumps at the treatment plant and pumping stations. This pressure is used to fill the elevated storage tanks. The water in the tanks will not be delivered into the system until the demand upon the system exceeds the capacity of the pumps. The tanks are equipped with automatic "altitude" valves which keep the tanks from overflowing and systems are typically set up so that tanks contribute some water each day. This keeps fresh water circulating through the tanks.

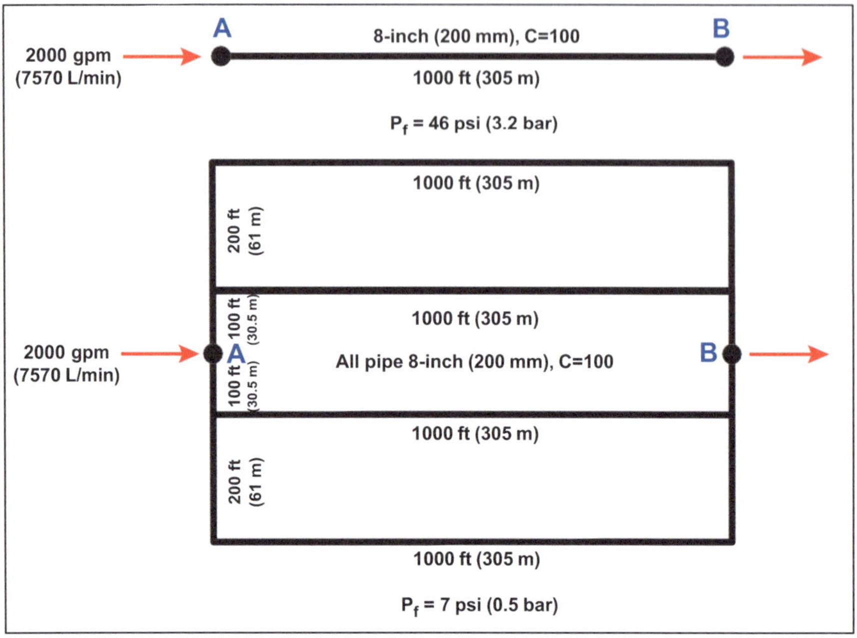

Fig. 7.5 Comparison of a nongridded to a gridded section of a water system

Storage Tanks

The storage capacity on a system is an important fire protection feature. For very large fires, the consumption on the system plus the fire flow demand may exceed the capacity of the treatment plant. If sufficient storage is available, all the water needs can be satisfied. It is often difficult to assess the fire flow contribution of elevated and ground-level storage. One method is to assume that the storage will be completely depleted during the time of fire flow demand. Let's look at an example.

Example 7.3

If an elevated tank has a minimum daily volume of 0.3 million gallons (1.14 million liters), what flow rate in gpm (L/min) and MGD (MLD) could be contributed by this tank over a 2-h fire flow duration when the tank is at its minimum level?

$$Q = \frac{0.3\,\text{million gallons}}{(2\,\text{h})(60\,\text{min/h})} = 2500\,\text{gpm}$$

And:

$$Q = (2500)(60\,\text{min/h})(24\,\text{h/day}) = 3.6\,\text{MGD}$$

Thus for a period of two hours (assuming the entire contents of the tank are depleted) the projected contribution from the tank would be 2500 gpm, or 3.6 MGD.

In S.I. Units the solution is:

$$Q = (1.14\,\text{million liters}) \div (2\,\text{h})(60\,\text{min/h}) = 9500\,\text{L/min}$$

And:

$$\text{Q} = (9500\,\text{L/min})(60\,\text{min/h})(24\,\text{h/day}) = 13.68\,\text{MLD}$$

So, for a period of two hours (assuming the entire tank contents are depleted) the projected contribution of the tank would be 9500 L/min or 13.68 MLD.

Example 7.4

If a town has a population of 150,000, a maximum daily consumption (MDC) rate of 100 gallons (380 L) per person, a maximum fire flow (FF) demand of 1000 gpm (3780 L/min) for two hours and a total minimum elevated storage capacity of 750,000 gallons (2.84 million liters), what treatment plant capacity in MGD (MLD) is needed to meet the maximum consumption and fire protection demands?

Solution:

$$\text{Needed Plant Capacity} = \text{MDC} + \text{FF} - \text{Storage}$$

$$\begin{aligned}\text{MDC} &= (150{,}000\,\text{people})(100\text{gal per person per day}) = 15\,\text{MGDFF}\\ &= (1000\,\text{gpm})(60\,\text{min/h})(24\,\text{h/day}) = 1.44\,\text{MGD}\end{aligned}$$

$$\begin{aligned}\text{Storage} &= (0.75\,\text{million gal}) \div [(2\,\text{h})(60\,\text{min/h})] = 6250\,\text{gpm}\\ \text{Storage} &= (6350\,\text{gpm})(60\,\text{min/h})((24\,\text{h/day}) = 9\,\text{MGD}\end{aligned}$$

$$\text{Needed Capacity} = 15\,\text{MGD} + 1.44\,\text{MGD} - 9\,\text{MGD} = 7.44\,\text{MGD}$$

In S.I. Units the solution is:

$$\text{Needed Plant Capacity} = \text{MDC} + \text{FF} - \text{Storage}$$

$$\begin{aligned}\text{MDC} &= (150{,}000\,\text{people})(380\,\text{liters per person per day}) = 57\,\text{MLDFF}\\ &= (3780\,\text{L/min})(60\,\text{min/h})(24\,\text{h/day}) = 5.44\,\text{MLD}\end{aligned}$$

$$\begin{aligned}\text{Storage} &= (2.84\,\text{million liters}) \div [(2\,\text{h})(60\,\text{min/h})] = 23{,}667\,\text{L/min}\\ \text{Storage} &= (23{,}667\text{L/min}60\text{min/hr})(24\text{h/day}) = 34.08\,\text{MLD}\end{aligned}$$

$$\text{Needed Capacity} = 57\,\text{MLD} + 5.44\,\text{MLD} - 34.08\,\text{MLD} = 28.36\,\text{MLD}$$

Valves

Valves are placed on water distribution systems to provide the ability to sectionalize the system and isolate portions of the system. This is a particularly important reliability feature. When a water main breaks, there needs to be sufficient valves so that only a small portion of the system must be put out of service to repair the break. These are typically underground valves that require a special wrench, often called a "key," to operate the valve (Fig. 7.6). These should ideally be located on each branch at the intersection of mains and no more than 500 ft (150 m) apart. There should also be a valve on every branch feeding a hydrant so that each hydrant can be isolated during repair or replacement.

Since these valves are underground, access is only possible through metal valve plates located at ground level. Unfortunately, these access plates are often covered over with dirt or pavement, leaving valves inaccessible and their locations hidden. A good maintenance program will require good water maps indicating valve location and yearly valve operations to make sure the valves work properly and are completely open.

Hydrants

No matter how strong or reliable a water system is, it is of little value if the water is not accessible. It is through the fire hydrants that the fire department gains access to the water supply system. Fire hydrants are available in two basic types—dry barrel and wet barrel. The most common type of hydrant used in the United States is the dry barrel type (Fig. 7.7). As the name implies, there is no water inside the hydrant until the hydrant is turned on. This hydrant, therefore, has application in areas where freezing temperatures are expected. The valving in the hydrant keeps water out of the barrel until it is opened, and then drains located at the bottom of the hydrant allow the water to drain out when the hydrant is closed. When the hydrant is fully opened, the drain holes will be closed off. If the hydrant is only partially opened, the drain holes will be partially open permitting a pressurized stream of water to be discharged out the drains beneath the ground, eroding the area at the base of the hydrant. For

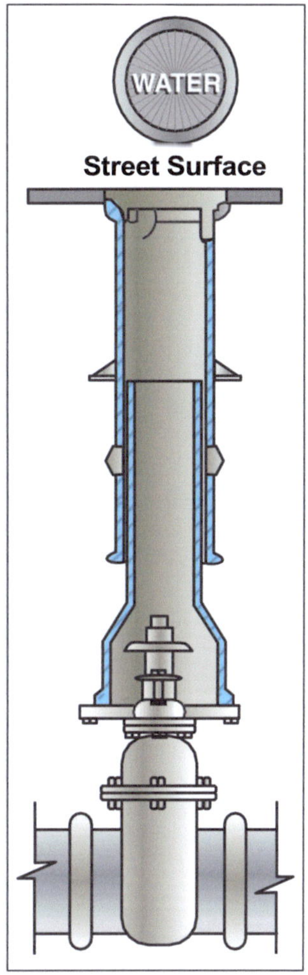

Fig. 7.6 A typical underground valve installation

this reason, dry barrel hydrants should always be completely open when in use. Dry barrel hydrants are easily recognized by the existence of the operating nut on the "bonnet" or top of the hydrant. Figure 7.8 shows a cutaway of a typical dry barrel hydrant.

The most common variety of dry barrel hydrant will have two 2 1/2-in. (65 mm) outlets and one 4 1/2-in. (115 mm) or 6-in. (150 mm) outlet. The larger outlet is often called a "pumper" outlet because it is common for the fire department to attach a suction line to this outlet to feed the pumping apparatus. It is also common to see dry barrel hydrants with only the two 2 1/2-in. (65 mm) outlets and no pumper outlet. This limits the capacity of the hydrant and these hydrants should not be used in industrial and commercial areas of a community. There are several other combinations of outlets that may be encountered, such as hydrants having only one

Fig. 7.7 A typical dry barrel hydrant

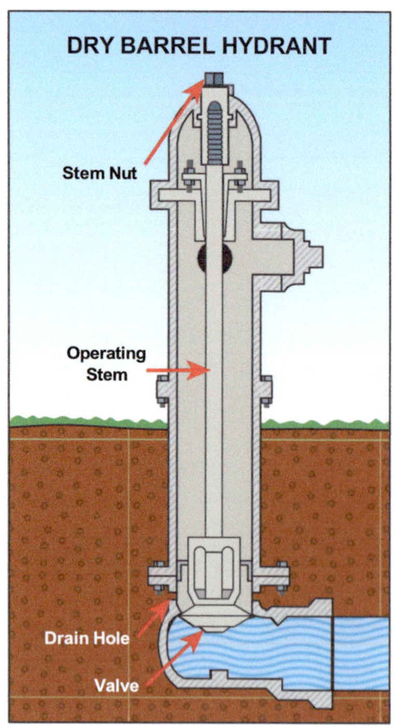

Fig. 7.8 This drawing shows the inner workings of a dry barrel hydrant

or two pumper outlets and no smaller outlets. However, the first two described above are by far the most common.

Wet barrel hydrants may be found in southern California, southern Florida, and other locations where freezing temperatures are not expected. Water is inside the hydrant at all times and would be subject to freezing in cold climates (Fig. 7.9). The operating nuts to turn the hydrant on generally extend through the side of the barrel, and there must be an operating nut for each outlet on the hydrant. This physically limits the number of outlets that can exist on a wet barrel hydrant. The most common variety will have a single 2 1/2-in. (65 mm) outlet and one pumper outlet. Figure 7.10 shows a cutaway of a typical wet barrel hydrant.

Whether wet barrel or dry barrel, the hydrants will be attached to the city main by what is typically a short piece of pipe called a "*branch*." This branch should be at least 6-in. (150 mm) in diameter. Branches 4-in. (100 mm) in diameter are common on older systems, but this limits the capacity of the hydrants. Each branch should also have a valve so that the hydrant can be isolated for repair or replacement without shutting down the city main to which it is attached.

Proper hydrant distribution and location is an important feature of an accessible water supply system. A good rule of thumb is that there should be a hydrant at every intersection, and hydrants should generally not be spaced farther than 500 ft (150 m) apart. However, the International Fire Code [2] bases hydrant spacing upon fire flow. It indicates that a 500-foot (150 m) separation is appropriate only where fire flow requirements are 1750 gpm (6600 L/min) or less. A spacing as little as 200 ft (60 m)

Fig. 7.9 A typical wet barrel hydrant

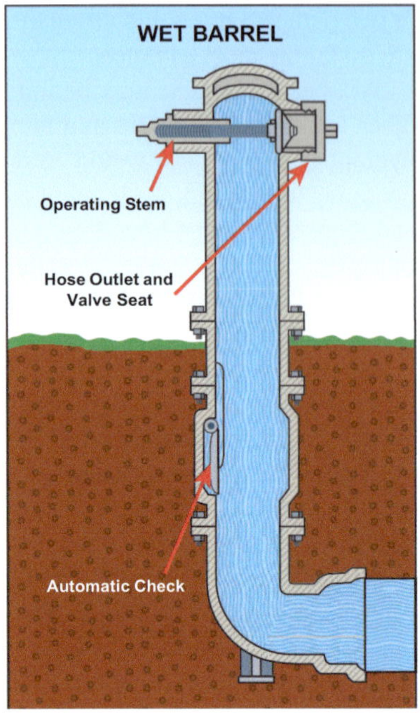

Fig. 7.10 This drawing shows the inner workings of a wet barrel hydrant

is indicated where the fire flow requirement is 7001 gpm (28,000 L/min) or more. The pumper outlet should be positioned facing the street and the hydrant should be set high enough that the hydrant wrench can be turned a full revolution when removing the cap from the lowest outlet (usually the pumper outlet).

From a maintenance standpoint, grass and bushes should be kept trimmed away from hydrants so that they remain visible and accessible. Hydrants should be flushed every year to make sure that they work and to clean rust and sedimentation from the pipes. It is also a good idea to measure the flow at each hydrant every year. By comparing the flows at each hydrant year by year, any deterioration of or obstruction to the water supply system can be easily detected.

In many communities, fire hydrants are color coded to reflect the gpm capacity of the fire hydrants. The system recommended in NFPA 291, *Recommended Practice for Fire Flow Testing and Marking of Hydrants* [3], is as follows:

Less than 500 gpm	Red
500–999 gpm	Orange
1000–1499 gpm	Green
1500 gpm or more	Light Blue

Each of the flows shown above would be available from the indicated hydrant at a residual pressure of 20 psi (1.4 bar).

The common components of a typical water supply system have now been identified. An important question is "How much water is available from a water system?" This question will be addressed in Chap. 8.

Summary of Important Concepts

1. Good water supply systems must be able to provide adequate flow and pressure to meet domestic and industrial consumption plus fire protection demands and be reliable.
2. Most water treatment systems include sedimentation, coagulation, filtration, and chlorination.
3. Reliability of water systems requires storage on the system and redundant system components.
4. Most piping in good water supply systems should be 8-in. (200 mm) or larger.
5. Adequate valving should be provided to minimize the impact of any break in a water main.
6. Fire hydrants will be either dry barrel or wet barrel, with the dry barrel provided in climates subject to freezing.
7. Hydrants in industrial and commercial areas should have a pumper outlet, a 6-in. (150 mm) branch, and be located not more than 500 ft (150 m) apart.
8. Many communities color code hydrants to indicate flow capacity.

Exercises

7.1 Change the following fire protection demands in gpm (L/min) into units of millions of gallons (millions of liters) per day.

   a. 500 gpm (1890 L/min)
   b. 1250 gpm (4700 L/min)
   c. 2000 gpm (7560 L/min)

7.2 Change the following fire protection demands in gpm (liters/min) into units of millions of gallons (millions of liters) per day.

   a. 750 gpm (2840 L/min)
   b. 1000 gpm (3780 L/min)
   c. 2250 gpm (8500 L/min)

7.3 Transform the following water treatment plant capacities in MGD (MLD) into flow rates in gpm (L/min).

   a. 7.3 MGD (27.63 MLD)

b. 9.7 MGD (36.72 MLD)
c. 18.4 MGD (69.65 MLD)

7.4 Transform the following water treatment plant capacities in MGD (MLD) into flow rates in gpm (L/min).

a. 11.2 MGD (42.4 MLD)
b. 17.8 MGD (67.38 MLD)
c. 20.4 MGD (77.22 MLD)

7.5 If a community has a population of 30,000 and consumes water at an average rate of 150 gallons (570 L) per person per day, what water treatment plant capacity in MGD (MLD) would be needed to meet the average daily consumption?

7.6 If a community has a population of 50,000 and consumes water at an average rate of 155 gallons (590 L) per person per day, what water treatment plant capacity in MGD (MLD) would be needed to meet the average daily consumption?

7.7 If a city of 500,000 people consumes water at a maximum rate in the summer of 130 gallons (490 L) per person per day, what water treatment plant capacity in MGD (MLD) would be needed to meet the maximum daily consumption?

7.8 If a city of 1,000,000 people consumes water at a maximum rate of 225 gallons (850 L) per person per day in August, what water treatment plant capacity in MGD (MLD) would be needed to meet the maximum daily consumption?

7.9 If a storage tank holds one million gallons (3.8 million liters), what contribution could be considered to be delivered from this tank in gpm (L/min) for a fire flow duration of two hours?

7.10 If an elevated tank holds 750,000 gallons (2.8 million liters), what fire flow in gpm (L/min) could be considered available from this tank for a fire duration of two hours?

7.11 If a tank has a capacity of 40,000 gallons (150,000 L), but is often only half full, what fire flow in gpm (L/min) could be considered available from this tank for a fire duration of one hour?

7.12 If a tank has a capacity of 100,000 gallons (380,000 L), but is often only 60% full, what fire flow in gpm (L/min) could be considered available from this tank for a fire duration of one hour?

7.13 If a community has a population of 80,000, a maximum daily consumption rate of 180 gallons (680 L) per person per day, a maximum fire flow demand of 3000 gpm (11,350 L/min) for two hours and two tanks each with 1,000,000 gallon (3.8 million liters) capacity that are always at least two-thirds full, what treatment plant capacity in MGD (MLD) is needed to meet the maximum consumption and fire protection demands?

7.14 If a community has a population of 210,000 people with a maximum daily consumption of 200 gallons (760 L) per person per day, a maximum fire flow demand of 3500 gpm (13,250 L/min) for two hours, and storage tanks having a minimum daily capacity of 3 million gallons (11.35 million liters), what treatment plant capacity in MGD (MLD) is needed to meet the maximum consumption and fire protection demands?

7.15 If a city has a population of one million with a maximum daily water consumption of 135 gallons (510 L) per person per day, a maximum fire flow demand of 6000 gpm (22,700 L/min) for two hours, and a storage capacity on the system of at least 20 million gallons (76 million liters) at all times, what treatment plant capacity in MGD (MLD) is needed to meet the maximum consumption and fire flow demands?

7.16 If a city has a population of 750,000 with an average daily consumption of 150 gallons (570 L) per person per day, a required fire flow of 3500 gpm (13,250 L/min) for two hours, and an average storage capacity of 2.5 million gallons (9.5 million liters), what treatment plant capacity in MGD (MLD) is needed to meet the average daily consumption rate plus the required fire flow?

7.17 If a rapid sand filter has a surface area of 400 ft^2 (37 m^2) and is operating at the top of the range for rapid sand filters, how many gpm (L/min) will be passing through the filter?

7.18 If a rapid sand filter has a surface area of 625 ft^2 (58 m^2) and is operating at the bottom of the range for rapid sand filters, how many gpm (L/min) will be passing through the filter?

Selected answers found in Appendix K.

References

1. Peavy, H.S. and Rowe, D.R, *Environmental Engineering* (New York: McGraw-Hill, Inc., 1985) pp. 177-178.
2. *International Fire Code*, International Fire Code Council, 2024, Appendix C.
3. National Fire Protection Association. NFPA 291, *Recommended Practice for Fire Flow Testing and Marking of Hydrants*. National Fire Protection Association: Quincy, MA, 2025 Edition

Supplemental Readings

James M. Montgomery, Consulting Engineers, Inc., *Water Treatment Principles and Design* (New York: John Wiley and Sons, 1985)

Peavy, H.S. and Rowe, D.R., *Environmental Engineering* (New York: McGraw- Hill, Inc., 1985)

Chapter 8
Testing and Analysis of Water Supply Systems

Abstract This chapter introduces the concepts of flow testing of hydrants. Flow testing is the best way to determine the available water supply at a given location. Flow tests must result in a static pressure, residual pressure, and measured flow. Test results may be analyzed graphically or mathematically to determine the adequacy of existing water supplies. The results may also be modified with proper corrections for friction loss and changes in elevation to other locations.

Tests conducted to establish the capabilities of water supply systems are often referred to as *flow tests* because they involve flowing fire hydrants. The objective of a flow test is to establish quantity (gallons or liters per minute) and pressures available at a specific location on a particular water supply system.

Flow Testing

To accomplish this objective mathematically is tedious and yields only approximations, even when using the most sophisticated computer programs. This is because the actual internal condition of the pipes, effective diameters of the pipes, and existence of partially or completely closed valves or other obstructions can only be estimated. Therefore, actual flow testing is the best method to establish the quantities and pressures available at a particular location.

A single test will yield results only good for the immediate vicinity of the test hydrant. This is because the water supply available from point to point on a system will vary due to elevation and friction loss. Higher parts of a community will typically have lower pressures available than low lying areas. Also, those areas near pumping stations or those areas supplied by large-diameter circulating mains will have better water supplies than remote areas supplied by small diameter dead-end piping.

The magnitude of water consumption nearby can also have an impact. If a factory is using large volumes of water in a manufacturing process, the pressures and flows remaining for fire protection purposes may be affected. Not only will the water

P. Brock and V. Charter, *Fire Protection Hydraulics and Water Supply Analysis*, The Society of Fire Protection Engineers Series,
https://doi.org/10.1007/978-3-032-09877-1_8

supply be different from location to location, but it also may change over time at the same location. This can result from population increases where greater domestic and industrial consumption leaves reduced amounts and pressures of water available. The supply can also change throughout the year due to environmental conditions. Most testing will yield higher flows and pressures during the cool, wet months of the year than in the hot, dry months. Even during the course of a single day the water supply at the same location may vary.

Tests conducted at 10:00 a.m. often result in higher values than tests conducted at 6:00 p.m., when the local population has returned home from their daily activities and domestic consumption is higher. For these reasons, the knowledgeable fire protection specialist understands that water supply testing yields approximate results which change over time. It is impractical to wait until July or August at 6:00 p.m. to conduct the water flow tests that will yield the most conservative results. It is for this reason that many fire marshals impose additional safety factors upon water flow test results used in the design of fire sprinkler systems.

Piping corrosion and deterioration can also result in greater pressure losses to friction and reduced carrying capacity of the pipe. In addition, changes can occur in the system. Many of these changes can be for the better, such as replacement of old mains or looping and gridding of dead-ends. However, any changes may have an impact upon the water supplies available. Because of these factors, the most practical method of establishing the actual water supply available at a particular location and time is to conduct hydrant flow tests.

Equipment for Flow Testing

The following is a list of the typical equipment needed to conduct flow tests:

- Pitot tube and gauge
- Tapped hydrant cap and gauge
- Ruler
- Hydrant and spanner wrenches
- Elbow
- Smoothbore nozzle or playpipe.

Most of this equipment is available from dealers who sell fire protection equipment, or it can be constructed. One of the most essential items on the list is the pitot tube and gauge, which come in several different styles (Fig. 8.1). However, all styles have a thin blade with a small diameter opening in the blade. The blade is inserted in a flowing stream, and the water velocity pressure is reflected on the gauge. Some pitot tubes have an air chamber extending past the gauge to form a handle. This air chamber serves as a shock absorber and reduces needle vibration on the gauge. Use of a liquid-filled gauge can accomplish the same thing.

Most pitot tubes are handheld; however, some can be attached to the water flow device or hydrant itself. This type comes with various orifice sizes to allow increasing system velocities for better pitot readings.

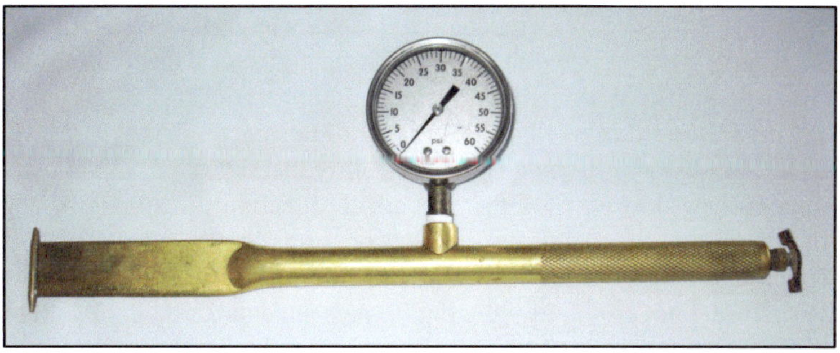

Fig. 8.1 A typical pitot tube

A hydrant cap that has been tapped so that a standard test gauge can be screwed into it is also essential (Fig. 8.2). This cap should have a small valve installed. This valve permits evacuating the air when the hydrant is being turned on and equalizing the suction pressure when a dry barrel hydrant is draining back after being turned off. It is important to know that the thread design on fire hydrants in different communities may be different. Most communities in the United States use National Standard Hose (NSH) threads on their fire hydrants, but many communities have their own customized hose thread design. Before attempting to conduct a flow test, the local thread design must be determined to make sure the cap and gauge can be attached. If the threads on the hydrant are different from the cap and gauge, adapters are generally available. The range of the pressure gauge is also important. If possible, select a gauge that will result in pressure readings near mid-range on the gauge.

A small ruler is needed to measure the actual diameter of hydrant outlets. Practically any ruler will work. Pocket cards with hydraulic tables are even available with one edge printed as a ruler.

On a public water system, it is the responsibility of the water utility or the fire department to open the hydrants. Fire hydrants belonging to the city or water utility company should only be opened by their personnel. In these situations, you will

Fig. 8.2 A hydrant cap and gauge

probably not need the hydrant wrenches or spanner wrenches. However, these tools must be available for a private system with private hydrants if the hydrants are to be tested. Tools must also be available for firefighting operations.

Having a 2 1/2-in. (65 mm) elbow fitting will allow directing the stream away from the hydrant. During the test, there is the possibility that thousands of gallons (liters) per minute will be flowing onto the ground. Redirecting the stream is important where drainage, erosion, or traffic conditions pose a problem (Fig. 8.3).

One point of difficulty involving the elbow is flow calculation. The basic equation used to transform pitot pressures into gpm (as presented in Chap. 4, Flow Measuring Devices) requires the use of a specific mathematical coefficient depending on whether the flow is from a hydrant outlet or a nozzle. There is no coefficient for an elbow. Therefore, in conjunction with an elbow, a straight stream nozzle or an Underwriters Playpipe must be used (Fig. 8.4).

It should also be noted that the flow directly from a hydrant outlet is very turbulent. If the flow is from a strong hydrant delivering a high velocity, it can be difficult (even

Fig. 8.3 The elbow may be used to redirect the flow of water

Fig. 8.4 The Underwriters playpipe used in conjunction with an elbow

Fig. 8.5 A stacked tip nozzle may be used in order to reduce the outlet diameter

for an experienced person) to obtain a good and consistent reading on the pitot tube. Therefore, another advantage of using a smooth bore nozzle or playpipe is to straighten the stream and make it easier to get a good pitot tube reading.

Also, when the flow from a hydrant is so poor that no pressure registers on the pitot tube gauge, reduction of the diameter of the outlet with a nozzle can increase the stream velocity enough to provide a pressure reading. Installing a smaller nozzle tip accomplishes this (Fig. 8.5). With the required test equipment available, we can look at the general test procedure.

General Test Procedure

Choose a hydrant as close as possible to the location in question and place the tapped cap with gauge on the hydrant and open it. In this text, this hydrant is called the *test* hydrant. However, in current publications of the National Fire Protection Association, this hydrant is called the *residual* hydrant. It is appropriate to call it the test hydrant because all of the test data taken will be relevant at this location. Making the test results relevant at other, nearby locations may require additional friction loss and elevation calculations.

Once the gauge is in place and the hydrant turned on, read and record the pressure. This pressure is called the *static* pressure, since no test water is flowing at this time.

Next, at one or more nearby hydrants, remove a cap and open the hydrant. These are called *flow* hydrants. Use the pitot tube to measure the velocity pressure of the flowing stream. The pitot pressure recorded can be converted to gpm (L/min) by the formula described in Chap. 4. While the pitot reading is being taken at the flow hydrant, the pressure at the test hydrant must also be recorded. The pressure at the test hydrant while water is flowing is called the *residual pressure* because it represents the pressure remaining in the system while the test water is flowing.

With the recording of the residual pressure and the pitot pressures, three pieces of data have been obtained: the static pressure, the residual pressure, and the pitot pressures. To have any real meaning, a flow test must result in these three pieces of information. Once obtained, this data can be used for mathematical or graphical analysis of the water supply at the test hydrant location. Before looking at more specific testing techniques or the analysis method, let's look at converting pitot pressures into the number of gpm (L/min) flowing. The equations used for this conversion are the ones presented in Chap. 4.

$$Q = (29.83)(C_\mathrm{d})(D)^2\left(\sqrt{P}\right) \tag{8.1}$$

In S.I. Units the equation is:

$$Q = (0.666)(C_\mathrm{d})(D)^2\left(\sqrt{P}\right) \tag{8.1a}$$

This formula allows computation of gpm (L/min) flowing from a nozzle, hydrant outlet, or orifice. As indicated in Chap. 4, application of the formula requires that you know **D** and $\mathbf{C_d}$ and have measured **P**. Using a calculator that can perform square roots, the flow can be easily computed. Should you choose not to make the calculations yourself, numerous tables and charts are available that convert pitot readings directly to gpm (L/min). Tables of this type are included in Appendix G. Once this conversion is made, the data is complete for analysis of the water supplies. Before we get to that, however, let's look at a few specific test techniques.

Specific Test Techniques

Establishing Hydrant Coefficients

First, you should always check the flow hydrants to determine what coefficient of discharge to use in the formula. This is not important on the test hydrant since no water flow is measured there.

Fire hydrant outlets are traditionally described as coming in three basic designs (Fig. 8.6). In most modern designs, the shoulders are rounded where the hydrant nozzle or outlet connects to the barrel. This is the most favorable design, and the coefficient of discharge for this hydrant is 0.9. If the shoulders are squared off, the coefficient is 0.8. In some rare cases, the nozzle may actually protrude into the barrel. This design would receive a coefficient of 0.7. You can see that ignoring the coefficient of discharge in the equation could introduce error as high as 30%. Coefficients for straight stream hose nozzles are about 0.99 with the Underwriters Playpipe being about 0.97.

Determining the coefficient of discharge of a hydrant does not involve a very sophisticated technique. In fact, you simply stick your fingers into the opening and feel it. People with large hands or short fingers may find it helpful to remove the

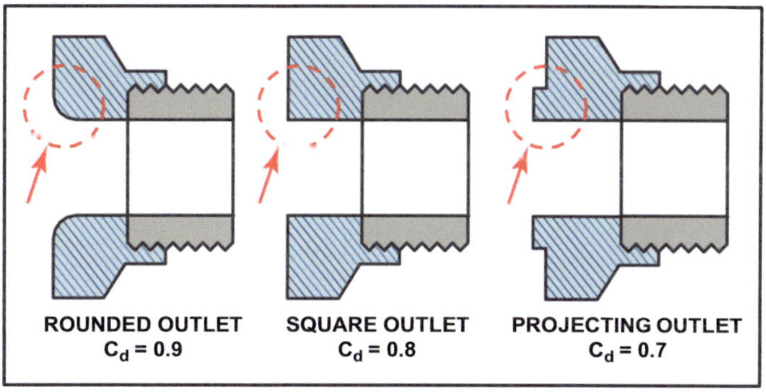

Fig. 8.6 The three most common designs of hydrant outlets

cap from the larger pumper outlet to gain access to the inside of the 2 1/2-in. (65 mm) opening. In a few cases, the coefficient of discharge may be stamped onto the hydrant. Of course, this manual technique will not work with the wet barrel hydrants found in warm climates. Since these hydrants are constantly under pressure, you cannot get your hand inside them unless you are very strong and willing to get very wet. With wet barrel hydrants it is best to depend upon manufacturer's information. However, using a value of 0.8 is generally a good estimate.

Measuring the Diameter of the Hydrant Outlet

Another important technique is to measure the diameter of the hydrant opening or nozzle tip to the nearest sixteenth of an inch (1.0 mm). This is important because even an error of one-sixteenth of an inch (1.0 mm) can result in more than a 5% error in calculated results. Also, the 2 1/2-in. (65 mm) outlet may actually be as small as 2 1/4 in. (56 mm) or as large as 2 9/16 (64 mm). It sounds strange, but you cannot assume a 2 1/2-in. (65 mm) outlet will be 2 1/2 in. (65 mm) in diameter.

Using the Pitot Tube

Another important technique is to hold the opening in the blade of the pitot tube one-half the diameter of the nozzle opening away from the nozzle (Fig. 8.7). This point is where the stream size is smallest and the stream velocity is greatest. The formula for **Q** assumes that the pitot pressure will be measured at this point. Holding the pitot tube at any other location in the stream will result in less accurate calculations. However, after witnessing and conducting several thousand tests, it is clear that eyeball accuracy is sufficient. Holding the blade 8 in. (20.3 cm) too far from the outlet makes a big difference. Holding the blade 1/8 of an inch (0.3 cm) too far or too close has no significant impact.

Fig. 8.7 Taking a pitot tube reading

It is also important to hold the tube so that the air chamber is slightly elevated and to make sure that the water has been drained out of the tube prior to each use. This will permit the air chamber to work as a shock absorber and help reduce gauge needle vibration.

Selection of Hydrant Outlets

The formula for gpm (L/min) also assumes that the stream being used will be a full stream at the point where the pitot reading is taken. However, when water flows from the large pumper outlets, there are actually void spaces in the stream. For this reason the 2 1/2-in. (65 mm) outlets should always be used when possible.

If the large pumper outlets must be used, conversion factors must be used to improve the accuracy of the results. A table of these conversion factors is contained in Appendix H. The results calculated from Eqs. 8.1 or 8.1a must be multiplied by these conversion factors.

Choosing Hydrants

Another basic principle is to flow hydrants downstream of the test hydrant where possible. This ensures that the test results are truly valid at the test hydrant. Where the water main is dead-end, the test hydrant should be the one closest to the building in question. In the situation illustrated in Fig. 8.8, the obvious choice for the flow hydrant would be the one to the right, since it is downstream. The only way that the water in this situation could get to the flow hydrant is to pass by the test hydrant. Thus, we know that any flow available at the flow hydrant is also available at the test hydrant and at the pressures indicated on the test hydrant gauge.

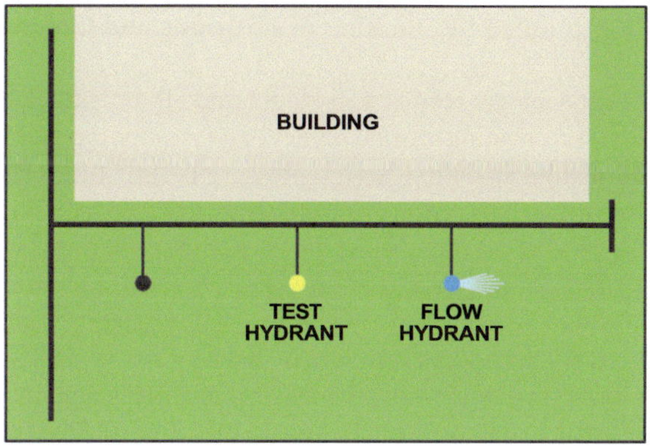

Fig. 8.8 Correct locations of test and flow hydrants on a dead-end main

However, if the left or upstream hydrant were used as the flow hydrant, no conclusion could be made regarding the flow and pressure available at the test hydrant without additional friction loss calculations and elevation corrections.

In the dead-end situation shown in Fig. 8.8, identifying the downstream hydrant is quite simple. However, how would you identify the downstream hydrant in Fig. 8.9, where the main is supplied from both ends? Here the trick is to consider the flow as coming from the side supplied by the larger mains. Indeed, most of the water usually will flow from the larger supply mains. If both mains are the same size, either or both hydrants may be used without significantly affecting the results.

Since establishing which hydrants are upstream or downstream requires knowing something about piping arrangements and diameters, this points out an obvious problem. How can you know anything about piping buried under the ground? This requires access to and study of water maps of the area where tests are being conducted. Good maps will show all hydrant and valve locations plus pipe layout and diameters. Conducting flow tests without first consulting water maps may result in erroneous test data. These maps are available from the local water utility and should normally be accessible to the public. The fire department also usually has copies of local water maps.

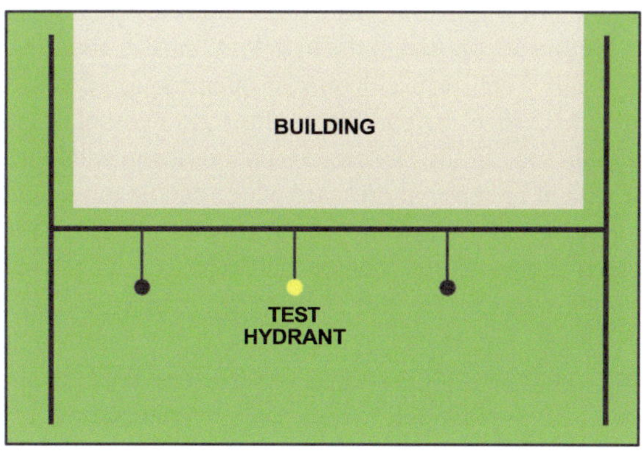

Fig. 8.9 Correct location of the test hydrant on a gridded system

Determining the Number of Hydrants and Outlets to Open

More accurate results will be obtained if a significant pressure drop from the static pressure reading can be achieved when one or more hydrants are opened. A 25% drop from static to residual pressure is sometimes recommended. Others may look for at least a 15 psi (1.0 bar) difference between the static and residual pressures. There is no standard value, but larger pressure drops generally give better test results.

Thus, if the static pressure is 80 psi (5.51 bar), a residual pressure no higher than 60 psi (4.1 bar) is desirable. If the static pressure reading is 60 psi (4.1 bar), it is good to obtain a residual pressure no higher than 45 psi (3.1 bar). Having a pressure drop of this magnitude minimizes the impact of gauge inaccuracy and human error. It is difficult to defend the significance of a 3 psi (0.2 bar) pressure drop where gauge error might be as high as 2 psi (0.14 bar), and sight accuracy is only good to plus or minus 1 psi (0.07 bar). So, the bigger the pressure drop, the better. To achieve a significant pressure drop may require opening more than one outlet on the flow hydrant or even three or four hydrants. If multiple outlets are used or multiple hydrants, they must all be opened simultaneously, and the flows from each outlet measured at the same time and added together for the total flow. The residual pressure must also be recorded while all the outlets are flowing.

Test Precautions

Anticipate where the water will drain. In one instance, the water from a flow test drained over a portion of newly poured concrete pavement. In another case, test water drained into an open manhole where a person was working. These are just two examples of what can happen. Plan ahead to avoid problems.

It was mentioned previously that operation of fire hydrants should be left to the fire department or water department, with the exception of privately owned industrial hydrants. The reason for this is that hydrants will break and can be damaged. Should one be damaged during a test, it is better from a liability standpoint if properly authorized personnel were responsible. Also, in some communities it is a violation of the law for unauthorized individuals to open fire hydrants.

If you are operating the hydrant yourself, count the turns it takes to fully open the hydrant. Then, when closing, if operating properly, it should take the same number of turns. Also, be careful to close the hydrant very slowly. Stopping any flowing stream too rapidly can cause a phenomenon called *water hammer*. Water hammer is a violent increase in pressure which can be large enough to rupture the piping.

After a dry barrel hydrant has been turned off, it should always be checked to see if it is draining. One of the simplest tests is to place your hand firmly over the outlet for a few seconds. A properly draining hydrant will create enough suction as the water drains that you can easily feel and hear it as you remove your hand. If the hydrant is not draining, repairs are called for. If repairs are impossible before freezing weather, the water should be pumped out of the barrel.

There are relatively few dangers involved with conducting flow tests, but there are a few safety precautions that should be followed. Horror stories are told about hydrant stems rocketing up out of the top of the hydrant. Although the possibility of this is remote, it is wise not to position your head and torso above the hydrant when operating it. Also, make sure that the caps on outlets which are not being used are secure and will not blow off. Take care in handling the pitot tubes. The blades on the tubes are often razor sharp and will cause injury if handled carelessly. Remember that a 2 1/2-in. (65 mm) stream has considerable impact and can knock people off their feet and even cause automobile accidents. Testing in freezing weather should be avoided to prevent ice from forming on streets and sidewalks.

Sometimes the water mains supplying the test and flow hydrants will not have been flushed for years. As a result, corrosion and particulate matter will have accumulated in the mains. You should not be surprised to find red or even black water flow initially from a hydrant. Because of this, even the test hydrant should be flushed before attaching the cap and gauge. Never insert the pitot tube into a dirty stream. To do so can result in damage to your equipment.

The homeowners and businesses in the area should be warned, if possible, that some water discoloration is likely to occur. This may prevent people from doing laundry during the flow tests and ending up with rust-colored sheets. Occasionally, the news media can even be used to publicize the test a few days in advance.

With these test techniques and precautions in mind, it is time to recall that a flow test must result in obtaining three pieces of data: the static pressure, the residual pressure, and the flow (gpm or L/min) during the test. With this information, analysis of the water supply available can be made. Analysis can be made mathematically using formulas. However, the easiest and most versatile analysis involves drawing a simple graph.

Graphical Analysis of Flow Test Results

Graphical analysis involves using special hydraulic graph paper. This is a special semi-logarithmic paper which allows the test results to be plotted as a straight line. Several blank copies are contained in Appendix J. The test results are plotted as follows:

1. The static pressure is plotted above 0 gpm (L/min), since no test water was flowing when the static pressure was measured.
2. Plot the residual pressure directly above the flow that was measured during the test. When these two points are connected by a straight line, that line represents the water supply available in the water main at the test hydrant location.

Any questions regarding pressure and flow at the test hydrant location can then be answered.

Example 8.1

Suppose a flow test yielded the following results:

- Static pressure of 86 psi (5.9 bar)
- Residual pressure of 42 psi (2.9 bar)
- Pitot pressure of 28 psi (1.9 bar).

If the pitot reading was taken on a single stream from a 2 1/2-in. (65 mm) outlet with a coefficient of discharge of 0.8, the formula gives the following:

Solution:

$$Q = (29.83)(C_{\mathrm{d}})(D)^2\left(\sqrt{P}\right)$$
$$Q = (29.83)(0.8)(2.5)^2\left(\sqrt{28}\right)$$
$$Q = 789\,\mathrm{gpm}$$

In S.I. Units the Solution is:

$$Q = (0.666)(C_{\mathrm{d}})(D)^2\left(\sqrt{P}\right)$$
$$Q = (0.666)(0.8)(65)^2\left(\sqrt{1.9}\right)$$
$$Q = 3103\,\mathrm{L/\,min}$$

To draw the graph, the static pressure of 86 psi (5.9 bar) is plotted above 0 gpm (liters/min) (Fig. 8.10a illustrates the graph in English units with S.I. units shown in Fig. 8.10b). The residual pressure is plotted next directly above the 789 gpm (3103 L/min) calculated flow. These two points are connected with a straight line, which is generally extended through the second point and down to at least 20 psi (1.4 bar). Now, any question about the water supply can be answered.

For example, the flow available at 20 psi (1.4 bar) can be read immediately below the intersection of the curve and the 20 psi (1.4 bar) line. It turns out to be about 980 gpm (3750 L/min). Also, the pressure available at 500 gpm (1890 L/min) is read immediately above 500 (1890) to be about 67 psi (4.9 bar). Suppose a sprinkler demand of 750 gpm (2800 L/min) at 36 psi (2.5 bar) is needed. At 750 gpm (2800 L/min), this water supply can provide about 46 psi (3.5 bar), and therefore would be more than adequate to meet the demand.

A quick way to establish the adequacy of a water supply with respect to various demands is to note the location of the demand point relative to the water supply line. If the point falls on or below the line, the water supply is adequate. If the demand point falls above the line, the water supply is not capable of meeting that demand.

(a)

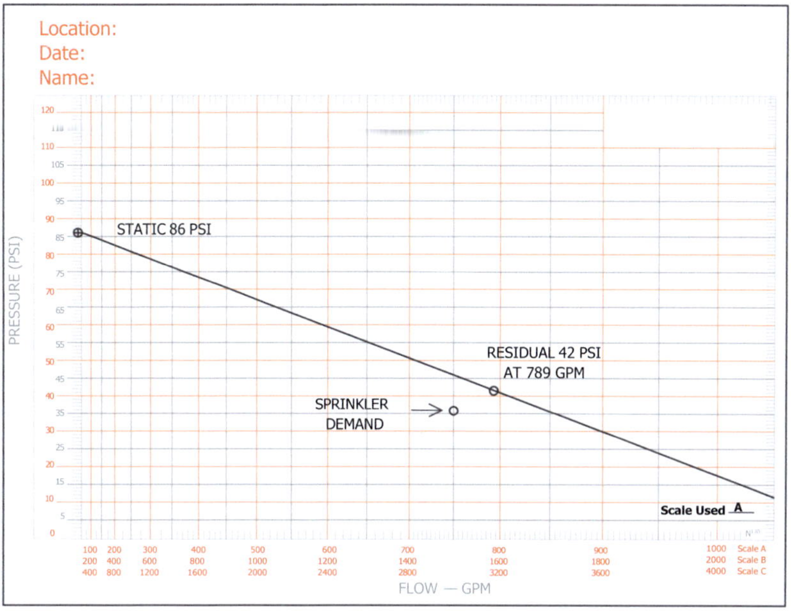

(b)

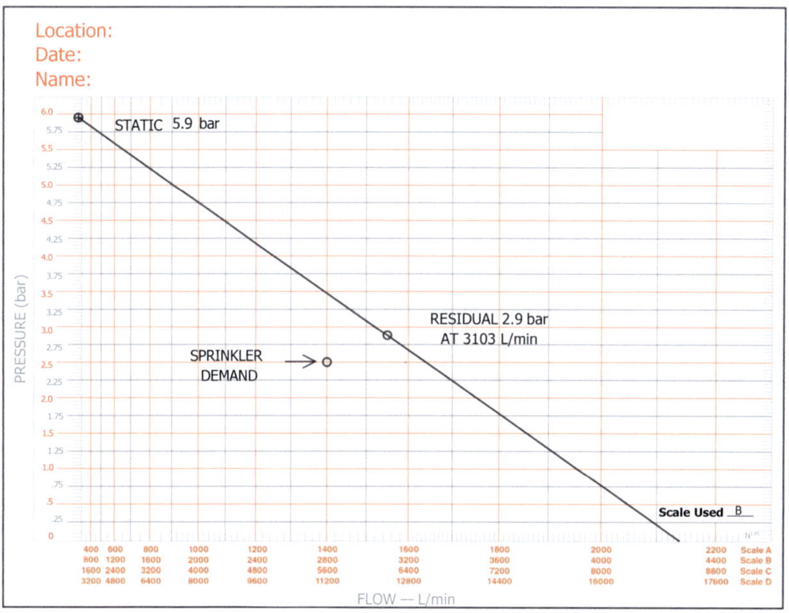

Fig. 8.10 **a** Test flow graph for Example 8.1 **b** S.I. unit flow test graph for Example 8.1

The comparison of a demand point to the supply curve is illustrated in Fig. 8.10a and b, where the demand point falls below the curve indicating an adequate supply.

One thing to note about the graph paper is that most hydraulic graph paper comes with at least three different scales along the bottom. They are simply multiples of each other to allow plotting a broader range of flows. The scale used should be clearly marked. Which scale to use depends upon the flow. Typically, the scale with the smallest range possible that will allow the curve to pass below 20 psi (1.4 bar) should be used. Twenty pounds per square inch (1.4 bar) is a historical minimum pressure for supplying fire department pumpers and industrial fire pumps.

Example 8.2

Suppose a static pressure of 72 psi (5.0 bar) and a residual pressure of 37 psi (2.6 bar) were recorded during a flow test. However, in this case, the residual pressure was measured while two outlets on each of two hydrants were flowing (a total of four outlets). If each of the hydrant outlets was 2 1/2 in. (65 mm) in diameter with a coefficient of discharge of 0.9, and pitot pressures were two at 24 psi (1.7 bar) and two at 36 psi (2.5 bar), the total flow is calculated as follows:

Solution:

$$Q_1 = (29.83)(0.9)(2.5)^2\left(\sqrt{24}\right) = 822\,\text{gpm}$$

$$Q_2 = (29.83)(0.9)(2.5)^2\left(\sqrt{24}\right) = 822\,\text{gpm}$$

$$Q_3 = (29.83)(0.9)(2.5)^2\left(\sqrt{36}\right) = 1007\,\text{gpm}$$

$$Q_4 = (29.83)(0.9)(2.5)^2\left(\sqrt{36}\right) = 1007\,\text{gpm}$$

$$\text{Total}\,Q = Q_1 + Q_2 + Q_3 + Q_4 = 3658\,\text{gpm}$$

In S.I. Units the Solution is:

$$Q_1 = (0.666)(0.9)(65)^2\left(\sqrt{1.7}\right) = 3302\,\text{L/min}$$

$$Q_2 = (0.666)(0.9)(65)^2\left(\sqrt{1.7}\right) = 3302\,\text{L/min}$$

$$Q_3 = (0.666)(0.9)(65)^2\left(\sqrt{2.5}\right) = 4004\,\text{L/min}$$

$$Q_4 = (0.666)(0.9)(65)^2\left(\sqrt{2.5}\right) = 4004\,\text{L/min}$$

$$\text{Total}\,Q = Q_1 + Q_2 + Q_3 + Q_4 = 14,612\,\text{L/min}$$

Note that the flow from each outlet must be calculated separately, then all four added. It is not proper to add or average pitot readings. The results of this test are plotted in Fig. 8.11a and b.

Such a large flow during the test dictates that scale C (scale D for S.I. units) be used in plotting the results. Once again the static pressure is plotted above 0 gpm (L/min). The residual pressure of 37 psi (2.6 bar) is plotted above the total flow of 3658 gpm (14,612 L/min). Connecting the two points illustrates the water supply available. At 20 psi (1.4 bar), the fire department would have a little over 4500 gpm (17,000 L/min) available from the water supply for fire fighting purposes. This water supply can also provide over 45 psi (3.1 bar) at 3000 gpm (11,350 L/min).

The technique for plotting other test results will be the same. After the test line is drawn, comparing the water supply needed with the supply available permits evaluation of the adequacy of the water supply.

Mathematical Analysis of Test Results

With the aid of an exponential calculator or for computer applications, mathematical analysis is also possible. The following equation is used:

$$Q_{\mathrm{A}} = (Q_{\mathrm{T}})\left(\frac{P_{\mathrm{S}} - P_{\mathrm{A}}}{P_{\mathrm{S}} - P_{\mathrm{R}}}\right)^{0.54} \tag{8.2}$$

where

$\boldsymbol{Q_{\mathrm{A}}}$ = the flow available at some residual pressure P_{A}.

$\mathbf{Q_T}$ = the actual flow measured during the test.

$\mathbf{P_S}$ = the measured static pressure.

$\mathbf{P_R}$ = the measured residual pressure.

$\mathbf{P_A}$ = the pressure of interest.

This formula can be used to determine what flow would be available at any residual pressure as indicated by the actual test results. Also, algebraic manipulation of the formula will permit solving for any one of the other variables. Equation 8.2 is applicable using both English and S.I. units.

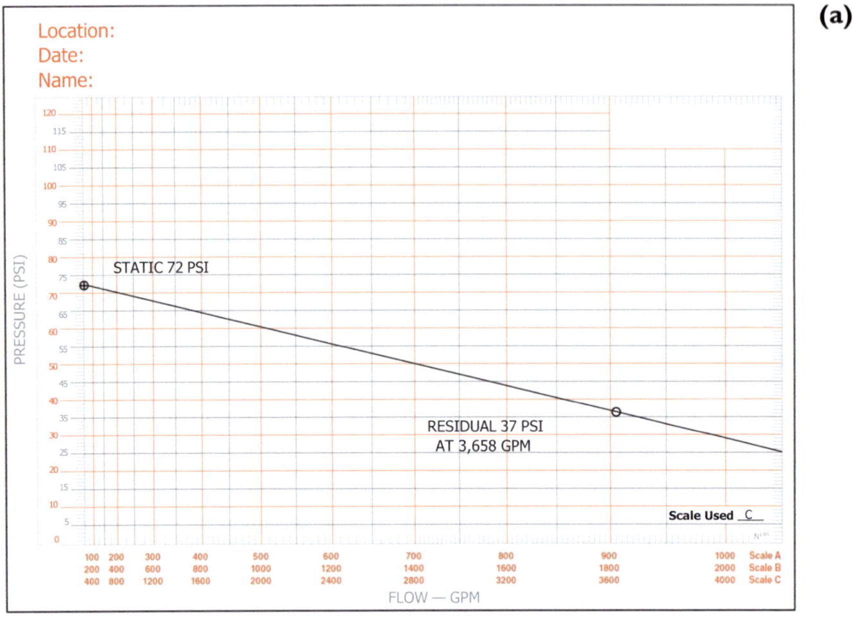

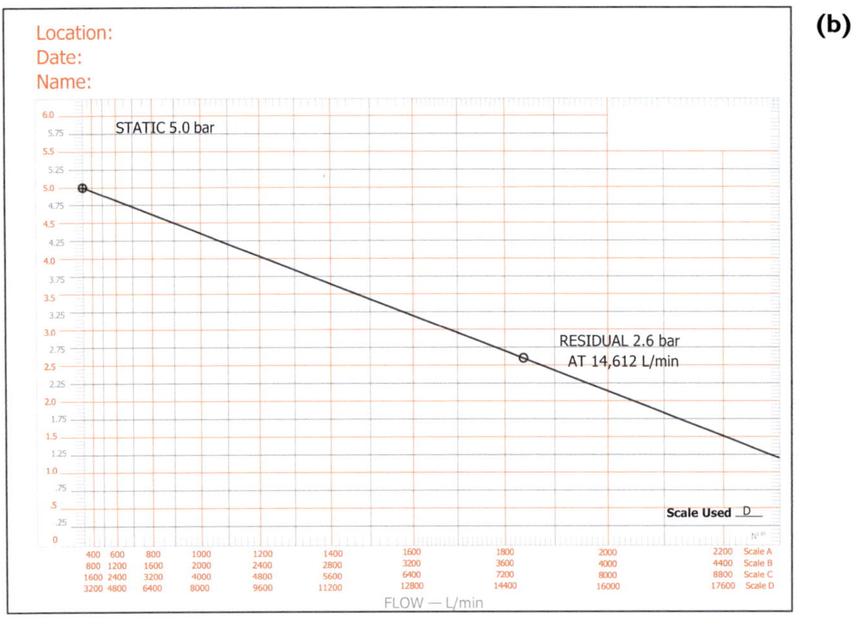

Fig. 8.11 **a** Test flow graph for Example 8.2**b** S.I. unit flow test graph for Example 8.2

Example 8.3

Suppose it is desired to know the flow available at 20 psi (1.4 bar) residual pressure at the test location described in Example 8.1. Recalling that the static pressure was 86 psi (5.9 bar), the residual was 42 psi (2.9 bar), and the measured flow was 789 gpm (3103 L/min), the solution becomes the following:

Solution:

$$Q_A = (Q_T)\left(\frac{P_S - P_A}{P_S - P_R}\right)^{0.54}$$

$$Q_A = (789)\left(\frac{86 - 20}{86 - 42}\right)^{0.54} = 982\,\text{gpm}$$

In S.I. Units the solution is:

$$Q_A = Q_T\left(\frac{P_S - P_A}{P_S - P_R}\right)^{0.54}$$

$$Q_A = (3103)\left(\frac{5.9 - 1.4}{5.9 - 2.9}\right)^{0.54} = 3863\,\text{L/ min}$$

You can see that the results are practically identical to those obtained through graphical analysis. Now, suppose it is also desired to know what residual pressure would be available at a sprinkler demand of 750 gpm (2840 L/min). Manipulation of Eq. 8.2 gives the following:

$$P_A = P_S - (P_S - P_R)\left(\frac{Q_A}{Q_T}\right)^{1.85} \tag{8.3}$$

Substituting the test results provides the solution:

$$P_A = 86 - (86 - 42)\frac{(750)^{1.85}}{789} = 46\,\text{psi}$$

And, in S.I. Units:

$$P_A = (5.9) - (5.9 - 2.9)\frac{(2840)^{1.85}}{3103} = 3.4\,\text{bar}$$

Note once again the almost identical results to those obtained through graphical analysis.

Example 8.4

Using the same test results from Example 8.2, determine mathematically the flow available at a 20 psi (1.4 bar) residual pressure. Also, determine the residual pressure that would be available at a flow of 1500 gpm (5680 L/min).

Solution:

At 20 psi:

$$Q_{\mathrm{A}} = (Q_{\mathrm{T}})\left(\frac{P_{\mathrm{S}} - P_{\mathrm{A}}}{P_{\mathrm{S}} - P_{\mathrm{R}}}\right)^{0.54}$$

$$Q_{\mathrm{A}} = (3658)\left(\frac{72 - 20}{72 - 37}\right)^{0.54} = 4530\,\mathrm{gpm}$$

At 1500 gpm:

$$P_{\mathrm{A}} = P_{\mathrm{S}} - (P_{\mathrm{S}} - P_{\mathrm{R}})\left(\frac{Q_{\mathrm{A}}}{Q_{\mathrm{T}}}\right)^{1.85}$$

$$P_{\mathrm{A}} = 72 - (72 - 37)\left(\frac{1500}{3658}\right)^{1.85} = 65\,\mathrm{psi}$$

In S.I. Units the Solution is:

At 1.4 bar:

$$Q_{\mathrm{A}} = (Q_{\mathrm{T}})\left(\frac{P_{\mathrm{S}} - P_{\mathrm{A}}}{P_{\mathrm{S}} - P_{\mathrm{R}}}\right)^{0.54}$$

$$Q_{\mathrm{A}} = (14,612)\left(\frac{5.0 - 1.4}{5.0 - 2.6}\right)^{0.54} = 18,189\,\mathrm{L/\,min}$$

At 5680 L/min:

$$P_{\mathrm{A}} = (P_{\mathrm{S}}) - (P_{\mathrm{S}} - P_{\mathrm{R}})\left(\frac{Q_{\mathrm{A}}}{Q_{\mathrm{T}}}\right)^{1.85}$$

$$P_{\mathrm{A}} = 5.0 - (5.0 - 2.6)\left(\frac{5680}{14,612}\right)^{1.85} = 4.6\,\mathrm{bar}$$

It is important to understand that the test results are only valid at or near the test hydrant. This is particularly true if the water mains upon which the test was conducted are 6 in. or smaller. For larger mains, the test results would typically be valid for several hundred feet in each direction. Also, for large-diameter mains where the piping is well looped and gridded, the test results may have valid application over an area of several blocks. The reason that the results might not be valid at all locations nearby is friction loss. Elevation differences will also have an impact. In the next section, the impact of elevation and friction loss on water supplies will be addressed.

Adjusting Flow Test Results for Friction Loss and Elevation Differences

Perhaps the most significant factor affecting available pressures in a water system is friction loss. It is friction loss that makes 4- and 6 -in. (100 and 150 mm) pipe of limited value in fire protection applications. It is because of friction loss that the residual pressure in a flow test is less than the static pressure. The pressure lost between the supply source and the test hydrant is dependent on the amount of pressure depleted by friction as water flows between these points. As indicated in Chap. 5, Friction Loss In Piping Systems, the amount of pressure that will be lost to friction depends upon the quantity flowing (gpm or L/min), pipe roughness, pipe diameter, and pipe length. There is no friction loss if there is no water flow. That is why, at the same elevation, the static pressure on the fire protection system will be found the same throughout a facility. However, when water begins to flow, the resulting pressure may be different at different locations. The greater quantity of water that flows, the greater the friction loss will be. Also, the longer the pipe, the greater the friction loss. As you might expect, friction loss is also greater when the pipe is rough than when smooth. Also, the smaller the pipe, the greater the friction loss. This should explain why larger pipes are always more desirable.

Let’s look at how friction loss can affect flow test results where the location of interest is not close to the test hydrant. In Fig. 8.12, a sprinklered building is located 800 ft (244 m) away from the test hydrant and supplied by a 6-in. (150 mm) main. If the flow test results show a static pressure of 69 psi (4.8 bar) and a residual pressure of 36 psi (2.5 bar) available with 750 gpm (2840 L/min) flowing, what will the water supply be at the building in question? Answering this question will require adjusting the test curve.

The technique of adjusting water supply curves for friction loss is quite simple. Note again that with no water flowing, there is no friction loss. Therefore, the static pressure remains unchanged. However, as water begins to flow, the friction loss will begin to have an impact. The adjustment to the original curve is accomplished through choosing any flow and computing the friction loss for that flow between the test hydrant and the point of interest. The pressure lost to friction is then subtracted from the original curve at the flow from which the friction loss was calculated. The

Fig. 8.12 Flow test results are for a location remote from the building of interest

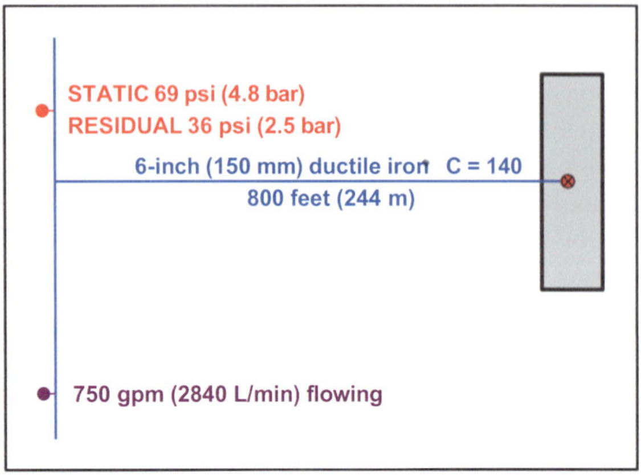

new line is then drawn through the original static pressure and this new pressure point. This is the adjusted curve.

In this example, if the measured flow of 750 gpm (2840 L/min) is chosen for the calculation of friction loss through the 800 ft (244 m) of enamel-lined, 6-in. (150 mm) cast-iron pipe with $C = 140$, the resulting friction loss is approximately 13 psi (0.9 bar). The original pressure existing at 750 gpm (2840 L/min) is 36 psi (2.5 bar). Subtracting the friction loss of 13 psi (0.9 bar) from the 36 psi (2.5 bar) yields a new pressure of 23 psi (1.6 bar). Thus, the adjusted curve would again have a static pressure of 69 psi (4.8 bar), but the pressure available at 750 gpm (2840 L/min) is now only 23 psi (1.6 bar).

Figures 8.13a and b illustrate the water supply adjusted for friction loss. It can be seen that the original water supply was adequate to deliver a sprinkler demand of 51 psi (3.5 bar) and 500 gpm (1900 L/min). However, the water supply at the building 800 ft (244 m) away from the original test location is not adequate to meet the sprinkler demand.

It is important to understand that the same adjusted curve would have been created if 500 gpm (1900 L/min), 800 gpm (3030 L/min), or any other flow was used to calculate the friction loss. The different friction from the different flows would simply give a different point on the same adjusted curve.

Another adjustment that may have to be made in transforming water supply tests from one location to another is an adjustment due to elevation. In Fig. 8.14, it might be assumed, due to large piping and the gridded design, that friction loss would have a minor impact in comparing the water supply available at Points A and B. This would be true, but what if Point B is 40 ft (12.19 m) higher than Point A?

Suppose the water supply at Point A is illustrated by the solid lines in Fig. 8.15a and b, with a static pressure of 59 psi (4.1 bar) and a residual pressure of 40 psi (2.8 bar) at 1100 gpm (4165 L/min). To adjust for the elevation difference between Points A and B involves first calculating the elevation correction. As should be second nature by now, the elevation correction is as follows:

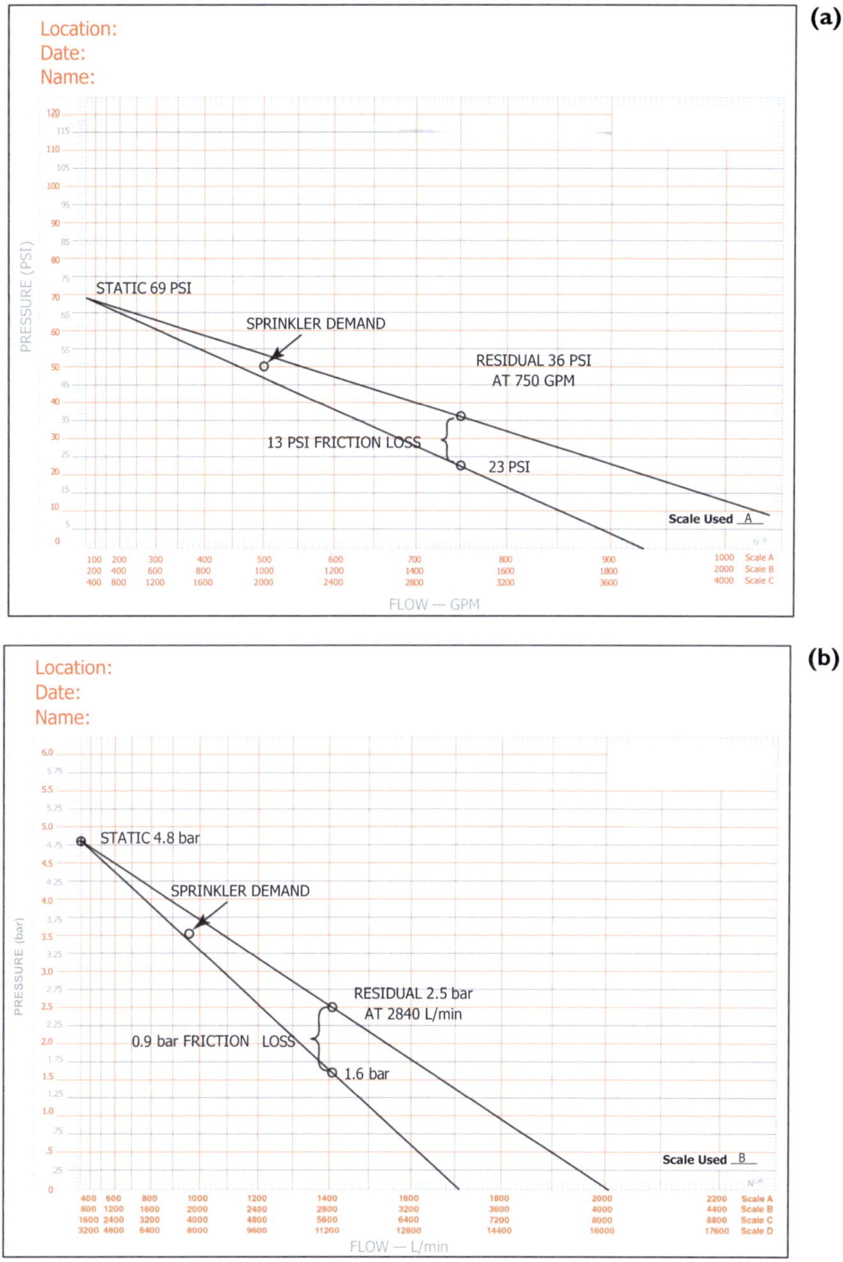

Fig. 8.13 **a** Flow test curve adjusted for friction loss. **b** Flow test curve adjusted for friction loss in S.I. units

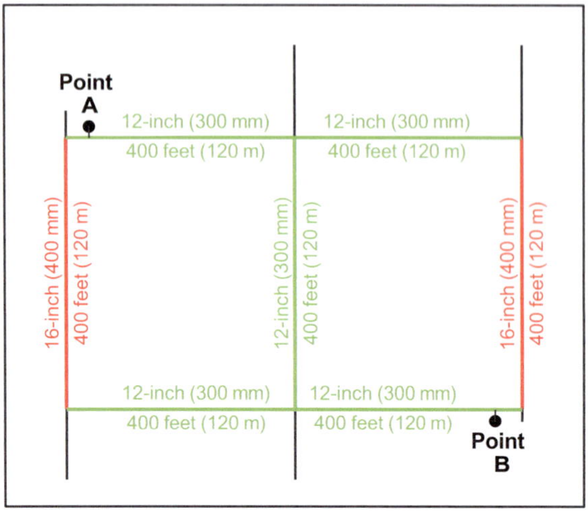

Fig. 8.14 A gridded water supply system

$$P_e = (0.433)(\mathrm{h})$$

where

P_e = pressure in psi.

h = the height difference in feet.

Or, in S.I. Units:

$$\boldsymbol{P_e} = (0.098)(\mathrm{h})$$

With

$\mathbf{P_e}$ = pressure in bar.

h = height in meters.

then

$$\boldsymbol{P_e} = (0.433\,\mathrm{psi/ft})(40\,\mathrm{ft}) = 17.3\,\mathrm{psi}$$

And

$$\boldsymbol{P_e} = (0.433\,\mathrm{psi/ft})(40\,\mathrm{ft}) = 17.3\,\mathrm{psi}$$
$$\boldsymbol{P_e} = (0.098)(12.19\,\mathrm{m}) = 1.2\,\mathrm{bar}$$

(a)

(b)

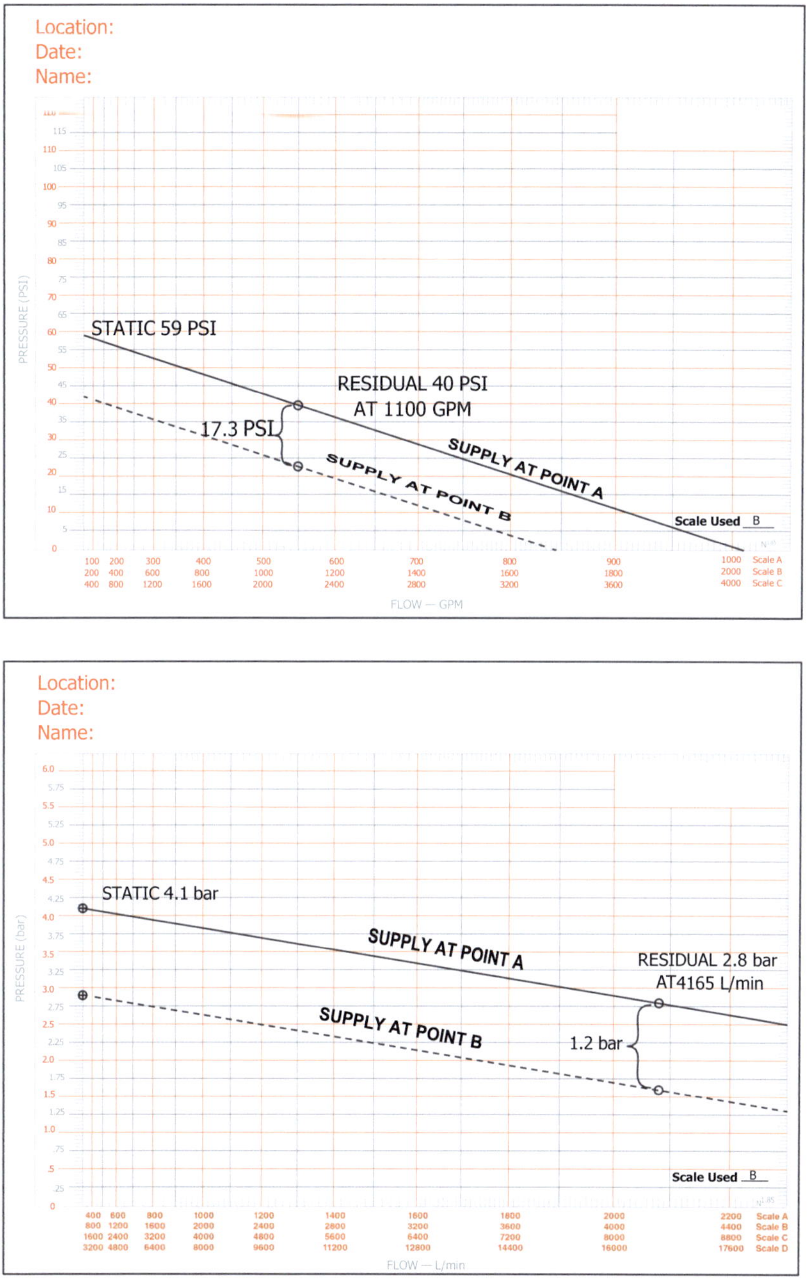

Fig. 8.15 **a** A water supply test curve corrected for elevation difference **b** A water supply test curve corrected for elevation difference in S.I. units

Since elevation corrections have nothing to do with pipe size, the amount of water flowing or any of the other factors affecting friction loss, the correction of 17.3 psi (1.2 bar) must be applied to every pressure point. Therefore, a curve corrected for elevation will always be parallel to the unadjusted curve. In this case, since Point B is at a higher elevation than Point A, the adjusted curve will fall below the unadjusted curve. This is illustrated by the dotted line in Fig. 8.15a and b, which is 17.3 psi (1.2 bar) below the unadjusted curve at every point.

Of course, it is possible that both an adjustment for friction loss and an adjustment for elevation may be necessary. Either may be done first as long as the corrections are made as previously described.

Some Final Cautions

In summary, a few warnings are in order. Flow test results should never be considered more than approximations of the water supply available. There are several things that contribute to this. Pressures existing on a water system change from month to month, from day to day, and even from hour to hour. Pressures tend to be higher in winter months, lower on weekends, and higher in the early morning hours, all depending upon the consumption demand on the system. Also, tests conducted by different people will yield different results. Five different people may read five different pitot pressures on the same hydrant at the same time. In addition to human inconsistency, gauge accuracy also has an effect. The use of nozzles or playpipes will also give different (and better) results than tests where pitot readings are taken directly from the hydrant.

With all of the above contributing to possible variation in results, it should be clear why flow test results must be treated as approximations. It should also be clear why many fire marshals and plan reviewers are nervous about sprinkler system demand points that fall right on the supply curve; a curve that they know is only an approximation.

A few words should also be said about the currency of flow test results. Flow test results at a specific location will change over time. This is due to aging and deterioration of the piping, increase in water consumption in the community, and changes in the water system. For these reasons, flow test results more than three years old should be considered invalid. In a community that is growing rapidly, test results more than one year old may even be questionable.

Invalid test techniques can also yield invalid results. Proper techniques have already been described. However, one of the most common questionable techniques involves the use of only a single fire hydrant to obtain all of the test data. Using one hydrant to get the static, residual, and pitot pressures yield results inconsistent with the method using two or more hydrants. It can be pointed out, however, that using a pressure gauge on the opposite side of the hydrant from a flowing outlet yields pressures almost identical to a pitot tube placed in the open stream (Fig. 8.16). This

Fig. 8.16 A hydrant cap and gauge attached to the flow hydrant will give approximately the same reading as a pitot tube

is easily explained by applying Bernoulli's Equation and remembering that all of the energy in the open stream is kinetic.

Finally, it should be emphasized that flow test results permit the approximation of the water supply available *inside* the water main at the test hydrant location. People often mistakenly believe the test results indicate the flows available from the test hydrant itself. Referring back to Example 8.2, the test results indicated a flow of more than 4500 gpm (17,000 L/min) available at 20 psi (1.4 bar). It is unlikely that any single hydrant can deliver 4500 gpm (17,000 L/min). To make full use of the water supply available at this location would require opening several hydrants.

Summary of Important Concepts

1. Flow testing of hydrants is the best way to determine the available water supply at any location.
2. Flow tests must result in three pieces of information:
   - Static pressure
   - Residual pressure
   - Measured flow
3. The test hydrant, rather than the flow hydrant, is the location of test relevancy.
4. Flow test results may be analyzed graphically or mathematically to determine the adequacy of existing water supplies.
5. Flow test results may be transferred to different locations with proper corrections for friction loss and changes in elevation.

Exercises

8.1 Plot the following flow test results:

a. static pressure: 108 psi (7.4 bar) residual pressure: 48 psi (3.3 bar) pitot pressure: 36 psi (2.5 bar) $C = 0.8$d = 2 1/2 in. (65 mm)
b. static pressure: 61 psi (4.2 bar) residual pressure: 42 psi (2.9 bar) pitot pressures: 31, 32, 24, and 24 psi (2.1, 2.2, 1.7 and 1.7 bar) $C = 0.9$ $d = 2$ 1/2-in. (65 mm)
c. static pressure: 83 psi (5.7 bar) residual pressure: 39 psi (2.7 bar) measured flow: 1185 gpm (4485 L/min)
d. static pressure: 76 psi (5.2 bar) residual pressure: 52 psi (3.6 bar) measured flow: 1930 gpm (7300 L/min)

8.2 Plot the following flow test results:

a. static pressure: 74 psi (5.1 bar) residual pressure: 34 psi (2.3 bar) pitot pressure: 28 psi (1.9 bar) $C = 0.97$ $d = 1.75$ in. (44 mm)
b. static pressure: 90 psi (6.2 bar) residual pressure: 47 psi (3.2 bar) pitot pressure: 26, 26, 22 and 22 psi (1.8, 1.8, 1.5 and 1.5 bar) $C = 0.8$ $d = 2$ 1/ 2 in. (65 mm)
c. static pressure: 48 psi (3.3 bar) residual pressure: 36 psi (2.5 bar) measured flow: 2300 gpm (8700 L/min)
d. static pressure: 66 psi (4.6 bar) residual pressure: 39 psi (2.7 bar) measured flow: 1060 gpm (4000 L/min)

8.3 Using the flow test results from Exercise 8.1, mathematically determine the following and compare with the results from the graphs.

a. The flow available at a residual pressure of 20 psi (1.4 bar)
b. The residual pressure available at a flow of 1000 gpm (3785 L/min)

8.4 Using the flow test results from Exercise 8.2, mathematically determine the following and compare with the results from the graphs:

a. The flow available at a residual pressure of 20 psi (1.4 bar)
b. The residual pressure available at a flow of 1000 gpm (3785 L/min)

8.5 Will a sprinkler system demand of 850 gpm (3220 L/min) at 48 psi (3.3 bar) be available at a location where the flow test results are as follows: static pressure: 75 psi (5.2 bar) residual pressure: 35 psi (2.4 bar) measured flow: 1130 gpm (4280 L/min)

8.6 Will a fire pump demand of 2250 gpm (8500 L/min) at 20 psi (1.4 bar) be available at a location where the flow test results are as follows: static pressure: 93 psi (6.4 bar) residual pressure: 47 psi (3.2 bar) measured flow: 1460 gpm (5525 L/min)

8.7 At Point A, the following flow test results were obtained: static pressure: 68 psi (4.7 bar) residual pressure: 47 psi (3.2 bar) measured flow: 1320 gpm (5000 L/min)

Graphically illustrate the water supply available at Point B which is 1450 ft (440 m) away from, but connected to, Point A by a dead-end, cement-lined, 8-in. (200 mm) cast-iron main. Both A and B are at the same elevation. What flow is available at Point B with a residual pressure of 20 psi (1.4 bar)?

8.8 At Hydrant A, the following flow test results were obtained:

static pressure: 82 psi (5.7 bar)
residual pressure: 53 psi (3.7 bar)
measured flow: 980 gpm (3710 L/min)

Graphically illustrate the water supply available at Hydrant B which is 400 ft (120 m) away from, but connected to, Hydrant A by a 4-in. (100 mm) unlined, cast-iron, dead-end main with $C = 100$. Both A and B are at the same elevation. What flow is available at Hydrant B with a residual pressure of 20 psi (1.4 bar)?

8.9 If Point B from Exercise 8.7 is 26 ft (8 m) higher in elevation than Point A, graphically depict the water supply available at Point B. Will the water supply at Point B provide a fire pump demand of 1125 gpm (4260 L/min) at 20 psi (1.4 bar)?

8.10 If Hydrant B from Exercise 8.8 is 38 ft (12 m) lower in elevation than Hydrant A, graphically depict the water supply available at Hydrant B. Will the water supply at Hydrant B provide a sprinkler system demand of 750 gpm (2840 L/min) at 68 psi (4.7 bar)?

Selected Answers Found in Appendix K.

Supplemental Readings

Alliance of American Insurers. *Simplified Water Supply Testing*. Alliance of American Insurers: Chicago, IL, 1982.

National Fire Protection Association. *The Fire Protection Handbook*. National Fire Protection Association: Quincy, MA, 21st Edition, Sec. 15, Ch. 2.

National Fire Protection Association. NFPA 291, *Recommended Practice for Fire Flow Testing and Marking of Hydrants*. National Fire Protection Association: Quincy, MA, 2025 Edition.

Chapter 9
Industrial Fire Pumps and Pump Installation Components

Abstract This chapter introduces commonly used industrial and commercial fire pump installations, focusing mostly on horizontal split-case pumps and vertical shaft turbine pumps. Pump drivers are also presented, including electric motors, diesel engines, and steam turbines. Standard pump installations are introduced including the valves, piping, gauges, pump controllers, and testing equipment.

This chapter addresses the types of fire pumps commonly installed in business and industrial facilities to support existing water supplies. A discussion of the valves, piping, components, and accessories that make up a standard pump installation is also included. The requirements for an installation to qualify as standard are those listed in NFPA 20, *Standard for the Installation of Stationary Pumps for Fire Protection* [1].

Common Fire Pump Types

The earliest fire pumps were positive-displacement types and used either rotary gears or pistons. However, modern industrial fire pumps are centrifugal pumps. Pressure is added to the water by the centrifugal force of a rotating wheel or the rotating vanes of a turbine. That is why they are called centrifugal pumps.

Several basic types of centrifugal fire pumps exist: end-suction pumps, in-line pumps, horizontal split-case pumps, vertically mounted split-case pumps, and vertical-shaft turbine pumps. This chapter is limited to the discussion of the most common: horizontal split-case pumps, vertical-shaft turbine pumps, and overhung impeller in-line pumps.

P. Brock and V. Charter, *Fire Protection Hydraulics and Water Supply Analysis*, The Society of Fire Protection Engineers Series,
https://doi.org/10.1007/978-3-032-09877-1_9

Horizontal Split-Case Pumps

These pumps are referred to as split-case pumps. This is because the casing in which the shaft and impeller rotates is split in the middle and can be separated, exposing the shaft, bearings, and impeller. This provides easy access for repair or replacement of internal components (Fig. 9.1).

Two types of split-case pumps exist: the horizontal and the vertically mounted. The horizontal split-case centrifugal fire pump is the more common of the two types and is acceptable for use where water can be supplied to the pump under some pressure. This is because the pump is not self-priming. The pump must be supplied by a public water supply system or a tank located above the level of the pump-intake port. Otherwise, priming of the pump is necessary, which is unacceptable where automatic pump operation is required.

In this pump, the water pressure is increased due to the rotation of a wheel inside the pump casing. This wheel is called an *impeller*. Water is fed into the center, or eye, of the impeller from one or both sides and then thrown to the outer edges by the rotation of the impeller. As it is slung to the outer edge of the impeller, the energy or pressure increase depends upon how fast the impeller is turning and the diameter and design of the impeller. The impeller turns on a shaft, which is usually driven by an electric motor or a diesel engine. Though relatively rare, you may occasionally encounter a pump driven by a steam turbine.

Pumps are often referred to as single-stage or multiple-stage pumps. A single-stage pump is one with a single impeller. Most horizontal split-case pumps are single-stage pumps. Multiple-stage pumps have the ability to deliver higher pressures and are commonly used for standpipe water supplies in high-rise buildings. Historically, most pumps provided on fire department pumper trucks have been two-stage pumps.

Fig. 9.1 A horizontal, split-case fire pump

UL list as standard, fire pumps rated as small as 25 gpm (95 L/min). However, you will not often encounter pumps rated at less than 500 gpm (1892 L/min). From a practical standpoint, the only time less than 500 gpm (1892 L/min) is adequate for fire protection purposes is with the Class II standpipe and hose system, which is composed of hose cabinets and 1–1/2 in. (38 mm) hoselines, and with small, light-hazard sprinkler systems. The most common pump sizes are 500, 750, 1000, and 1500 gpm (1892, 2839, 3785 and 5677 L/min), but pumps are manufactured with ratings of 5000 gpm (18,925 L/min) or larger. No standard pressure rating exists, but single-stage horizontal split-case centrifugal pumps may be purchased with pressure ratings as low as 40 psi (2.7 bar) and as high as 290 psi (20 bar) or higher. No standard fire pump is permitted to have a pressure rating less than 40 psi (2.7 bar).

As its name implies, the horizontal split-case pump has its shaft oriented horizontally. A grease fitting is provided for the bearing at each end of the shaft, allowing proper lubrication; and lubrication at regular intervals is essential. There are also fiber packings on each end of the shaft; these are inserted in the packing gland (which is a kind of sleeve). These fiber packings are squeezed into the packing gland on either side of the impeller. Their function is to seal the shaft and prevent excess water leakage around the shaft. They are intended to be water cooled and lubricated, so a small leakage is required. About one drop per second is generally considered adequate. Too much or too little leakage requires tightening or loosening of the packing gland.

Additional important components for pump maintenance are case wearing rings. These are located at the interface between the pump casing and the impeller on each side of the impeller. Their function is to prevent wearing of the pump casing. Since these wearing rings suffer all the wear, they will have to be checked regularly and replaced periodically.

Every fire pump must have a driver to turn the pump impeller and an electric controller to turn the pump on and off. The fire pump must be installed in a system of piping and valves. More attention will be given to these various components in the following sections.

Vertically Mounted Split-Case Pumps

Even though the name implies that a horizontal split-case pump will be installed with the drive shaft in the horizontal position, there is a modern variation. In some modern installations, where the pump is driven by an electric motor, the motor may be installed on top of the pump (Fig. 9.2). This is a standard installation technique and saves floor space. This is the type of pump referred to as the vertically mounted split-case pump.

This pump is basically the same as the horizontal split-case and has identical applications. The impeller rotates on a shaft within the pump casing. The water is discharged in a direction perpendicular to the shaft, and the shaft rotates on ball-type bearings at each end.

Fig. 9.2 A split-case fire pump mounted vertically

In-Line Pumps

In-line pumps are a variety of centrifugal pump. With this pump, the driver (usually an electric motor) is supported by the pump, with the pump shaft positioned vertically with respect to the ground. The pump discharge and suction openings are approximately on the same centerline (Fig. 9.3). These pumps are also described as having "overhung impellers" and the motor can be coupled to the impeller shaft in several different coupling methods. The application of the in-line pumps is identical to the horizontal and vertically mounted split case pumps. However, the range of gpm (L/min) ratings for the in-line pumps is not as great as for the horizontal split-case pumps. An important advantage of the in-line pump is conservation of floor space.

Vertical-Shaft Turbine Pump

The vertical-shaft turbine centrifugal fire pump was originally designed to pump water from wells. Presently, they still have application where the water supply is from a non-pressurized source. This includes wells, ponds, rivers, and underground storage tanks. Vertical-shaft pumps never require priming because the rotating turbines are positioned down inside the water source. Vertical-shaft pumps ordinarily have more than one turbine and are therefore multistage pumps. The number of turbines present is determined by the pressure rating; increase the number of impellers and a greater pressure will be developed.

Figure 9.4 illustrates the essentials of the vertical-shaft pump. Note that each turbine sits in a slight enlargement of the casing (referred to as a bowl). Each bowl has a wear ring to prevent wearing of the bowl itself. The shaft can be either water

Fig. 9.3 An inline pump driven by an electric motor

lubricated or oil lubricated. A strainer is located at the bottom to keep fish, snails, leaves, and other objects out of the pump.

The shaft is turned by either an electric motor, a steam turbine, or a diesel engine. There is also a fiber packing at the top of the shaft to seal the shaft against water leakage. The fiber packing is contained inside a packing gland. The gland can be either tightened or loosened to obtain the proper packing lubrication.

Though not common, the vertical-shaft turbine pump may also be used to support pressurized water supplies. It is commonly referred to as a “can-type” installation.

Essentially, a cylinder or canister is constructed, and the city water supply feeds this canister. The vertical-shaft pump is simply set into this canister and is then able to boost the pressure of the incoming water supply (Fig. 9.5).

The available ratings of vertical-shaft fire pumps range from 250 gpm (946 L/min) to 5000 gpm (18,925 L/min). Again, the most common are the 500, 750, 1000, and 1500 gpm (1892, 2839, 3785 and 5677 L/min) pumps. As with the horizontal-shaft pumps, there is no standard pressure rating; however, manufacturers may have models ranging in pressure from 40 psi (2.7 bar) to over 500 psi (34 bar). The reason that vertical-shaft pumps are available with higher pressure ratings than horizontal-shaft pumps is that some of the pressure developed by the vertical-shaft pump must be expended in simply getting the water up to ground level. How much pressure is lost in bringing the water to the surface depends upon how far below grade the impellers are located.

Even though the impellers of a vertical-shaft pump are located below ground level inside the water source, the driver and control panel are accessible. Vertical-shaft pumps are usually driven by an electric hollow-shaft motor mounted above the pump or by a diesel engine through a right-angle gear drive.

Both horizontal- and vertical-shaft pumps may have a pressure rating in terms of feet (meters) of head rather than in pounds per square inch (psi) or bar. Therefore,

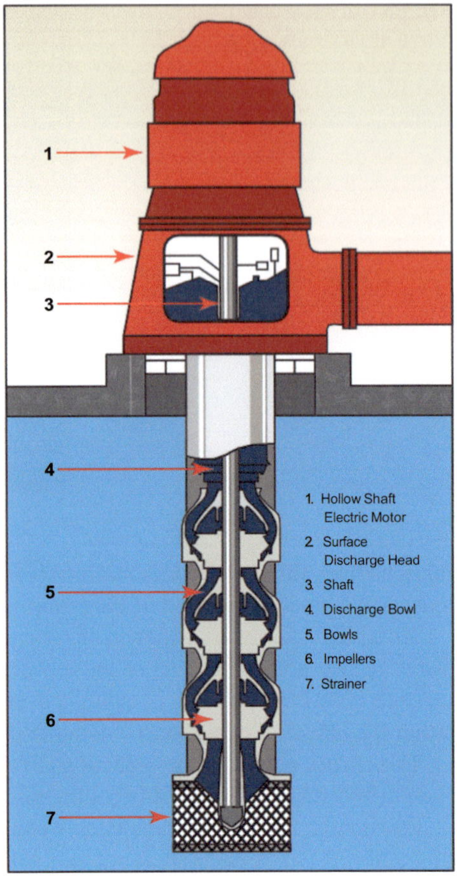

Fig. 9.4 A vertical-shaft turbine fire pump

a pump that is rated at 231 ft (70 m) of head is equivalent to a pump that is rated at 100 psi (6.9 bar), since 231 ft × 0.433 psi/ft is equal to 100 psi (70 m × 0.098 bar/m = 6.9 bar). Some pumps are designed to pump seawater. If these pumps are rated in feet (meters) of head, 0.445 psi/ft (0.101 bar/m) should be used to convert to psi (bar).

Industrial fire pumps are tested and listed by third party organizations such as UL Solutions or FM. The companies that manufacture fire pumps and descriptions of their equipment are contained in lists published by these two organizations.

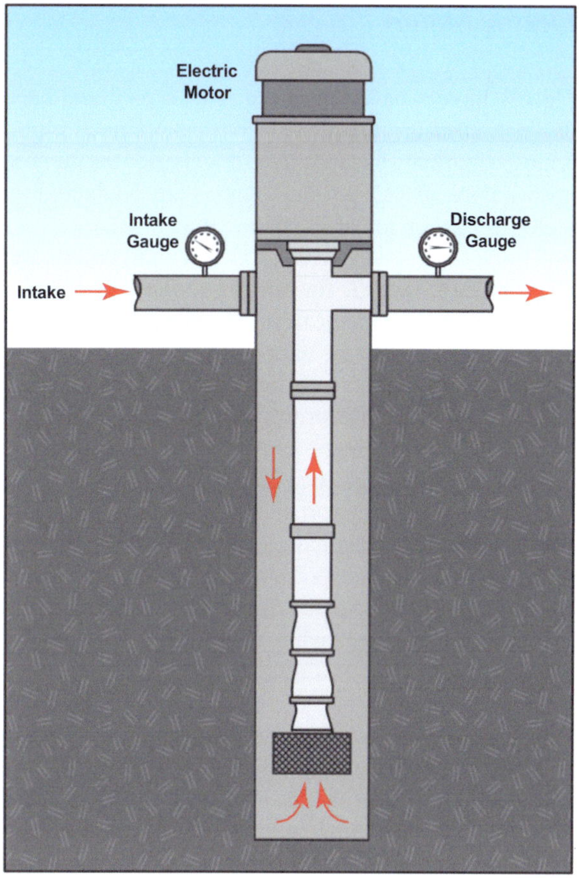

Fig. 9.5 A can-type vertical-shaft fire pump

Pump Drivers

The pump is just one component of many in an industrial fire pump installation. In the next two sections, we shall look at two of the most important components, the drivers and controllers. The drivers are the engines or motors used to turn the pump. The controllers are the electric control panels used to switch the pumps on and off and to control their operation.

Currently, three types of drivers are acceptable for use with industrial fire pumps: electric motors, diesel engines, and steam turbines. Before 1974, other kinds of internal combustion engines, such as gasoline engines and natural gas or propane engines, were permitted. These engines used a spark plug ignition system and were found less reliable in starting than the diesel engine. You may still see some of these kinds of engines on older installations, but modern pump installations should have diesel engines if an internal combustion driver is used.

Electric Motors

Electric motors have long been an accepted and dependable source of power for driving centrifugal fire pumps. For the horizontal split-case pump, the electric motor is most commonly located on the floor beside the pump as shown in Fig. 9.6. However, to save floor space, split-case pumps are available for vertical installation with the motor located on top as illustrated in Fig. 9.2.

The electric motors driving vertical-shaft pumps are mounted vertically on top of the pump turbine shaft. They must be hollow-shaft motors with an antireverse ratchet, which prevents the motor from turning backwards. If not equipped with this feature, the motor can turn in the wrong direction when water drains down the column after the pump is turned off or if the motor is wired incorrectly. The motor must also be equipped with thrust bearings to carry the weight of the pump plus withstand the thrust of the water (Fig. 9.7). An in-line pump driven by an electric motor is shown in Fig. 9.3. With an in-line pump, the pump must support the weight of the motor.

One of the most important considerations is the size of the motor relative to the pump. The motor must have adequate power to turn the pump at its rated speed so that the pump can deliver its rated flow and pressure. A motor of ample size must be provided. The adequacy of the size of the electric motor will become apparent during acceptance tests when the motor is required to turn the pump under various load conditions.

A decided advantage of the electric motor is the relatively small amount of maintenance required. Proper lubrication of the motor bearings, in accordance with manufacturers' instructions, is all that is required.

A disadvantage of the electric motor involves reliability. Storms, fires, or other accidents involving power lines, transformers, or substations can leave motors without power and fire pumps useless. Wiring installations providing power for the electric motors and the controllers are required to comply with the provisions of NFPA 70, *The National Electrical Code*®. NFPA 20 does not normally require a back-up power supply unless the primary power supply is not considered reliable. However, an alternate power supply is required in buildings that are so tall that fire

Fig. 9.6 An electrically driven, horizontal split-case fire pump

Fig. 9.7 An electrically driven, vertical-shaft type fire pump

department pumpers cannot deliver adequate fire protection water supplies to the top floor. Other codes and standards also require emergency generators for high-rise buildings, hospitals, and research laboratories. Whenever an emergency generator is provided for a building, the fire pump should also be connected to this alternate power supply through an automatic transfer switch.

It is beyond the scope of this text to discuss the details of the electrical power system. However, a few points are worth noting. The pump motor should be wired so that shutting off the power to the plant or facility can be accomplished without shutting off the power to the pump motor. This can be accomplished by providing separate power services or by locating the fire pump service connection ahead of all plant service equipment. Also, power lines supplying electricity to the pump should not run through the protected building unless inside fire-resistive enclosures.

Voltages in excess of 600 should not be used for fire pump service unless approved by the authority having jurisdiction. Each electric motor should have an information plate giving the current and voltage ratings, horsepower, RPM, and service factor. This data will be useful in determining the acceptability of pump performance during the tests which will be described later in Chap. 10. Motors are required to bear a NEMA (National Electrical Manufacturers Association) label and be specifically listed for fire pump service.

Electric-motor-driven pumps are often neglected. This is partly due to their simplicity and the fact that very little maintenance is required. However, those pumps should be turned on at least monthly and permitted to run for at least 10 min.

Diesel Engine Drivers

The diesel engine is a common and reliable method of driving industrial fire pumps (Fig. 9.8). The diesel engine has proven to be the most dependable of all of the internal combustion engines and is presently the only kind of engine considered acceptable for fire protection applications. Diesel engines, like the pumps, are listed by UL Solutions and FM for fire protection applications. This means that not all engines are acceptable for driving fire pumps. Look for the label—it is the manufacturer's responsibility to properly match the motor to the pump.

Engine Power

The diesel engine must have enough power to turn the pump at rated speed under all required load conditions. The engine must turn the pump at rated speed when pumping at churn and also when pumping at 150% of its rated capacity. The horsepower ratings of the engine will commonly be found on an information plate on the engine. The horsepower varies according to the speed at which the engine is operating. For this reason, it is very important to keep an engine well maintained and properly adjusted. Otherwise, the pump cannot be expected to meet its performance specifications.

If the engine should operate at speeds that are too high, the pump could develop excessive water pressures and the engine and pump could also be damaged. If the engine operates too slowly, the pump will not develop rated pressures. The engine is therefore required to have an adjustable governor to maintain engine speed within a 10% range. The governor is required to be set to maintain the rated pump speed at maximum pump load. The engine is also required to have an over-speed shutdown

Fig. 9.8 A diesel engine driven fire pump

device. If the governor fails to limit pump speed, this device will shut down the engine at a speed of 10–20% above the rated engine speed. The shutdown device is required to be manually reset and supervised to send a trouble signal to the control panel until it is reset.

Gear Drives

The diesel engine is connected to a horizontal split-case pump by a flexible coupling. Typically there is no gearing between the engine and the horizontal split-case pump, and the pump speed is identical to the engine speed.

The vertical-shaft pump is driven by a diesel engine through a right-angle gear drive with universal joints. It is a geared connection and the primary purpose is to change drive direction. However, it is common to find a gear ratio that will provide a pump speed different from the operating speed of the engine. Proper evaluation of pump performance requires that you check carefully for a gear ratio or take pump speed measurements on the pump side of the gear drive.

The gear drives must also be equipped with the anti-reverse ratchet feature and the thrust bearings previously described for the electric motor driven vertical-shaft pumps.

Engine Requirements

The following instruments should be placed on a panel securely fastened to the engine at an accessible location.

- Tachometer—to indicate revolutions per minute (rpm) of the engine
- Oil pressure gauge—to indicate the pressure of the lubricating oil
- Temperature gauge—to indicate the temperature of the engine coolant in the engine jacket.

A speed-sensitive switch signals when the engine is running. Engines may be started electrically by storage batteries, by compressed air, or even hydraulically. Batteries are the most common method to start an engine. For reliability, each engine is required to have at least two battery units. Each battery unit must have sufficient capacity to maintain engine cranking speed through six consecutive cranking cycles of 15 s cranking and 15 s resting.

Two means of recharging the batteries must be provided. One means of recharging is to use the generator or alternator that comes with the engine. The other method is to use an automatic charger that takes power from an alternating current source. The battery charger is incorporated into the design of the controller and must be capable of fully recharging the batteries within 24 h. The batteries should be located so that they are not subject to flooding, mechanical damage, high or low temperatures, or vibration. They should be readily accessible for easy servicing and must not be located in front of the instrument panel or engine controls.

If a diesel engine is located in an environment subject to flammable vapors, starting the engine through the ordinary electric starting motor could create a fire or explosion hazard. For this reason, pneumatic and hydraulic starting is available. This is accomplished by forcing a compressed gas or water through a turbine which will turn the engine.

Cooling System

Diesel engines are water cooled and most commonly make use of closed-circuit types of cooling systems. The basic components of the system include a water pump driven by the engine, a heat exchanger, and a reliable device for regulating the water temperature in the engine jacket. The heat exchanger works by taking water from the discharge side of the pump to cool the water in the engine. The exchange of heat from the engine water to the pump discharge water takes place in the heat exchanger. The provision of water from the discharge side of the pump requires a special piping arrangement including bypass line, valves, pressure regulator, and strainers. NFPA 20 contains very specific requirements for the piping feeding and discharging from the heat exchanger.

Fuel Storage

Fuel is an important consideration for diesel engines. Safe storage, transmission, and adequate quantities must be provided. Diesel fuel is not as volatile as gasoline, but can be dangerous and must be handled carefully. Diesel fuel tanks are required to be located above ground and preferably inside of the pump room. The tank should be large enough to provide at least 1 gallon of fuel per horsepower (5.07 L per kW) plus 10% for expansion and sump. Therefore, a 200 horsepower (150 kW) engine should be provided with fuel storage capable of holding at least 220 gallons (836 L) of diesel fuel. Tanks are also required to be maintained at least two-thirds full at all times. Any exposed fuel lines must be protected against mechanical damage. Fuel piping should be rigid except where the fuel line connects to the engine. At that point, flame-resistant flexible hose is required.

For environmental protection, containment must be provided to prevent the runoff of any possible leakage from the tank. This is commonly accomplished through construction of a small containment dike around the base of the tank. The containment volume must be at least large enough to hold the entire contents of the tank. Other containment methods are also possible, including sloping the floor to direct leaks to a containment basin or placing a sill around the pump room.

For personnel safety, the exhaust fumes from the diesel engines must be piped outside the building. The integrity of the exhaust system should be continually inspected for leaks inside the pump room. The exhaust pipe should not be located close to combustible materials. Also, diesel engines are noisy, and hearing conservation practices should be in place when engines are running.

Maintenance

Diesel engines require more care and maintenance than electric motors. To ensure reliability, the engine must be started and operated for 30 min every week and should run smoothly at the rated speed. The engine should be kept clean, dry, and well lubricated. The manufacturer's instructions should be followed in scheduling oil changes, but the interval between changes should not exceed one year.

The batteries must be examined and tested regularly to make sure they are fully charged. With the automatic battery charger, it is easy to assume that the batteries will always be in good shape. Do not let yourself fall into this trap of negligence. If more water is needed in the batteries, use only distilled water and make sure the plates are kept submerged at all times. Take care to ensure that fuel tanks are kept free of water and foreign materials.

As the temperature drops, diesel engines become increasingly difficult to start. For this reason, the current standard requires that pump rooms be maintained at least at the minimum temperatures recommended by the manufacturer. If these maintenance procedures are consistently followed, the diesel-driven fire pump can provide a reliable source of fire protection water supplies.

Steam Turbines

Although not common, you may occasionally encounter a fire pump driven by steam. NFPA 20 lists steam turbines as one of the acceptable driver types. However, some of the older installations used piston-type reciprocating steam engines. Both horizontal-shaft and the vertical-shaft pumps can be driven by the steam turbine. The only feasible application of the steam-driven pump is when steam is already available in sufficient quantities and sufficient pressure. Otherwise, economic considerations would dictate use of the electric or diesel-driven equipment.

Pump Controllers

The controller governs the operation of the pump—it turns the pump on and off. The controller can be designed to operate the pump automatically by use of microprocessors, simple electronic circuits, or manual operation. The controller is usually in a large red box that is against or hangs on the wall. High voltage wiring exists inside the controller. Work inside the controller should be done only by experienced electricians or pump maintenance personnel. Mistakes there can be fatal. Let's look in detail at the functions of the controller and begin with the controllers for electric motor driven pumps.

Controllers for Electric Motor Driven Pumps

The controller should be tested and listed for fire protection use by one of the nationally recognized testing laboratories. The controller should be located inside the pump room and as close to the pump as possible. It should be protected from water discharge, and all current-carrying parts of the controller must be at least 12 in. (30.5 cm) off the floor. Figure 9.9 is a drawing of a typical automatic fire pump controller. Every automatic controller will not be exactly like this one; however, they all should have the basic components shown here.

Number 1 in Fig. 9.9 is the *circuit breaker*. The circuit breaker provides overcurrent protection and opens if too many amps are being drawn. The circuit breaker must be accessible and operable from the outside of the controller. Proper load rating of the circuit breaker is important. It must allow at least 115% of the rated full load current without tripping and also permit normal starting of the motor without tripping.

Number 2 is the *isolation switch*. This switch is located between the power supply and the circuit breaker. It is required that the circuit breaker always be opened before the isolation switch is operated. In some controllers, the isolation switch and the circuit breaker is interlocked, so the isolation switch cannot be operated with the

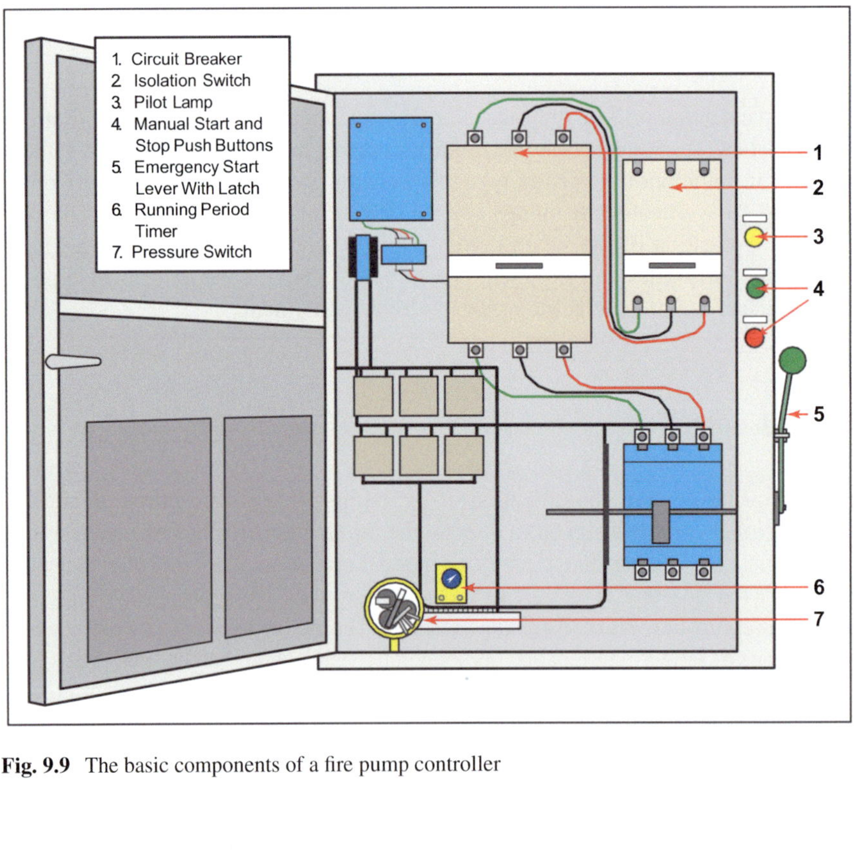

Fig. 9.9 The basic components of a fire pump controller

circuit breaker closed. In other cases, the operating handle of the isolating switch is equipped with a spring latch that requires the use of both hands to operate. The isolation switch lever is always located outside the control panel.

Number 3 is a *pilot lamp*. It may be in various locations; however, every controller should have one. It should be visible without getting inside of the controller. Its purpose is to indicate when power is available to the pump control panel.

Number 4 is the manual *start* and *stop buttons*. These are standard controller components and located on the outside of the controller enclosure. Manual starting and stopping of the pump can be accomplished by using these buttons.

Number 5 is an *emergency start lever with latch*. This switch can be latched in the operating position and provides for continuous non-automatic operation. This operation is independent of the timer or automatic starting mechanisms.

Number 6 is the *running period timer*. The timer is used to shut off the motor after starting causes have returned to normal. The running period timer is required to be set for at least 10 min. Therefore, if a pump is automatically controlled, whenever it comes on it operates for at least 10 min unless manually shut down or wired for manual shut down only.

Number 7 is the *pressure switch*. The most common method of automatically turning on industrial fire pumps is by using a pressure switch connected to the fire protection water supply system on the discharge side of the pump. The pressure switch is set to close the circuitry and turn on the fire pump in the event water pressure on the system drops below the switch setting. Such a pressure drop is ordinarily caused by a sprinkler opening on a sprinkler system or the use of hoselines somewhere on the system.

The pressure switch is adjustable with independent high- and low-pressure settings. When the pressure on the system drops to the low-pressure setting, the pump turns on. When the pressure is restored to normal, the high-pressure setting turns the pump off (if wired to do so). Ordinarily, the high-pressure setting should be about 10 psi (0.7 bar) below shut-off or churn pressure, and the low-pressure setting about 20 psi (1.4 bar) below the rated pump pressure. Consistent pressure fluctuations of sufficient magnitude on the fire protection system may require the use of a jockey pump to prevent false pump starts. The jockey pump is discussed later. It should be noted, however, that if the pump is the sole source of fire protection water supply, the pump must be set up for automatic start and manual shutdown.

Historically, pump controllers have used mercury switches to turn the pumps on and off (Fig. 9.10). In modern controllers, however, pressure sensing and control is through electronic pressure transducers (Fig. 9.11) and is programmable through microprocessors with digital readouts (Fig. 9.12).

Also, in modern controllers, provision for measuring current and voltage are provided as a part of the control panel. Measuring current and voltage is an integral part of pump testing, as demonstrated in Chap. 10, Fire Pump Applications, Testing, and Maintenance. Some modern fire pump controllers may also be wired for "soft" starting. This allows the motor to take up to 10 s to come up to full speed, which reduces the load on the power supply while also reducing the water hammer created by pump start up.

Fig. 9.10 A Mercoid type pressure switch

Fig. 9.11 Modern pressure switch

Sometimes a pump is used to supply a deluge system or other fire protection system where turning the pump on independent of a pressure drop is needed. In this case, relays can be used to turn the pump on when a signal is received from a fire- or smoke-detection device. If two or more pumps are connected in parallel, the controller must be capable, through provision of sequential-timing devices, to prevent the motors from starting at the same time. If the demand becomes great enough that both pumps are needed, the controller is required to provide intervals of 5–10 s between starting of additional pumps.

If the pump installation is the only source of water supply for a sprinkler system or standpipe system, the automatic start controller is required to be wired for manual

Fig. 9.12 Modern electronic control panel

shutdown. This is to ensure that the pump will be turned off only after a fire emergency is over. Also, if the pump room is not constantly attended, audible or visual alarms must be transmitted to an attended location to signal that the pump is running or to signal that the power supply to the pump has been interrupted.

Some pumps may not be equipped with automatic controllers. If manual controllers are used, there will be both a manually operated electric switch to turn the pump on and off, plus a mechanical control consisting of a lever or handle which can be latched in the ON position.

Diesel Engine Controllers

The controllers for the diesel engines are not interchangeable with the controllers for electric motors. The electric motor controllers were designed to start the pumps by closing high voltage circuits. On the diesel controller, the main function is to close the circuit between the storage batteries and the engine starter motor.

One feature to notice is the alarm and signal devices located on the controller itself. If the controller is an automatic controller, a pilot light indicates when the controller is in the AUTOMATIC position. Separate lights and a common audible alarm are also required to indicate the following:

- Low engine oil pressure
- High engine coolant temperature
- Failure of engine to start automatically
- Engine shut down due to overspeed
- Battery failure
- Low pressure in the storage tanks when air or hydraulic starting is used.

In addition, the controller requires separate lights for battery charger failure, but this does not require the provision of the audible alarm.

If the pump room is not constantly attended, alarm signals must be transmitted to a constantly attended location. The alarm indicates when the engine has started, when the controller has been turned off or turned to manual operation, and when trouble exists with the engine or controller.

The controller is required to be equipped with a pressure recording device. The recording is required to run continuously for at least seven days without resetting or rewinding. The recorder's chart drive must be spring wound, AC electric with spring-driven backup, or air powered.

The provisions for automatic starting of the engine due to water pressure or fire protection devices are similar to those for the electric motor controller. If the pump installation is the sole supply to standpipes or sprinkler systems, the automatic start controller must be wired for manual shutdown. Also, the automatic controller is required to be arranged to automatically start the engine every week to assure reliability in engine starting. If the controller is designed to automatically shut down the engine when system conditions return to normal, it must provide a running time of at least 30 min before shutting the engine off. However, if the overspeed governor operates, the controller is to shut off the engine without time delay.

Modern engine-driven pump installations can also be provided with variable speed pressure limiting control. This control automatically adjusts the engine speed to reduce the possibility that fire system components will be subjected to pressures high enough to cause damage. Chapter 10 introduces the pump affinity laws. These mathematical relationships demonstrate that pump pressures are related to pump speed. Thus controlling pump speed can also control the pressure developed by the pump.

Pump Installation Components and Accessories

The following components fit together to make a complete fire pump installation: pipe, fittings, power supply, relief valves, test equipment, jockey pumps, gauges, alarms, and the pump house.

Pipe and Fittings

The aboveground pipe in a pump installation is required to be of steel construction. The underground pipe may be of any material complying with the requirements of NFPA 24, *The Standard for the Installation of Private Fire Service Mains and Their Appurtenances* [2]. These other materials include plastic, asbestos cement, and ductile iron. The above ground steel pipe can be joined together by screwed or flanged fittings or with grooved fittings. The pipe can also be welded together. However, cutting and welding in the pump house to modify or repair piping is permitted only if done in compliance with NFPA 51B, *Standard for Fire Prevention During Welding, Cutting, and Other Hot Work.*

Suction pipe is to be installed and tested in accordance with NFPA 24 [2]. The specifics of the testing is discussed in Chap. 10. An OS&Y type indicating control valve is required to be located in the suction line for control of the water supply. Butterfly valves are not permitted in the suction line due to the greater turbulence created by these types of valves, unless located at least 50 ft away from the pump.

The discharge pipe is to be hydrostatically tested in accordance with NFPA 20 [1]. A check valve is required in the discharge piping. This check valve permits water to flow in one direction only. This keeps water from flowing backwards through the pump. An indicating control valve is also required in the discharge line. It is to be located on the system or downstream side of the check valve. This control valve may be either an OS&Y type or a butterfly type.

The proper sizing of both suction and discharge pipe is important. With pipe sizing too small, the pump cannot perform at its rated capacity. NFPA 20 dictates what appropriate pipe sizing should be. The size depends upon the rating of the pump. The information from the table in NFPA 20 for the most common pump sizes is summarized in Table 9.1. The table indicates 5-in. (125 mm) pipe is the acceptable minimum for a 500 gpm (1892 L/min) pump. Since 5-in. (125 mm) pipe is not commonly available, 6-in. (150 mm) pipe is the smallest you will find with any pump rated at 500 gpm (1892 L/min) or larger.

In most cases, the minimum sizes are the same for both the suction line and the discharge line. For the 1000 gpm (3785 L/min) pump, the minimum for the suction is larger than the discharge. Although this is not a consistent requirement, in practice it is common to see the suction pipe larger than the discharge pipe. This is particularly true where a horizontal-shaft pump is receiving its supply from an aboveground tank. The larger piping reduces the friction loss and increases the pressure available at the pump.

When the size of the suction pipe is different from the inlet port of the pump, this means that a reducer is installed in the suction line. The reducer is required to be of the eccentric type as shown on the right in Fig. 9.13, not concentric as shown on the left. It is required to be an eccentric type in order to eliminate the possibility of

Table 9.1 Minimum pipe size

Pump rating		Suction pipe		Discharge pipe	
(GPM)	(L/min)	(Inches)	(mm)	(Inches)	(mm)
500	1892	5	125	5	125
750	2839	6	150	6	150
1000	3785	8	200	6	150
1250	4731	8	200	8	200
1500	5677	8	200	8	200
2000	7570	10	250	10	250

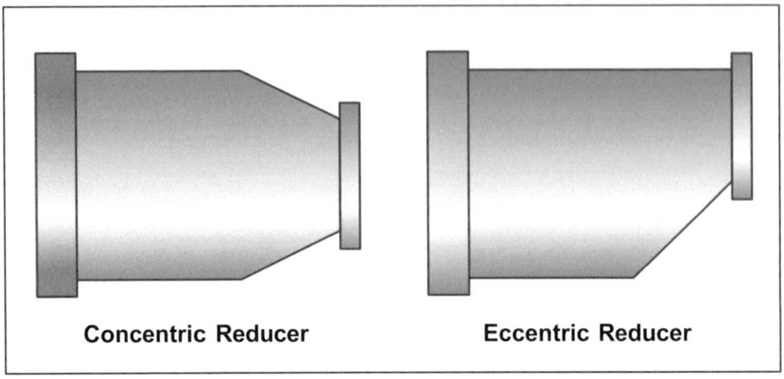

Fig. 9.13 Eccentric and concentric reducers

air pockets. Air being entrained in the water stream can reduce the efficiency of the pump and can even damage the pump due to a phenomenon known as cavitation. The eccentric reducer is to be installed with the flat side on top.

NFPA 20 also requires that no device that will restrict the flow be installed on the suction side of a horizontal split-case pump. This includes backflow prevention devices which are a common component on modern fire systems. Their proper location, therefore, would be on the discharge side of the pump. However, the standard permits the local authority to place the backflow preventer in the suction line. Another component that may be found on the suction side of a pump is a strainer. This device is most likely to exist where non-potable water is supplied to a split-case pump from an aboveground tank. Strainers can also restrict the flow to the pump, especially if undersized or dirty.

Relief Valves

Historically, if a pump was driven by an adjustable-speed driver, such as a diesel engine, a pressure relief valve would be required in the installation. More recent versions of NFPA 20 require these large relief valves only if the pressure at churn is high enough to damage system components. The purpose of this relief valve is to prevent pressures that are high enough to damage system piping or fittings. It is possible for an internal combustion engine to get out of adjustment and develop excess speeds. Pump pressure is related to the square of the pump speed. That means doubling the revolutions per minute increases the pressure developed by four times. Therefore, the relief valve is provided to open up and discharge water should the pressure become excessive. The relief valve should be located between the pump and the discharge check valve. It is supposed to discharge to an open drain or in some manner so that water flow can be visually detected.

The size of the relief valve and its discharge line depends upon the rating of the pump. If the pump was chosen with a proper pressure rating, the relief valve should be set to open up a little above the normal discharge pressure when the pump is operating at churn.

Circulation Relief Valve

Even though the large relief valve is ordinarily not present on electrically driven pump installations, a small relief valve is present. This is called a circulation relief valve or casing relief valve. It is supposed to open up and provide enough water flow into and out of the pump to prevent the pump from overheating when it is operating at churn against a closed system. This relief valve and the piping are to be 3/4-in. (19 mm) in size for pumps up to 2500 gpm (9462 L/min) and 1 in. (25 mm) for larger pumps. All pumps are required to have this circulation relief valve, except for the diesel engine-driven pump where engine cooling water is taken from the pump discharge.

The circulation relief valve should be set to flow a full stream when the pump is operating at churn, and then close off when water begins to flow in the system. The discharge piping should be installed to discharge into a drain where the flow from the line can be inspected.

Test Equipment

Every pump is required to be provided with components that allow testing of the installation. The most prominent of these components is the test manifold. The test manifold piping should be connected to the pump discharge line between the check valve and the indicating control valve. There should also be an indicating control valve in the test piping which is required to be supervised in the closed position. This piping should terminate in a hose valve header located outside of the building. The hose valve header should be equipped with 2-1/2 in. (65 mm) hose connections with a shut-off valve for each connection. This manifold and test header will permit flowing water from the pump installation through hoselines and nozzles for test purposes. The flow from the nozzles will be measured using pitot tubes.

The required size of the hose header supply pipe and the number of hose valves required depends upon the rating of the pump. It is often possible to estimate the rating of the pump by counting the hose connections. For the 500, 750, 1000 and 1500 gpm (1892, 2839, 3785 and 5677 L/min) pumps, there will be one 2 1/2-in. (65 mm) hose connection for each 250 gpm (946 L/min) of pump rating. For example, a 500 gpm (1892 L/min) pump will most often have two hose connections and a 750 gpm (2839 L/min) pump will have three hose connections. As you can see from Table 9.2, the method is not foolproof and breaks down with the 1250 and 2000 gpm (4731 and 7570 L/min) pumps.

Table 9.2 Required number of test valves

Pump rating		Number of hose valves	Test pipe size	
(GPM)	(L/min)		(Inches)	(mm)
500	1892	2	4	100
750	2839	3	6	150
1000	3785	4	6	150
1250	4731	6	8	200
1500	5677	6	8	200
2000	7570	6	8	200

Fig. 9.14 Flow from the fire pump may be measured by a flow meter

Some more recent pump installations will not have the test headers and hose valves. Instead, they will be equipped with a metering device which can be used to measure the gpm delivered by the pump (Fig. 9.14). This is acceptable, but the meter line should discharge to the outside or back to the tank if a tank is used for the supply source. Circulating back to the suction side of the pump is permitted but does not allow evaluating the condition of the suction supply.

Jockey Pumps

In some cases, there is enough leakage in the fire protection system or enough fluctuation in the pressure of the water supply to the pump to cause the automatic controller to turn the pump on periodically in nonemergency situations. This can be a serious bother if it happens often, particularly since many installations require manual shutdown after automatic operation. A jockey pump is used to prevent these false starts.

A *jockey pump* is a small-capacity, high-pressure pump used to maintain constant pressures on the fire protection system. This pump takes suction from the fire pump suction line and discharges into the fire pump discharge line on the system side or downstream side of the indicating control valve. The jockey pump should have adequate capacity to keep up with any leaks in the system. However, it should be small enough so that any demand on the fire protection system, even a single sprinkler opening, will result in fire pump operation.

The pressure rating of the jockey pump should be high enough to maintain the desired fire protection system pressure. The pressure switch in the fire pump controller should be set to correspond to the system pressure maintained by the jockey pump. Thus, when the jockey pump cannot maintain the system pressure due to a demand on the system, the fire pump controller will activate the fire pump. The jockey pump installation requires the provision of a check valve in the discharge pipe from the jockey pump, as well as indicating control valves. In some cases, pressure relief valves are even required.

Jockey pumps have an additional benefit. By maintaining a relatively high pressure on the fire system, the impact of the water hammer created when the pump starts is diminished. Thus, the potential for damage to the fire system is diminished.

Air Release Valves

Pumps which are automatically controlled are required to be provided with an automatic air release. This allows the release of any air that might be trapped inside the pump. This air release is a float-type device that automatically closes when the pump casing fills with water. It has been pointed out previously that air inside the pump can reduce the pump efficiency and even damage the pump.

Gauges

On a horizontal split-case pump installation, there should be two gauges. One of the gauges should be located near the discharge port of the pump and the other near the intake. The gauge on the discharge side should be at least 3-1/2 in. (89 mm) in diameter and be capable of registering pressures of at least 200 psi (13.8 bar) or twice the rated pressure of the pump, whichever is larger. However, you will probably encounter gauges of all sizes and with various pressure ranges.

The suction side gauge is also required to be 3-1/2 in. (89 mm) in diameter and have a pressure range of two times the maximum suction pressure. If suction pressure is anticipated to be below 20 psi (1.4 bar), a compound pressure gauge is required which reads both positive and negative pressures (Fig. 9.15).

Fig. 9.15 Compound pressure gauge

A vertical-shaft pump is required to have only a single gauge on the discharge pipe. Special attention should be given to pressure gauges. Inaccurate gauges can invalidate test results and give no idea how well a pump is performing. Often during the testing of a pump, the gauge needle may vibrate so widely that an accurate reading is very difficult. If this is the case, the use of a liquid-filled gauge can reduce the vibration problem.

Component Arrangement

The major components of a pump installation have now been identified. Shown in Fig. 9.16 is a line drawing of a typical fire pump installation using a horizontal split-case pump. Note that the indicating control valves on the supply sides of the fire pump and jockey pump must be of the OS&Y type. The other three control valves may be of the OS&Y or butterfly type. Note the locations of the check valves and the test header connection.

In some instances, the supply for the pump may be taken from a city water supply system that has pressures high enough to be of value, even without the pump. In this case, an additional bypass line should be provided around the pump (Fig. 9.17). Ordinarily, this bypass line should be as large in diameter as the pump discharge line. It may seem to you that there are a lot of indicating control valves required in a pump installation. The reason for so many is to be able to isolate any component, such as the jockey pump or the check valve in the bypass line. This allows working on or even removing that component without completely shutting off the pump installation.

Figure 9.18 further illustrates a horizontal-shaft pump installation taking suction from an aboveground storage tank. The items identified are as follows:

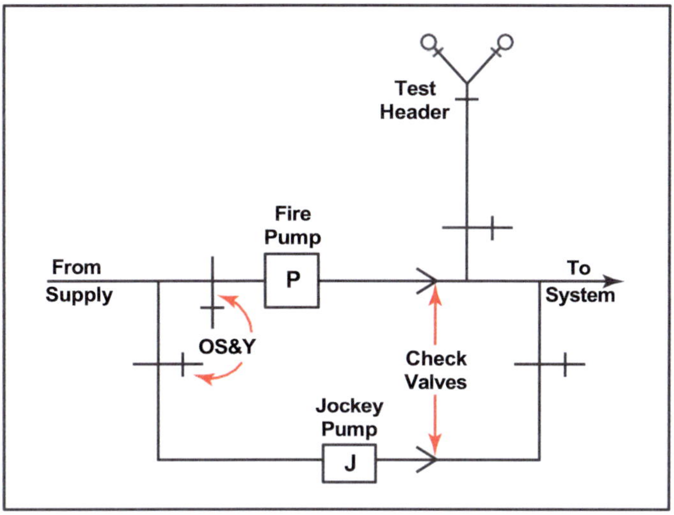

Fig. 9.16 The standard arrangement for a fire pump installation

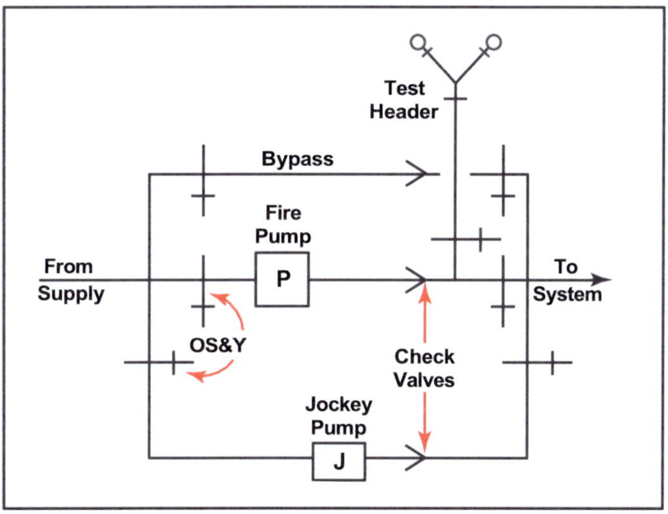

Fig. 9.17 The arrangement for a fire pump installation that is equipped with a bypass

#1—Storage tank
#2—Entrance elbow
#3—Suction pipe protected from freezing by the frost-proof casing (#4)
#4—Frost-proof casing
#5—Flexible coupling for strain relief
#6—OS&Y valve controlling the suction line

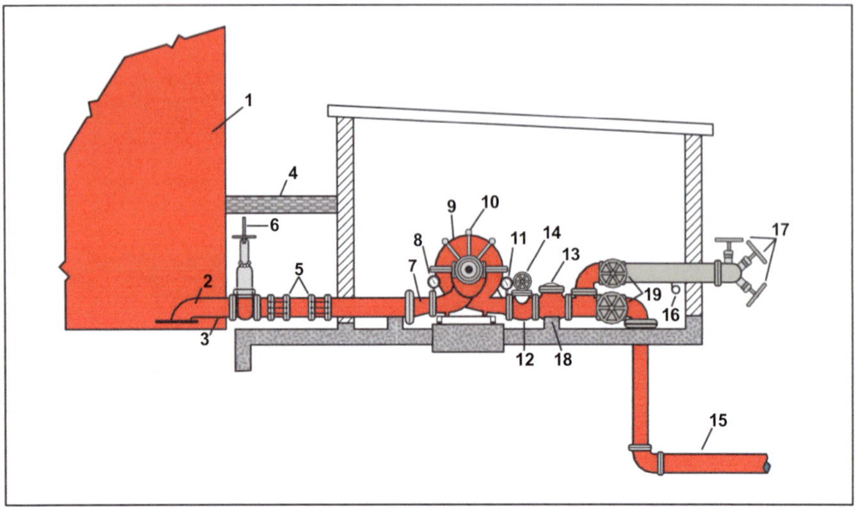

Fig. 9.18 A horizontal split-case fire pump taking suction from a tank

#7—Eccentric reducer (Note again how it should be oriented with the flat side on top.)
#8—Suction pressure gauge
#9—The pump itself
#10—Automatic air release
#11—Discharge pressure gauge
#12—A reducing tee (A reducing tee will be used if the discharge pipe is larger than the discharge port of the pump.)
#13—The check valve
#14—If a diesel engine is used for a driver, a relief valve may be needed
#15—The discharge pipe to the system
#16—The ball-drip valve installed to drain the test line which supplies the hose connections
#17—Hose connections
#18—Support for the piping
#19—Represents the indicating control valves in the discharge line and the test line.

The general piping arrangements and components of an installation using a vertical-shaft pump are shown in Fig. 9.19. Note that the components in the discharge side of the pump are essentially the same as they were for the horizontal pump. There is the relief valve, if needed, the check valve, and the test and discharge pipes with their indicating control valves. Notice the single pressure gauge in the discharge line.

A vertical-shaft pump often takes suction from a wet pit. This type of installation would ordinarily exist where the water supply source is a pond, lake, or river. An important feature here is the double screening required between the pond and pit

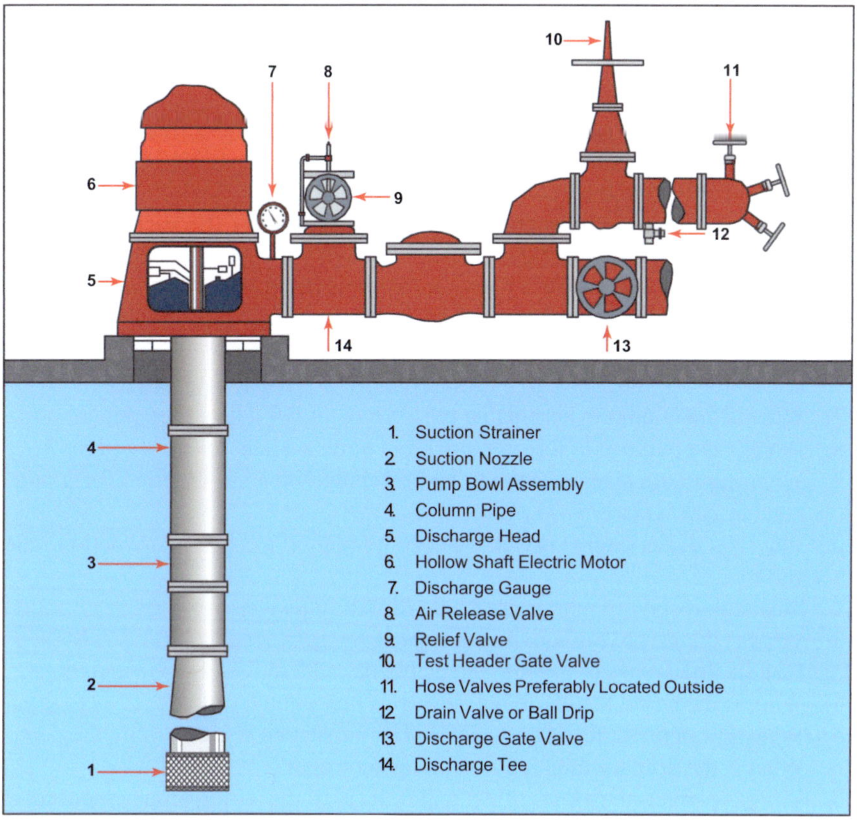

Fig. 9.19 A vertical-shaft, turbine tube fire pump

which is designed to keep foreign matter out of the pump. The screens should be removable for easy cleaning.

Summary of Important Concepts

1. Horizontal split-case and vertical-shaft turbine pumps are the types most commonly used in industrial and commercial installations.
2. Modern pumps may be driven by diesel engines, electric motors, or steam turbines.
3. Pump controllers are used to automatically turn pumps on.
4. Standard pump installations include valves, piping, gauges, and testing equipment.

Exercises

1. How many pressure gauges will normally be found on a vertical shaft pump installation?
2. What type of relief valve will be found on the electric motor driven pump but should not be found on the diesel-driven pump?
3. What type of indicating control valve is required on the suction side of a horizontal split-case pump?
4. In what situation is a bypass line around the horizontal split-case pump needed?
5. When should a vertical-shaft turbine pump be used?
6. How many 2-1/2 in. (65 mm) hose connections should be provided on a test header for a 1500 gpm (5677 L/min) pump?
7. What is the minimum acceptable pressure rating for a listed fire pump?
8. What is the purpose of the pressure relief valve on a diesel-driven pump?
9. What minimum sizes of suction and discharge pipe are required for a pump rated at 2000 gpm (7570 L/min)?
10. Why is a diesel engine-driven pump not required to have a circulation relief valve?
11. Most automatic fire pumps are set up to start in response to what?
12. What is the purpose of a jockey pump?
13. List the three types of drivers acceptable on modern fire pump installations?
14. What is the NFPA standard that addresses fire pump installations?
15. What type of pressure gauge is required on the suction side of a split-case pump?
16. What is the primary function of a pump controller?
17. What is perhaps the most critical requirement of the wiring arrangement for an electric motor driving a fire pump?
18. What is the lowest gpm (L/min) rating of any pump recognized by NFPA 20?
19. What temperature must be maintained in a pump room housing a diesel driven pump?
20. For automatic starting, what type of water supply is required for a split-case pump?

Selected answers are found in Appendix K.

References

1. NFPA 20, *Standard for Installation of Centrifugal Pumps* (Quincy: National Fire Protection Association, 2025).
2. NFPA 24, *Standard for the Installation of Private Fire Service Mains and Their Appurtenances* (Quincy: National Fire Protection Association, 2025).

Supplemental Readings

National Fire Protection Association. *The Fire Protection Handbook*. National Fire Protection Association. Quincy, MA, 21st edition, Sec. 15, Ch 6 and Ch. 7.

National Fire Protection Association. *Stationary Fire Pumps and Standpipe Systems Handbook*. National Fire Protection Association: Quincy, MA, 2022 edition.

Chapter 10
Fire Pump Applications, Testing, and Maintenance

Abstract This chapter introduces the basics of fire pump applications. This includes the three standard points of flows and pressures utilized for testing a fire pump. Fire pumps must be properly maintained and started on a regular basis. Additionally, a performance test should be conducted annually on each pump installation.

The most common type of water supply for fire protection comes from a public water supply system. These water supplies can be adequate in every respect and reasonably reliable. However, it is common to find industrial or commercial facilities in locations where there is no public water supply, or the public water supply is too weak to provide for fire protection needs. In such cases, adequate fire protection water supplies can still be provided by using a fire pump.

Two situations can be identified which require the use of fire pumps:

- The public supply can deliver enough gpm (L/min), but the pressure is too low to meet the fire protection demand.
- The public supply may or may not exist and can provide neither an adequate number of gpm (L/min) or adequate water pressure.

Let's begin by looking at the first of these situations.

Suppose a test conducted on a water system reveals that a static pressure of 58 psi (4.0 bar) exists. A residual pressure of 22 psi (1.5 bar) is available with 1450 gpm (5490 L/min) flowing; 1450 gpm (5490 L/min) is a lot of water, but 22 psi (1.5 bar) is not much pressure at all. If a building near this test location has a sprinkler system demand of 1200 gpm (4540 L/min) at a pressure of 48 psi (3.3 bar), the existing water supply is not adequate. The demand point falls well above the supply curve on the water supply graph (Fig. 10.1a and b).

A fire pump could be used in this situation to boost the pressure of the existing water supply to a point where it could be considered adequate. A fire pump used in this situation is sometimes referred to as a *booster pump*. If a 1000 gpm (3785 L/min) fire pump rated at 40 psi (2.7 bar) is installed on this water system, the resultant combination of a booster pump and public water system provides a water supply that graphs out as shown in Fig. 10.2a and b.

P. Brock and V. Charter, *Fire Protection Hydraulics and Water Supply Analysis*, The Society of Fire Protection Engineers Series,
https://doi.org/10.1007/978-3-032-09877-1_10

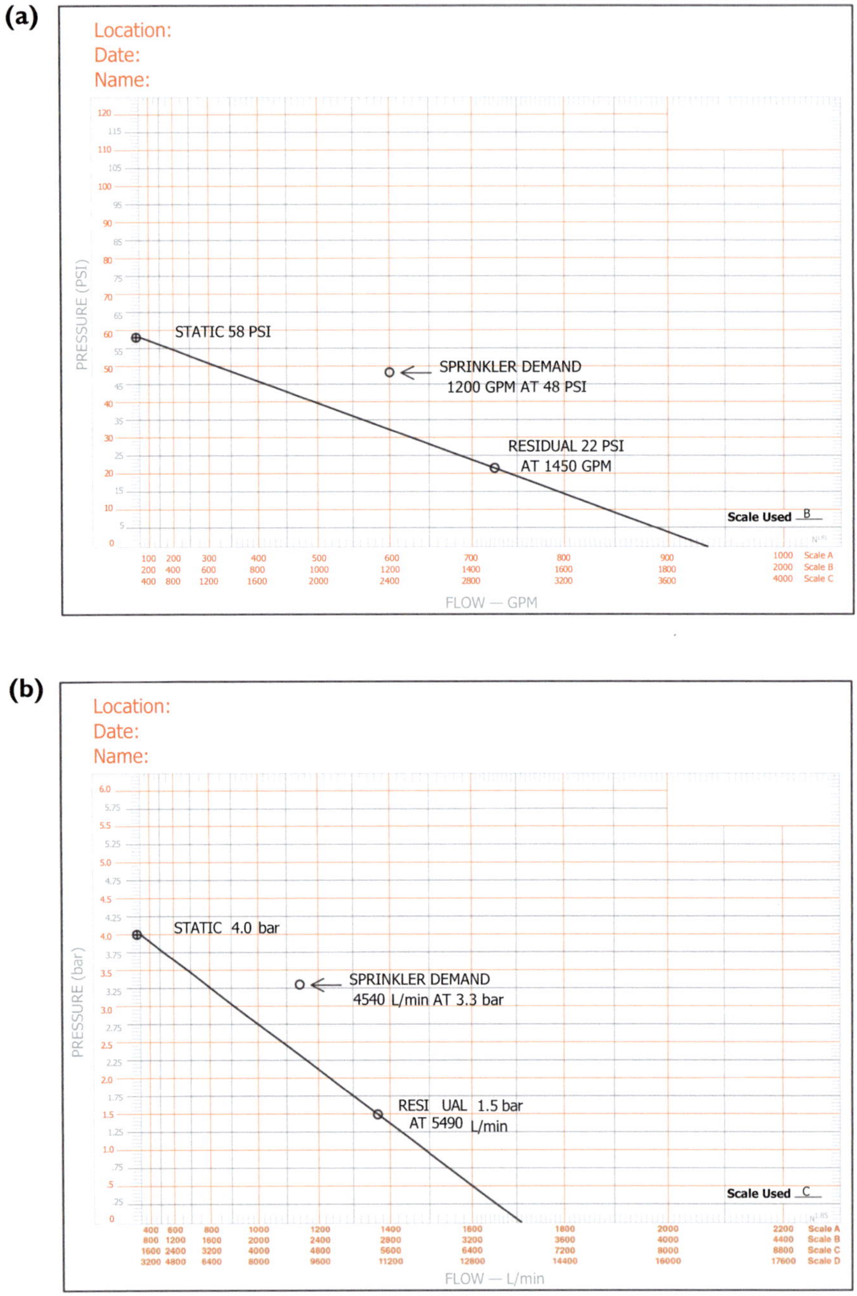

Fig. 10.1 A water supply that could benefit from a fire pump

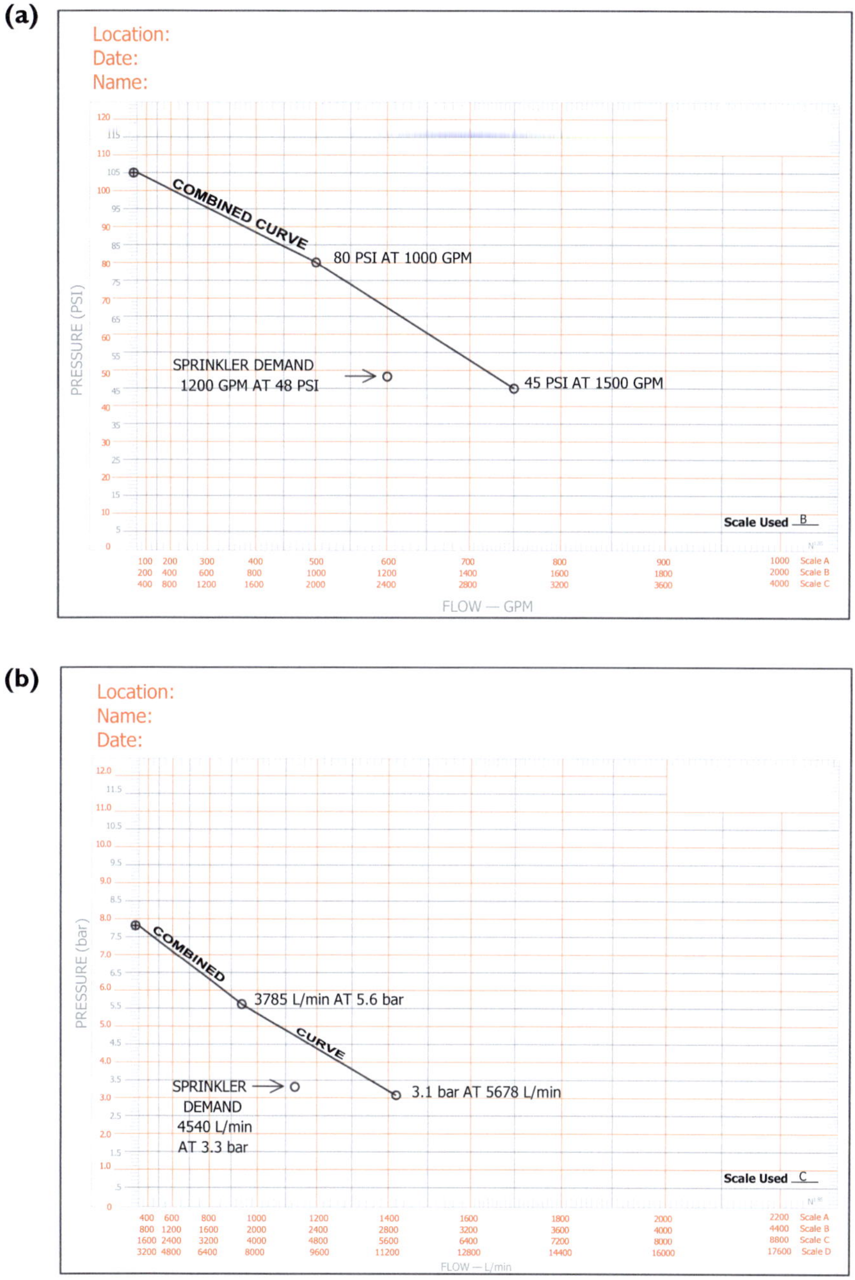

Fig. 10.2 The support of a water supply system by a fire pump

Now, the combined curve is well above the sprinkler demand point, and the water supply is considered adequate. More time will be spent on constructing water supply graphs later. For now, just understand that fire pumps do not manufacture water. They pump out only as much as is supplied to them. *They can, however, significantly increase the pressures available on an existing water supply system.*

In the second situation, where there is no public supply, or it is too weak in both capacity and pressure to be of value in fire protection, a fire pump can still be used. However, a tank or reservoir must be used to provide the number of gallons (liters) needed. The tank can be constructed aboveground or underground. Or, the water supply can be taken from a pond, well, lake, river, canal, or even the ocean. With any of these supply sources, a properly sized, installed, and maintained fire pump can provide adequate fire protection water supplies.

Standard Performance Specifications

The primary performance criteria for standard fire pumps are contained in NFPA 20, *Standard for the Installation of Stationary Pumps for Fire Protection* [1]. The FM data sheets also contain information on industrial fire pumps. For the most part, the FM rules are the same as those contained in NFPA 20. Let's look at some of the important performance requirements for industrial fire pumps. In order to be considered standard under the provisions of NFPA 20, a new industrial fire pump must be capable of satisfying three test points.

In the first test point, the pump must not develop more than 140% of its rated pressure when operating against a closed system (also referred to as churn or shut-off). However, it should be noted that all horizontal pumps manufactured before 1987 were required not to exceed 120% of the rated pressure at churn. These are maximum points.

Centrifugal fire pumps are designed so that the highest pressures will be obtained at the lowest flow. Thus, when operating against a closed system with no fire protection water actually flowing, the highest pressures should be expected. When a pump is operating against a closed system, the pump is said to be *churning* or *operating at churn*. In some cases it may be called *operating at shutoff*. Both expressions are referring to the same situation.

Specifying a maximum acceptable increase in pressure at churn offers a degree of protection against excess pressures developing which could damage system components. So, if a pump is rated at 100 psi (6.9 bar), no more than 140 psi (9.7 bar) should be developed at churn.

The second test point requires that a pump does at least what it is rated to do. That is, it must develop a minimum of 100% of its rated pressure while delivering its rated flow. Thus, a pump rated at 100 psi (6.9 bar) and 1000 gpm (3785 L/min) must develop at least 100 psi while delivering 1000 gpm (3785 L/ min).

The third test point requires that a pump have some excess capacity. Specifically, it must be capable of developing at least 65% of its rated pressure while delivering

150% of its rated flow. Therefore, the pump rated at 100 psi (6.9 bar) and 1000 gpm (3785 L/min) must be able to develop a pressure of at least 65 psi (4.5 bar) while delivering 1500 gpm (5678 L/min).

A new industrial fire pump bearing the UL or FM label can be expected to perform according to these specifications.

It is important to understand that the pressures referred to in the three test points are net pressures. In simple terms, the *net pressure* is the pressure added to the system by the pump. Therefore, if the city water supply is contributing some pressure, it must be subtracted out to get the net pressure of the pump.

For the horizontal and vertically mounted split-case and in-line pumps, the net pressure may be expressed as the discharge pressure minus the intake pressure. Both will have gauges near the intake port and the discharge port of the pump. For most practical applications, the difference in pressures indicated on these two gauges is the net pressure.

To be technically correct, it should be noted that if the discharge pipe of the pump differs in size from the intake pipe, there will be some pressure difference due to a change in velocity caused by changing pipe size. The proper equation for net pressure is shown in Eq. 10.1.

$$P_{\mathrm{N}} = (P_{\mathrm{D}} - P_{\mathrm{I}}) + (P_{\mathrm{DV}} - P_{\mathrm{IV}}) \tag{10.1}$$

where

P_{N} = net pressure in psi or bar.
P_{D} = discharge pressure in psi or bar.
P_{I} = intake pressure in psi or bar.
P_{DV} = velocity pressure on the discharge side of the pump in psi or bar.
P_{IV} = velocity pressure on the intake side of the pump in psi or bar.

The terms in the right-hand set of parentheses reflect the change in pressure that might exist if the discharge diameter of the pump is different from the intake diameter. If both intake and discharge diameters are the same, the terms in the right-hand parentheses can be ignored. Including the velocity pressures in the calculations will never change the results more than a couple of psi (bar). Let's look at an example.

Example 10.1

If the intake gauge on a pump reads 32 psi (2.2 bar) while the pump was delivering the rated flow and the discharge gauge reads 107 psi (7.4 bar), what was the net pressure developed by this pump? Assume that both discharge and intake ports are identical in size.

Solution:

$$P_{\mathrm{N}} = (P_{\mathrm{D}} - P_{\mathrm{I}}) + (P_{\mathrm{DV}} - P_{\mathrm{IV}})\, P_{\mathrm{N}} = 107 - 32 = 75\,\text{psi}$$

And, in S.I. Units:

$$P_N = 7.4 - 2.2 = 5.2\,\text{bar}$$

The computation of net pressures for vertical-shaft pumps is somewhat different. Note that on the typical vertical-shaft pump installation there will be only a single gauge that reads discharge pressures. There is no intake pressure gauge. This is because vertical-shaft pumps most often take water from unpressurized water sources such as wells or ponds. Also, the impellers and the intake port are down inside the water source.

For practical applications, the pressure developed by the vertical-shaft pump (located inside the water source) is equal to the discharge pressure (shown on the gauge), plus the pressure required to get the water up from the source to the level of the gauge. This elevation pressure is computed by multiplying the distance down to the water level while the pump is operating by 0.433 psi per foot (0.098 bar per meter). To be technically correct, the velocity pressure imparted to the water by the pump should not be ignored. The correct equation for net pressure is shown in Eqs. 10.2 and 10.2a.

$$P_N = P_D + (0.433)(H) + P_{DV} \tag{10.2}$$

where

P_N = net pressure in psi.
P_D = discharge pressure in psi.
P_{DV} = velocity pressure at the point where the discharge pressure gauge is connected in psi.
H = the distance from the discharge gauge down to the surface of the water in feet.

In S.I. Units the equation is as follows:

$$P_N = P_D + (0.098)(H) + P_{DV} \tag{10.2a}$$

where

P_N = net pressure in bar.
P_D = discharge pressure in bar.
P_{DV} = velocity pressure at the point where the discharge pressure gauge is connected in bar.
H = the distance from the discharge gauge down to the surface of the water in meters.

Let's make it clear why velocity pressure can be ignored in most cases. The velocity pressures in Eqs. 10.1, 10.2 and 10.2a can be calculated in the same manner that the kinetic energy component of Bernoulli's Equation was calculated in Chap. 3, only in units of psi (bar) rather than feet (meters) of head. After properly converting gpm (L/min) to velocity for existing pipe sizes, the velocity pressure may be calculated from:

$$P_V = \frac{(0.433\,\text{psi/ft})V^2}{2g} \tag{10.3}$$

where

P_V is velocity pressure in psi,
V is water velocity in fps,
g is acceleration due to gravity, 32.2 ft/s^2.

And, in S.I. Units:

$$P_V = \frac{(0.098\,\text{bar/m})V^2}{2g} \tag{10.3a}$$

where

P_V is velocity pressure in bar.
V is water velocity in meters per second.
g is acceleration due to gravity, 9.8 m/s^2.

Example 10.2

Suppose a vertical shaft pump is delivering 2000 gpm (7570 L/min) into a 6-in. (150 mm) Schedule 40 steel discharge line. The water velocity in the 6-in. (150 mm) pipe is:

$$V = \left(\frac{200\,\text{gpm}}{449\,\text{gpm/ft}^3/\text{s}}\right)\left(\frac{144\,\text{in}^2/ft^2}{\pi\,(3.0325\,\text{in})^2}\right) = 22.2\,\text{fps}$$

Then the velocity pressure calculated from Eq. 10.3 is:

$$P_V = \frac{(0.433\,\text{psi/ft})(22.2\,\text{fps})^2}{(2)(32.2\,\text{ft/s}^2)} = 3.3\,\text{psi}$$

And, in S.I. Units, the Velocity in the 150 mm Pipe is:

$$V = \left(\frac{7570\,\text{L/min}}{(\pi)(75\,\text{mm})^2}\right)\left(\frac{10^6\,\text{mm}^3}{\text{L}}\right)\left(\frac{1\,\text{m}}{1000\,\text{mm}}\right)\left(\frac{1\,\text{min}}{60\,\text{s}}\right)$$

$$V = 7.14\,\text{m/s}$$

$$P_{\text{V}} = \frac{(0.98\,\text{bar/m})(7.14\,\text{m/s})^2}{(2)\left(9.8\,\text{m/s}^2\right)} = 0.25\,\text{bar}$$

Therefore, the velocity pressure in this case will be 3.3 psi (0.25 bar). If larger pipe is used, which is required for a pump this large, the velocity pressure is even less. Actually, a discharge pipe smaller than 10 in. (250 mm) for a 2000 gpm (7570 L/min) pump is not permitted by NFPA 20. Table 10.1 gives a good idea of typical magnitudes of velocity pressures. Therefore, for practical field applications, velocity pressures can usually be ignored.

The following is an example of calculating the net pressure developed by a vertical-shaft pump.

Example 10.3

Suppose that while a vertical-shaft pump is pumping 1500 gpm (5667 L/min) into an 8-in. (200 mm) discharge line, the water level in the well drops to 65 ft (19.8 m) below the discharge pressure gauge which reads 107 psi (7.4 bar). Ignoring velocity pressure, what is the net pressure developed by this pump at a flow of 1500 gpm (5667 L/min)?

Solution:

$$P_{\text{N}} = P_{\text{D}} + (0.433)(H)$$

Table 10.1 Magnitudes of velocity pressures

Flow		Pipe diameter		Velocity pressure	
(GPM)	(L/min)	(Inches)	(mm)	(psi)	(bar)
1000	3785	4	100	4.3	0.3
1000	3785	6	150	0.8	0.06
1000	3785	8	200	0.3	0.02
1500	5677	6	150	1.9	0.13
1500	5677	8	200	0.6	0.04
1500	5677	10	250	0.2	0.01
2000	7570	6	150	3.3	0.23
2000	7570	8	200	1.1	0.08
2000	7570	10	250	0.4	0.03

$$P_N = 107 + (0.433)(65\,\text{ft})$$

$$P_N = 135\,\text{psi}$$

In S.I. Units the solution is:

$$P_N = P_D + (0.098)(H),$$
$$P_N = 7.4 + (0.098)(19.8\,\text{m}),$$
$$P_N = 9.3\,\text{bar}.$$

Had velocity pressure been included, the total pressure developed becomes 135.6 psi (9.34 bar) which is not a significant difference.

Let's return to the three required performance points for industrial fire pumps. Every pump bearing a UL or FM label is certified to perform at least up to these test points when the pump is new. This allows prediction of what pressures and flows will be available when a specific pump is used in combination with the existing water supply. A graph can be drawn predicting the water supply available. Let's look at an example for each type of pump.

Example 10.4

Suppose a vertical-shaft pump is to be purchased with a rating of 95 psi (6.5 bar) at 2000 gpm (7570 L/min) and is to take suction from an underground storage tank. The water surface is 10 ft (3 m) below the discharge gauge of the pump.

A graph can be drawn representing the pressures expected at the discharge pressure gauge while the pump is operating at various flows. This can be done by plotting the three test points and subtracting the elevation pressure lost in bringing the water up to ground level.

The first point for the vertical-shaft pump is the maximum point of 140% of rated pressure at shutoff; 140% of 95 (6.5) is 133 psi (9.1 bar). Subtracting the elevation pressure of 10 ft × 0.433 psi/foot (3 m × 0.098 bar/m) gives a ground level pressure of about 129 psi (8.8 bar). This point is plotted above 0 gpm (Fig. 10.3a and b). This point represents the maximum pressure that should exist at the discharge gauge at ground level when the pump is operating against a closed system. Higher pressure than this would not be acceptable for this pump. The actual pressure delivered by the pump will normally be lower than this, but is certified not to be higher.

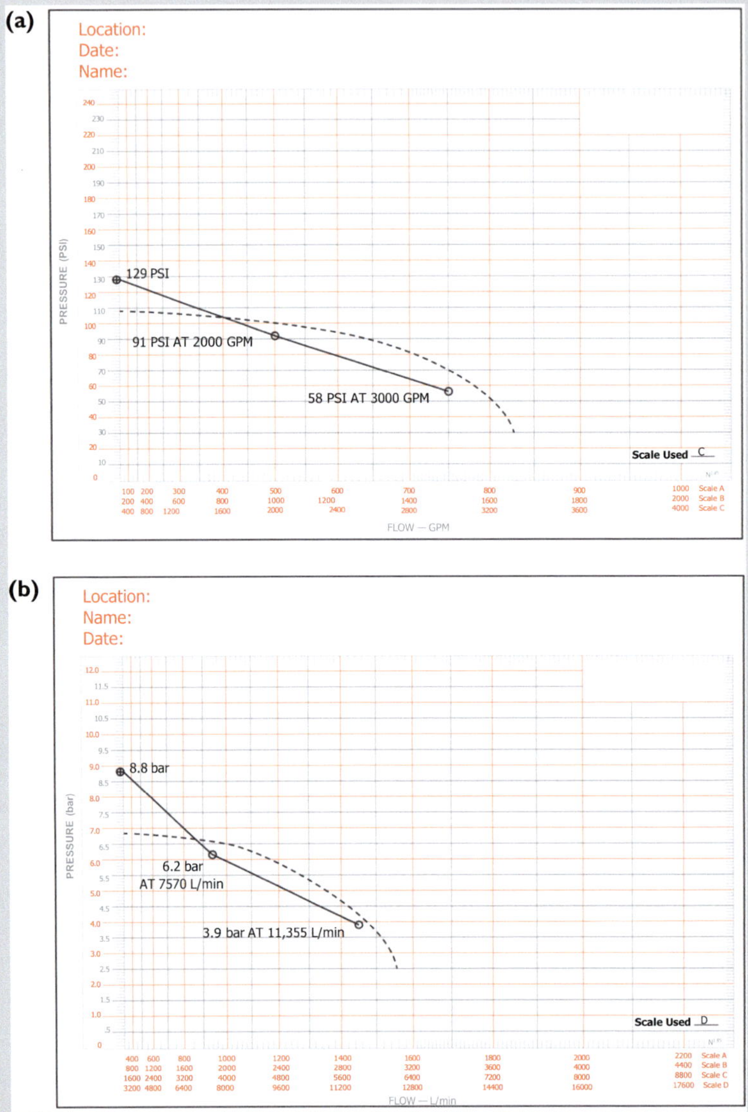

Fig. 10.3 The predicted performance of a vertical shaft pump

The second point is a minimum of 100% of rated capacity at 100% of rated pressure. After subtracting the 4 psi (0.3 bar) elevation pressure, the ground level pressure is found to be 91 psi (6.2 bar) while 2000 gpm (7570 L/min) are flowing. This point is plotted as shown. The pump may actually develop a little more than 91 psi (6.2 bar), but is certified to do at least that well when new.

The third point specifies 65% of rated pressure while delivering 150% of the rated number of gpm. This means the pump should be able to provide a pressure of at least 58 psi (3.9 bar) at ground level while delivering 3000 gpm (11,355 L/min). Now, by connecting these three points with straight lines, an approximation of the expected pump performance over all flows and pressures can be represented. An FM- or UL-listed pump rated at 95 psi (6.5 bar) and 2000 gpm (7570 L/min) should be expected to perform at least as well as represented by the solid line shown on the graph when the pump is new.

Actually, the pump will probably operate more in accordance with the dotted line shown. However, using the three specification points yields a safe approximation of performance. Any fire protection demand can now be compared to the predicted water supply made available by such a pump installation.

It should be noted, however, that in designing fire systems, no demand point should be allowed to exist in the triangular area at flows less than the rated pump gpm and pressures above the rated pump pressure. This is because the first standard point is a maximum point, and the pump is not likely to develop pressures near the 140% point. Thus, for flow demands less than the gpm rating of the pump, pressures no higher than the rated pressure of the pump should be assumed to be available.

Turning to the horizontal split-case pump, the three specification points are the same. An important difference, however, is that the split-case pump will be receiving water under some positive pressure. Remember that the net pressure is basically the difference between the intake pressure and the suction pressure. Therefore, the total water supply created by the use of a fire pump becomes the combined total of the pump performance plus the water system supplying the pump.

Example 10.5

Suppose we wish to use a pump rated at 60 psi (4.1 bar) and 500 gpm (1892 L/min) on the water supply graphically represented in Fig. 10.4a and b. This type of water supply system, where pressures are low, represents a typical situation for fire pump application. If the three specification points of the horizontal-shaft pump are plotted on the same chart and joined by straight lines, we have both the pump and the water supply characteristics (Fig. 10.5a and b).

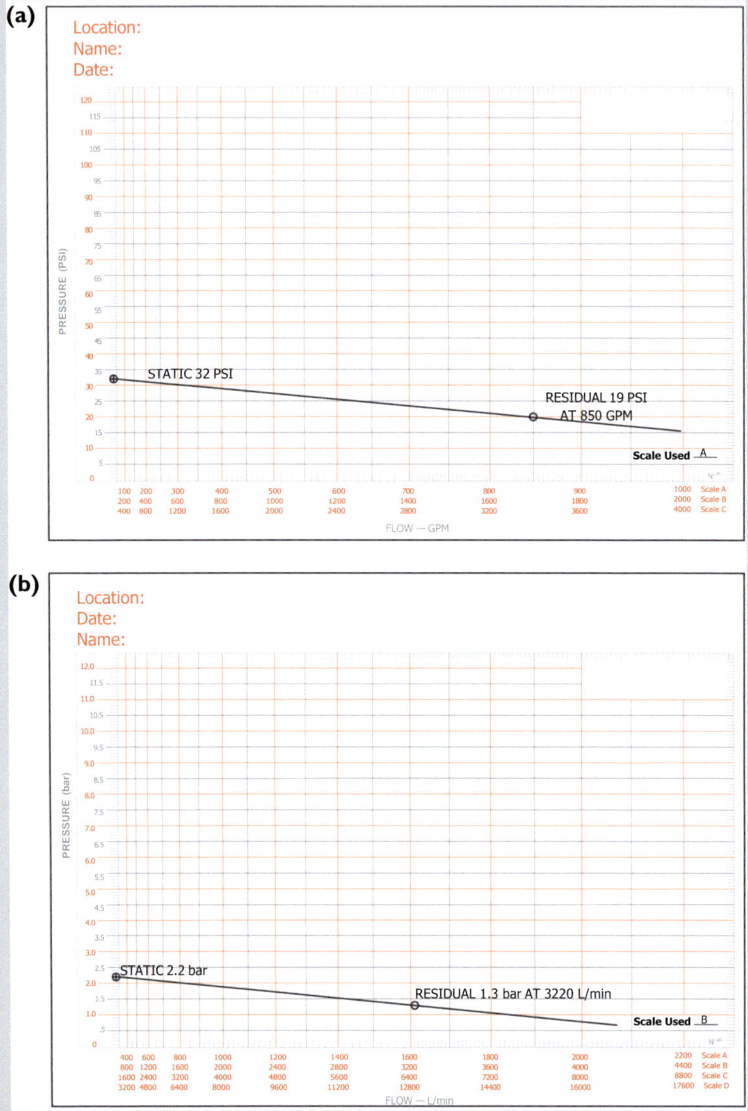

Fig. 10.4 A water supply that could benefit from a fire pump described in Example 10.5

(a)

(b)

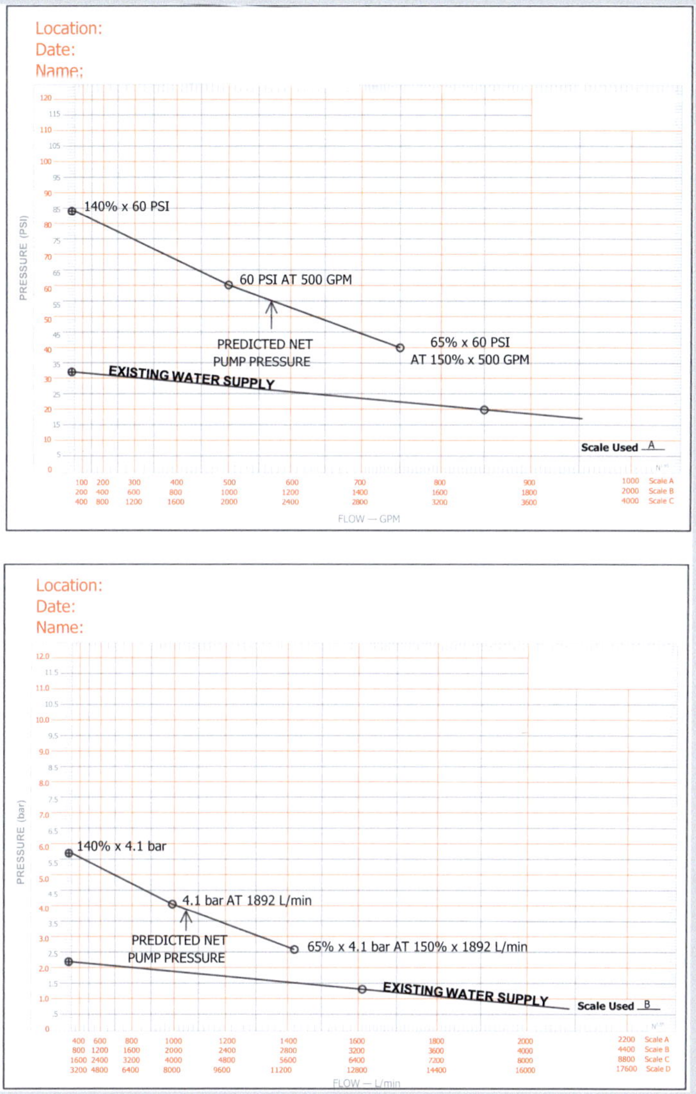

Fig. 10.5 The water supply curve and the predicted pump performance curve

Now, if the pressure developed by the pump at various flows is added to the pressure already existing in the water supply system at those flows, the combined water supply can be illustrated as shown in Fig. 10.6a and b.

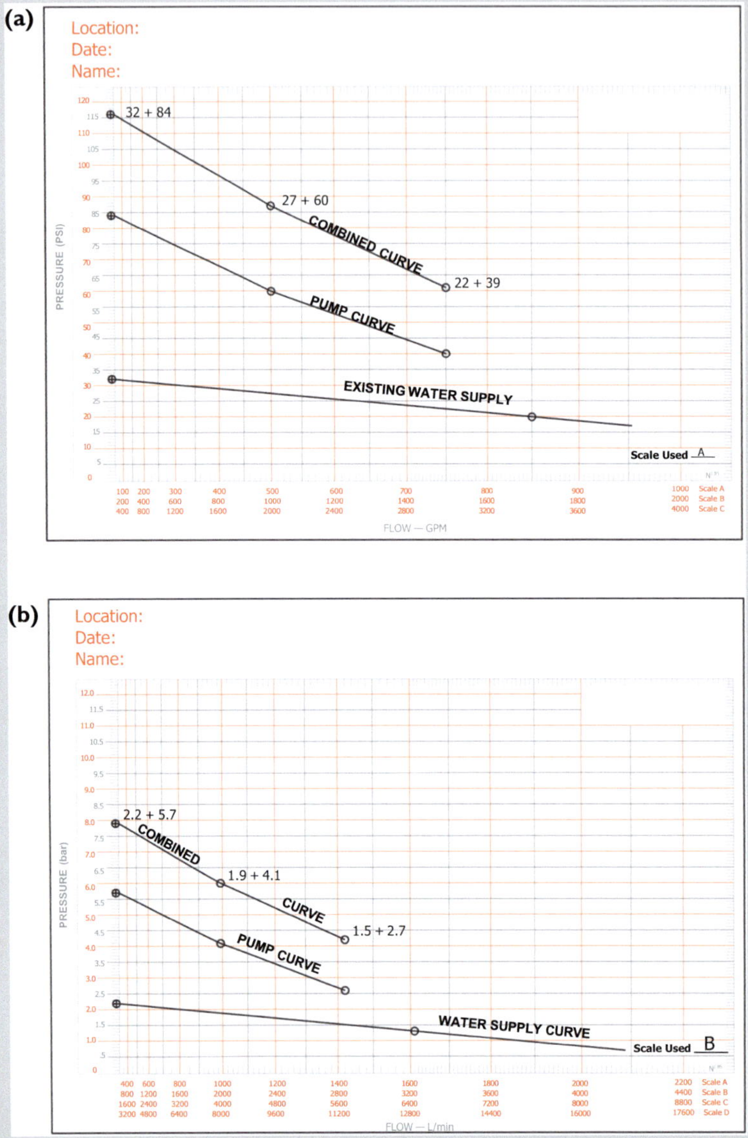

Fig. 10.6 The results of combining a water supply curve and the predicted pump performance curve

- The 84 psi (5.7 bar) maximum point of the pump is added to the 32 psi (2.2 bar) static water supply pressure to get the first point on the combined curve, which is 116 psi (7.9 bar) and is plotted at 0 gpm (L/min).

- At 500 gpm (1892 L/min), the rated pressure of 60 psi (4.1 bar) is added to the pressure delivered to the pump by the supply source. The combined total of 87 psi (6 bar) is plotted as a second point.
- A third point of 61 psi (4.2 bar) at 750 gpm (2838 L/min) is obtained by combining the two pressures available at that flow.

Other points can also be computed and plotted, but these three are really all you need to approximate the total combined water supply made available by the pump operating in conjunction with the supply source. Now, the combined supply can be compared to the fire protection water supply demand. Once again, the actual performance of the pump would probably be a little better than this, but for a safe approximation this method works well.

Again, exercise care with flow demands at points less than the gpm rating of the pump. Flow demands in this area of the graph should not be considered available at pressures exceeding the combination of the city pressure and the rated pressure of the pump. For example, in the combined supply illustrated by Fig. 10.6a and b at a flow of 300 gpm (1135 L/min), no more than 89 psi (29 + 60) or 6.1 bar (2.0 + 4.1) should be considered available for fire protection system design.

Water Supply Requirements

When specifying the size of a horizontal-shaft fire pump, there are several things to consider.

- The pump in combination with the supply source must produce a supply equal to or greater than the fire protection water supply demands. This means the demand point must fall on or below the combined curve. However, the local authority having jurisdiction should always be consulted. Many local authorities require additional safety factors in system design.
- Since listed pumps are required to be capable of delivering 150% of their rated capacity, it is possible to specify a pump that meets the flow demand when operating above rated capacity. This is acceptable as long as the demand point does not exceed 150% of the rated pump capacity. Thus, if 1500 gpm (5677 L/min) are needed, a 1000 gpm (3785 L/min) pump could be used. Again, it is always important to check with authorities having jurisdiction and insurance carriers to verify that this approach is accepted.

Ordinarily, city water supply pressures should never be reduced to less than 20 psi (1.4 bar). However, most pumps are capable of delivering 150% of the rated capacity, as long as the intake pressure remains positive. To drop the pressure lower than 20 psi (1.4 bar) introduces the possibility of collapsing old mains or even siphoning contaminated water into the public system. Low suction pressures can also damage pumps through a phenomenon called cavitation. For these reasons, a safe pressure to maintain is at least 20 psi (1.4 bar). This 20 psi (1.4 bar) pressure requirement should dictate the maximum pump size that can be used on a public system. There should be at least 20 psi (1.4 bar) available at 150% of the rated pump capacity. This pressure requirement allows the development of a simple equation for calculating the maximum gpm (L/min) rating of a fire pump used to support an existing water supply system. The equation is as follows:

$$Q_{\max} = \frac{Q_{20}}{1.5} \tag{10.4}$$

where

$Q_{\max}$ = the maximum gpm (L/min) rating of the pump.

Q_{20} = the number of gpm available at a residual pressure of 20 psi (1.4 bar).

After the computation, $Q_{\max}$ would then be rounded down to the nearest pump rating that is actually available.

Example 10.6

What maximum pump gpm (L/min) rating should be used on a water supply that can deliver 1230 gpm (4656 L/min) at a residual pressure of 20 psi (1.4 bar)?

$$Q_{\max} = \frac{1230}{1.5} = 820\,\text{gpm}$$

And, in S.I. Units:

$$Q_{\max} = (4656\,\text{L/min})/1.5 = 3104\,\text{L/min}$$

Since pumps are not rated at 820 gpm (3104 L/min), the largest pump that should be used in this situation would be a 750 gpm (2839 L/min) pump.

If a greater flow is needed than the city water system can safely supply at 20 psi (1.4 bar), a tank can be provided. The city supply can be used to fill the tank, and a fire pump can be used to take suction from the tank and provide the required flow and pressure. If taking suction from a tank, it is acceptable to drop the pressure on the suction side of the pump below 20 psi (1.4 bar) as long as the pressure remains positive.

Regarding water supplies for the vertical-shaft pumps, a primary problem is depending on sources that are affected by the weather. Wells are acceptable, but not if their water production capabilities are affected by dry weather. The same goes for ponds, streams, or lakes. It is important to remember that reliability is essential. Fires certainly can and do occur in hot dry weather.

If water supplies are provided from tanks, the most critical issue is proper capacity. Most of the NFPA standards have established flow duration requirements for water supply storage tanks. Tanks must be at least large enough to meet NFPA requirements.

To establish the exact appropriate sizing of the pump, the fire protection water supply demand must first be established. The pump size must be sufficient when coupled with existing water supplies to meet the fire protection demand. Determining water supply demands is addressed in Chap. 12, Establishing Water Supply Requirements.

Testing, Inspection, and Maintenance of Fire Pump Installations

The following sections look at the testing of new and existing installations, plus inspection and maintenance of industrial fire pumps. One of the most important things to be done is to operate pumps often and regularly. Fire protection literature is full of accounts of fire pumps failing to operate when needed.

The results of pump failure are most often catastrophic. The way to prevent this from happening is to continually make sure that pump installations are in good operating condition. The required frequency of pump operation is weekly for diesel driven pumps and monthly if driven by electric motors. Pumps should be operated from automatic starts, if so equipped, and brought up to full speed. It is not necessary to actually discharge water during the weekly start up. A diesel engine should be operated for at least 30 min and an electric motor for 10 min.

Testing Pump Installations

Both the suction and discharge piping is required to be hydrostatically tested in accordance with NFPA 20 [1]. Both are required to be hydrostatically pressurized for 2 h to 200 psi (13.8 bar) or 50 psi (3.4 bar) above the maximum static pressure, whichever is larger. For the above ground piping, any leakage constitutes failure. Underground pipe is permitted to leak a little depending upon the length and diameter of pipe.

The underground pipe is also required to be flushed out prior to connection to the fire protection system piping. The reason for this is to blow out all of the dead rabbits and rats, rocks, two-by-fours, and anything else that might accumulate in the pipe prior to installation. As might be expected, the required flow rate for flushing depends upon the diameter of the underground pipe. If the foreign materials are not blown out of the piping by flushing before connection to the fire protection system, these materials will end up inside the system piping. This can damage the pump and have a serious impact on the effectiveness of the fire protection system.

Before a pump installation is accepted from the installing contractor, the installation should be tested under the specifications of NFPA 20 [1]. The installation should not be accepted if it cannot meet the standard specifications. Before a pump is shipped by the manufacturer it is tested. The results of this test are plotted on graph paper. This plotted curve is called the *certified test curve* for the pump. The features of performance plotted are net pressure versus flow delivered, and horsepower delivered versus flow delivered. These characteristic curves are the manufacturer's guarantee of the new pump's capabilities. An important requirement of the acceptance test is that the pump operates at least as well as the pump characteristic curves.

The pump being tested must also meet the three standard performance points. At churn, not more than 140% of the rated net pressure may be developed. The pump must develop at least the rated net pressure while delivering the rated flow and must develop at least 65% of the rated net pressure while delivering 150% of the rated flow.

Equipment Needed for Pump Tests

Besides the pump installation itself, there is some basic equipment needed to conduct the test.

- One section of 2 1/2-in. (65 mm) or larger hose for each hose connection on the test header
- One Underwriters playpipe for each hoseline
- Method for safely securing playpipes
- Pitot tube and gauge
- Method of measuring pump speed
- Voltmeter
- Ammeter

If your system is equipped with a flow meter, then you do not need the first four items listed above. Otherwise, you will need at least one 50-foot (15 m) section of hose for each hose connection on the test header. This hose must have 2 1/2-in. (65 mm) couplings with threads compatible with the hose connections in the header and be at least 2 1/2 in. (65 mm) in diameter.

There should be an Underwriters playpipe for each hoseline and it is essential that the playpipes be firmly secured in place. The nozzle reaction on a playpipe connected to a 2 1/2-in. (65 mm) line can be tremendous. This reaction can actually lift a man off his feet. A 2 1/2-in. (65 mm) hoseline cannot be safely handled by inexperienced and untrained personnel. It is therefore necessary to provide a means of securing playpipes and hoselines. The playpipes can be tied to some substantial structure or special racks can be constructed.

A popular alternative to using playpipes, which eliminates the need of tying down the nozzles, is the use of listed diffusers such as the Hosemonster® (Fig. 10.7) and the Flowbuster® (Fig. 10.8). These devices eliminate the nozzle reaction and deliver a gentle flow which enhances safety and reduces damage to lawns and roadways caused by high velocity nozzle discharge.

The purpose of the hoselines, playpipes, and diffusers is to allow measurement of the flow by means of a pitot tube and gauge. Pitot tubes come in all shapes and sizes. A model with an air chamber for a handle or with a liquid-filled gauge will reduce needle vibration and give more accurate readings. For pump testing, gauges should be selected that will place the reading in the approximate center of the gauge range for the most accurate readings. For example if the desired pressure reading is 18 psi (1.2 bar), a gauge with a pressure range of 1–30 psi (0–2.1 bar) would be the best choice. When using the pitot tube, the blade of the tube is inserted into the flowing stream at a point one-half of the nozzle diameter away from the nozzle. The pressure registered on the gauge is the velocity pressure of the stream.

This velocity pressure is then converted to gpm (L/min) by use of the formulas presented in Chap. 4:

Fig. 10.7 A Hosemonster® type of diffuser

Fig. 10.8 A Flowbuster® type of diffuser

$$Q = (29.83)(C_{\mathrm{d}})(D^2)\left(\sqrt{P}\right) \tag{10.5}$$

Or, in S.I. Units:

$$\mathrm{Q} = (0.666)(\mathrm{C}_d)(\mathrm{D}^2)\left(\sqrt{\mathrm{P}}\right) \tag{10.5a}$$

However, as seen in the following section, an algebraic manipulation of Eqs. 10.5 and 10.5a is used to solve for the pitot pressure, which will represent the desired flow.

The rating of a pump is always for a specific pump speed. For example, a pump may be rated at 1000 gpm (3785 L/min) and 60 psi (4.1 bar) at 1770 revolutions per minute (rpm). The performance of such a pump will be certified at the speed of 1770 rpm. If the electric motor or diesel engine cannot turn the pump this fast, the pump cannot be expected to meet the performance specifications. For this reason, some means of measuring the rpm must be available. This can be accomplished by using either a hand-held revolution counter, a strobe-type tachometer, or by using the more modern reflecting-type digital tachometer.

A voltmeter and an ammeter are needed for testing of an electric motor driven pump unless the control panels are of a modern variety with digital display of voltage and current readings. A voltmeter and ammeter should only be used by a trained personnel equipped with proper personal protective equipment.

The inspection, testing, and maintenance of fire pump installations should be in compliance with the appropriate NFPA standard. Chapter 14 of NFPA 20 provides the requirements for acceptance testing of new pumps. Chapter 8 of NFPA 25, *Standard for Inspection Testing and Maintenance of Water-Based Fire Protection Systems* [2], gives the requirements for the annual testing of existing fire pump installation. The following discussion summarizes the important aspects of annual fire pump testing.

Steps for Annual Pump Testing

Before conducting the annual test, is it important to decide whether or not the fire protection systems will be isolated from the pump during testing. If systems are isolated, they will not be subjected to the water hammers that may result from starting and stopping the pumps. These pressure surges can sometimes cause leakage or damage to the piping systems. If the systems are isolated, no protection exists for the facility during testing plus weaknesses in the system may then only be discovered during a fire. There are advantages for each approach, but this is a decision that should be made by the facility owner.

Before an annual test is conducted, the pressure gauges should be evaluated for accuracy. On a split-case or in-line pump, if both gauges are giving the same reading before the pump starts, they may be considered acceptable. It is unlikely that two faulty gauges would be reflecting the same pressures. If the two gauges do not read the same, then one or both gauges need to be replaced. It is a common practice of many inspectors to carry a set of reliable gauges and routinely replace the existing gauges before conducting a test. In either case, spare reliable gauges should be on hand.

A final caution is that during a test on pumps taking suction from a municipal water supply system, the suction pressure should not be permitted to drop below 20 psi (1.4 bar). Should the pressure drop below 20 psi (1.4 bar) the test should be terminated.

Following are the basic steps in performing an annual fire pump flow test for a horizontal split-case pump driven by an electric motor.

Step 1: Calculate the expected pitot pressure for 100% and 150% of the rated flow by algebraically manipulating Eqs. 10.5 or 10.5a as follows.

$$P = \left(\frac{Q}{(29.83)(C_{\mathrm{d}})\left(D^{2}\right)}\right)^{2} \tag{10.6}$$

Or, for S.I. Units:

$$P = \left(\frac{Q}{(0.666)(C_{d})\left(D^{2}\right)}\right)^{2} \tag{10.6a}$$

Step 2: Connect all the hoselines and nozzles or diffusers. Make sure that all nozzles are securely fastened in place. All hose valves should be closed, and the control valve in the pipe to the test header should be closed.

Step 3: If a modern handheld reflecting-type revolution counter is being used to measure pump speed, a piece of reflective tape must be installed on the pump shaft. Proper attention to safety is critical with respect to any rotating equipment. Loose fitting clothing or ties should not be worn.

Step 4: With everything ready, the pump can be started. Initially, the pump should be operating against a closed system with no valves open and with no water flowing. This is the churn or shutoff point. The pump should be started by bleeding off the water pressure to determine if the pump will start automatically. Once the pump is operating, both the suction and discharge pressures should be read and recorded, the rpm measured, and voltage and current readings taken. It must be emphasized that electrical readings from inside the older generation control panels should only be taken by trained personnel equipped with proper personal protective equipment. While the pump is operating at churn, the circulation relief valve should have opened automatically and be flowing a solid stream of water. If water is not flowing, the relief valve must be adjusted until water begins to flow. Failure of the circulation relief valve to flow water at churn can result in pump damage.

Step 5: Open the control valve in the line leading to the test header and open the hose valves for the first flow measurement.
Step 6: Sufficient lines must be opened and adjusted so that the exact required pitot pressures for 100% of the rated flow is read on the pitot gauge. While the velocity pressures are being measured outside, the rpm, voltage, current, discharge, and suction pressures are to be measured inside the pump room. When the first line is opened, the circulation relief valve should again be checked. When the first line was opened, the relief valve should have closed off. If it continues to flow a solid stream, further adjustment is needed.
Step 7: When all readings are complete and recorded, the additional hoselines are opened and adjusted to the exact required pitot pressures for 150% of the rated flow. It usually requires flowing all of the hoselines to achieve 150% of the rated flow. It is important to calculate ahead of time what pitot pressure is needed at each nozzle to give both the 100% and the 150% points. It is convenient if the total flows required are divided by the number of hoselines to be used to determine the flow required from each line. This way the pitot reading required at each nozzle will be identical.

During the course of the test, attention should be paid to the temperature of the pump bearings and the pump itself. None of the components should become hot to the touch. A fiber packing on both sides of the pump shaft seals the shaft. This packing is water cooled and lubricated, and some water must drip from the shaft at both ends. If the packing gland is adjusted too tightly, it will prevent water from cooling the fiber packing and the packing will heat up. It can become hot enough to burn and smoke. During the initial operations of the pump, gradual adjustment of the packing gland is often necessary. Ultimately, you want to achieve an adjustment where about one drop every second is passing through the packing. More leakage than this requires tightening the packing and less leakage requires loosening. This adjustment must be done with the pump turned off to ensure safety.

Annual tests conducted on in-line, electrically driven pumps are identical to those conducted on the horizontal split-case pumps. On vertical-shaft turbine pumps the testing is almost the same, except that there is no suction pressure to read, and the pressure lost in bringing the water up to the level of the pressure gauge must be added to calculate the net pressure. When testing pumps driven by diesel engines, the procedure is essentially the same, except no voltage or current readings are taken.

Many modern fire pump installations are provided with a flow meter in a by-pass loop around the pump. This permits testing the pump by circulating water around the pump and eliminates the necessity of actually discharging water through hose lines and nozzles. Such testing can provide valid pump testing if the flow meter is calibrated and reliable. NFPA 25 permits testing in this manner two years out of every three. On the third year, actual discharge of water is required in order to test the availability of the water supply to the pump [2].

When the test has been completed, the data collected is used to construct performance curves, which are compared with the manufacturer's certified curves or the three standard performance points. In constructing the pressure versus flow curve, it is the "net pressure," as discussed previously, that is used. An increase or decrease in pipe size causes pressure changes because of the change in water velocity. But for most practical applications these pressure changes can be ignored.

NFPA 25 currently requires that a reduction in performance exceeding 5% calls for an investigation to determine the cause of the deterioration in performance [2]. The most common cause of this reduced performance will be wear on the pump. The pump must be repaired or replaced if it is no longer capable of meeting the fire system demands.

Pump speed significantly affects the performance of the pump; this is the reason pump speed measurements must be taken. If the pump is turning at a speed significantly different from the rated speed, any comparison with the certified performance or three standard points may not be legitimate. However, by use of mathematical relationships, called the "affinity laws," pump performance at any pump speed can be corrected to the rated pump speed. The affinity relationships are shown below:

$$\frac{Q_1}{Q_2} = \frac{\text{rpm}_1}{\text{rpm}_2} \tag{10.7}$$

$$\frac{P_1}{P_2} = \left(\frac{\text{rpm}_1}{\text{rpm}_2}\right)^2 \tag{10.8}$$

$$\frac{hp_1}{hp_2} = \left(\frac{\text{rpm}_1}{\text{rpm}_2}\right)^3 \tag{10.9}$$

Increasing the speed by 10% increases the number of gpm (L/min) delivered by 10%. The pressure, however, is related to the square of the pump speed. Thus, doubling the pump speed increases the pressure by four times, and increasing the speed by 10% increases the pressure by 21%.

Example 10.7

A pump rated at 1770 rpm delivers 800 gpm (3030 L/min) while developing a net pressure of 72 psi (5.0 bar). However, the pump speed is measured to be only 1650 rpm instead of 1770. These test results can be corrected to 1770 rpm by use of the affinity relationships as follows:

$$\frac{Q_1}{Q_2} = \frac{\text{rpm}_1}{\text{rpm}_2}$$

$$Q_1 = (Q)_2\left(\frac{\text{rpm}_1}{\text{rpm}_2}\right)$$

$$Q_1 = (800\,\text{gpm})\left(\frac{1770\,\text{rpm}}{1650\,\text{rpm}}\right)$$

$$Q_1 = 858\,\text{gpm}@1770\,\text{rpm}$$

And:

$$\frac{P_1}{P_2} = \left(\frac{\text{rpm}_1}{\text{rpm}_2}\right)^2$$

$$P_1 = (P_2)\left(\frac{\text{rpm}_1}{\text{rpm}_2}\right)^2$$

$$P_1 = (72\,\text{psi})\left(\frac{1770\,\text{rpm}}{1650\,\text{rpm}}\right)^2$$

$$P_1 = 83\,\text{psi}@1770\,\text{rpm}$$

And, in S.I. Units:

$$Q_1 = (Q)_2\left(\frac{\text{rpm}_1}{\text{rpm}_2}\right)$$

$$Q_1 = (3030\,\text{L/min})\left(\frac{1770\,\text{rpm}}{1650\,\text{rpm}}\right)$$

$$Q_1 = 3250\,\text{L/min}@1770\,\text{rpm}$$

And:

$$\frac{P_1}{P_2} = \left(\frac{\text{rpm}_1}{\text{rpm}_2}\right)^2$$

$$P_1 = (P)_2\left(\frac{\text{rpm}_1}{\text{rpm}_2}\right)^2$$

$$P_1 = (5.0\,\text{bar})\left(\frac{1770\,\text{rpm}}{1650\,\text{rpm}}\right)^2$$

$$P_1 = 5.8\,\text{bar}@1770\,\text{rpm}$$

If a motor or engine is not turning the pump at or near its rated speed, the motor or engine must be repaired or replaced. However, even with a poorly performing driver, a valid test may be performed on the pump by using the mathematical affinity law equations.

The voltage and current measured for the electrically driven pump are used to evaluate other acceptance criterion. An electric motor should have a nameplate on which various pieces of information are stamped, including the service factor, full-load current rating, and rated voltage.

During the annual test, the ratio of the products of the measured voltage times the measured current to the product of the rated voltage times the rated full-load current shall be less than or equal to the service factor. Finally, the measured voltage should never be more than 5% below or more than 10% above the rated voltage.

Another consideration is the large volume of water that will be discharged during a test. A 1000 gpm (3785 L/min) pump operated for one hour can discharge more than 60,000 gallons (227,000 L) of water. Attention must be given to where the water is flowing or draining. Care should be taken to avoid erosion or other property damage. Also, if possible, it is better if the hoselines can discharge back into the water source. This conserves water that is otherwise wasted onto the ground.

Routine Operation and Maintenance

The requirements for regular and routine inspection, testing, and maintenance are given in NFPA 25. This standard provides requirements for all water-based fire systems. NFPA 25 requires an annual flow test of the pump assembly be performed to determine its ability to continue to attain satisfactory performance at shutoff, rated flow, and peak loads. By doing this, the performance of the pump can be compared year by year.

On a quarterly basis, it is recommended that the bearings be re-lubricated. Oil or grease should be used depending upon the recommendations of the manufacturer. The packing should be adjusted monthly to obtain approximately one drop per second leakage. In addition to the packing and seals, it is a good practice to recheck the alignment regularly. Misalignment of the couplings can cause excessive vibration and pump damage.

As pointed out previously, the primary purpose of the weekly or monthly test is to make sure the pumps will start. Therefore, both diesel and electrically driven pumps should be started regularly, with the diesel driven pump started weekly and the electric motor driven pump started at least monthly. There is a tendency to use programmed timers to automatically start engines once per week. This is a bad practice if it encourages maintenance personnel to neglect frequent and regular visual inspection of pump facilities.

During the weekly inspections, it is also important to make sure the pump room is kept clean, dry, and free of combustible storage. The weekly inspection is a good time to ensure that all control valves supposed to be open are open and are supervised in the OPEN position by padlock and chain, electronic equipment, or tag and seal. This is also the appropriate time to make sure that water tanks and diesel fuel tanks are full.

During testing and inspection, you will eventually encounter a situation where the pump installation is not operating properly. Annex B of NFPA 20 provides a troubleshooting checklist that can help in identifying the causes of pump problems [1]. For example, excessive leakage at the stuffing box could be the result of improper tightening or the packing being improperly installed. Or, it might even be indicative of worn or scarred pump shafts. Thus, if the problem can be identified, several possible causes of the problem are suggested and possible corrective measures are outlined.

Fire pump installations are vital parts of a total fire protection system. Their proper selection, installation, and maintenance can be the difference between business as usual tomorrow, and a catastrophe that could destroy people, property, and jobs.

Summary of Important Concepts

1. Pumps may be used to provide water supplies where existing supplies are nonexistent or deficient.
2. Modern pumps must be able to meet the following three standard points:
   - A maximum of 140% of rated pressure at a flow of 0 gpm.
   - A minimum of 100% of rated pressure at 100% of rated flow.
   - A minimum of 65% of rated pressure at 150% of rated flow.
3. To provide dependable service, fire pumps must be properly maintained and started on a regular basis.
4. A complete performance test should be conducted on each pump installation every year.

Exercises

1. Use hydraulic graph paper to plot the following flow test results and determine the flow available at 20 psi (1.4 bar) for each:
   a. static pressure: 48 psi (3.3 bar) residual pressure: 16 psi (1.1 bar) measured flow: 850 gpm (3200 L/min)
   b. static pressure: 93 psi (6.4 bar) residual pressure:60 psi (4.1 bar) measured flow:1850 gpm (7000 L/min)
   c. static pressure:74 psi (5.1 bar) residual pressure: 36 psi (2.5 bar) measured flow: 775 (2930 L/min)
2. Use hydraulic graph paper to plot the following flow test results and determine the flow available at 20 psi (1.4 bar) for each:
   a. static pressure: 68 psi (4.7 bar) residual pressure: 26 psi (1.8 bar) measured flow: 850 gpm (3200 L/min)
   b. static pressure: 73 psi (5.0 bar) residual pressure: 52 psi (3.6 bar) measured flow: 1850 gpm (7000 L/min)
   c. static pressure: 54 psi (3.7 bar) residual pressure: 36 psi (2.5 bar) measured flow: 775 gpm (2930 L/min)
3. What is the maximum gpm (L/min) rating of a pump that should be used with each of the water supplies from Exercise 10.1?

4. What is the maximum gpm (L/min) rating of a pump that should be used with each of the water supplies from Exercise 10.2?
5. Plot the approximate net performance curve of a standard horizontal split-case pump rated at 50 psi (3.4 bar) and 1000 gpm (3785 L/min).
6. Plot the approximate net performance curve of a standard horizontal split-case pump rated at 80 psi (5.5 bar) and 1500 gpm (5677 L/min).
7. Plot the approximate net performance curve of a standard vertical-shaft pump rated at 90 psi (6.2 bar) and 1000 gpm (3785 L/min).
8. Plot the approximate net performance curve of a standard vertical-shaft pump rated at 120 psi (8.3 bar) and 500 gpm (1892 L/min).
9. Plot the combined performance of a 1500 gpm (5677 L/min), 75 psi (5.2 bar) horizontal split-case fire pump with the following water supply: static pressure: 42 psi (2.9 bar) residual pressure: 18 psi (1.2 bar) measured flow: 2400 gpm (9085 L/min)
10. Plot the supply curve, the approximate pump performance curve, and the combined water supply from the following information:

    WATER SUPPLY
    static pressure: 42 psi (2.9 bar)
    residual pressure: 8 psi (0.06 bar)
    measured flow: 900 gpm (3400 L/min)
    PUMP DATA
    horizontal split-case
    gpm rating: 500 gpm (1892 L/min)
    psi rating: 40 psi (2.8 bar)

11. Plot on graph paper the expected ground level performance of a vertical-shaft fire pump rated at 1500 gpm (5677 L/min) and 160 psi (11.0 bar) if the pump is taking suction from a well with a distance to the water surface of 60 ft (18.3 m) at full flow.
12. Plot on graph paper the expected ground level performance of a vertical-shaft fire pump rated at 1000 gpm (3785 L/min) and 80 psi (5.5 bar) if the pump is taking suction from a pit with a distance to the water surface of 20 ft (6.1 m) at full flow.
13. If a sprinkler system demand is 1200 gpm (4540 L/min) at 68 psi (4.7 bar), what minimum size horizontal split-case pump taking suction from the following water system will provide an adequate supply:

    static pressure: 49 psi (3.4 bar)
    residual pressure: 25 psi (1.7 bar)
    measured flow: 2000 gpm (7570 L/min)

14. If a sprinkler system demand is 620 gpm (2350 L/min) at 54 psi (3.7 bar), what minimum size horizontal split-case pump taking suction from the following water system will provide an adequate supply:

    static pressure: 73 psi (5.0 bar)

residual pressure: 25 psi (1.7 bar)
measured flow: 680 gpm (2570 L/min)

15. If a poorly tuned diesel engine turns a pump at 1400 rpm and the pump develops 73 psi (5.0 bar) while delivering 830 gpm (3140 L/min), what pressure and flow combination would result if the engine is properly tuned and operates at 1770 rpm?
16. If a poorly tuned diesel engine turns a pump at 1508 rpm and the pump develops 103 psi (7.2 bar) while delivering 1675 gpm (6340 L/min), what pressure and flow combination would result if the engine is properly tuned and operates at 1770 rpm?
17. If the fire system demand is 1000 gpm (3785 L/min) at 80 psi (5.5 bar), determine the minimum pressure rating required of both a 750 gpm (2839 L/min) and a 1000 gpm (3785 L/min) vertical-shaft pump if the water level is 36 ft (11 m) below the discharge gauge.
18. If the fire system demand is 2000 gpm (7570 L/min) at 100 psi (6.9 bar), determine the minimum pressure rating required of both a 1500 gpm (5677 L/min) and a 2000 gpm (7570 L/min) rated vertical-shaft pump if the pump takes suction from the ocean which is 22 ft (6.7 m)) below the discharge pressure gauge.
19. A fire system is 105 ft (32 m) higher than the discharge of a vertical-shaft pump and connected to the pump by 820 ft (250 m) of 6-in. (150 mm) unlined CL 52 ductile iron pipe. If the fire pump is rated at 1000 gpm (3785 L/min) and 150 psi (10.3 bar), and the water surface level is 15 ft (4.6 m) below the discharge line, what pressure will be available at the fire system at the system demand of 1250 gpm (4730 L/min)?
20. A fire system is 88 ft (26.8 m) higher than sea level and connected to a vertical shaft pump taking suction from the sea by 1600 ft (490 m) of 10-in. (250 mm) lined ($C = 140$) CL 50 ductile iron pipe. If the fire pump is rated at 2000 gpm (7570 L/min) and 125 psi (8.6 bar), what pressure will be available at the fire system demand of 2600 gpm (9840 L/min)?

Selected answers are found in Appendix K.

References

1. NFPA 20, *Standard for the Installation of Stationary Pumps for Fire Protection* (Quincy: National Fire Protection Association, 2025).
2. NFPA 25, *Standard for the Inspection, Testing and Maintenance of Water Based Fire Protection Systems* (Quincy: National Fire Protection Association, 2023).

Supplemental Readings

Hickey, Harry E. *Hydraulics for Fire Protection.* National Fire Protection Association: Quincy, MA 1980, pp.141–102.

National Fire Protection Association. *The Fire Protection Handbook*. National Fire Protection Association: Quincy, MA, 21st edition, Sec. 15, Ch. 6 and Ch. 7.

National Fire Protection Association. *Stationary Fire Pumps and Standpipe Systems Handbook*. National Fire Protection Association: Quincy, MA, 2022 edition.

Chapter 11
Elevated Storage

Abstract This chapter introduces the concept of water supplies use of elevated storage. The approach to understanding a purely statics source (i.e., tank) can be graphically illustrated through simple calculations since the difference between static and residual pressures is a result of friction loss. How to graphically represent a combined supply of a municipal system and a tank is also introduced. A conservative method of approximation is using the bottom of the tank as the elevation reference point.

How is pressure generated within a water supply system? A common response is that the pressure comes from elevated tanks. If this is true, how does the water get into those tanks? There are some communities in mountainous areas that generate their water pressure purely by gravity or elevation. However, in most cases, tanks are filled by the pressure already existing in the water system, and this pressure comes from pumps.

Elevated tanks are used to stabilize or balance the pressures on a water system at times of peak demand. As long as the pumps supplying a system can keep up with the demand, the water in the tanks will not be used. However, once the demand upon a system becomes so great that the pumps cannot keep up, the system pressure will begin to drop. When the pressure drops to a point where it cannot keep the tank full, then the tank will begin to add water to the system.

One of the greatest peak demand periods is when a major fire occurs in a community. For this reason, the existence of elevated storage becomes an important fire protection feature. This chapter addresses the analysis of combining an existing water supply with elevated storage to determine the total supply available. Chapter 12, Establishing Water Supply Requirements, will address the determination of how much storage is needed.

P. Brock and V. Charter, *Fire Protection Hydraulics and Water Supply Analysis*, The Society of Fire Protection Engineers Series,
https://doi.org/10.1007/978-3-032-09877-1_11

Plotting the Supply from Purely Static Sources

For the purpose of the initial concepts, the example of a supply taken from a large lake or reservoir will be used. The lake is to be large enough that no appreciable change in water surface elevation will occur when water is flowing to the system (Fig. 11.1).

It might be interesting to note at this point, that if Hydrant A or B in Fig. 11.1 were opened, the theoretical instantaneous velocity of the stream from the hydrant would be

$$V = \sqrt{2gh} \tag{11.1}$$

where:

$\boldsymbol{V}$ = stream velocity in fps (m/sec).

$\boldsymbol{G}$ = acceleration due to gravity, 32.2 ft/sec^2 (9.8 m/sec^2).

$\boldsymbol{h}$ = elevation of the water in ft (m).

In our example, where the water level is 160 ft (49 m) above the hydrants, the theoretical velocity would be:

$$V = \sqrt{((2)(32.2\,\text{ft/sec}^2)(160\,\text{ft}))}$$
$$V = 101.5\,\text{fps}$$

And in S.I. Units:

$$\boldsymbol{V} = \sqrt{((2)(9.8\,\text{m/sec}^2)(49\,\text{m}))}$$
$$\boldsymbol{V} = 31\text{m/sec}$$

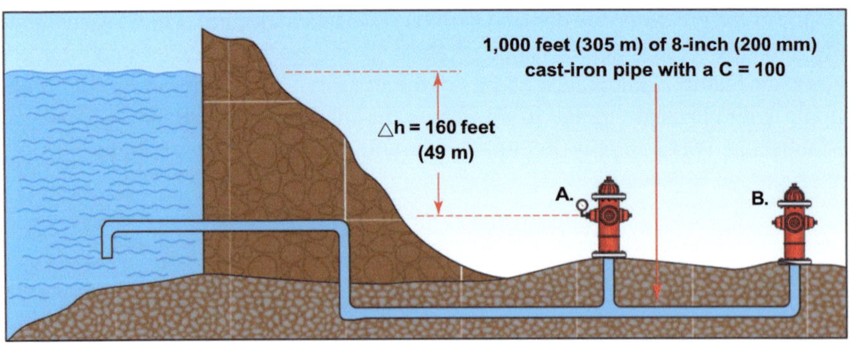

Fig. 11.1 An example of a static water supply

These velocities are theoretical, of course, because they do not recognize the impact of pressure losses due to friction as water travels through the pipe.

With no water flowing from the lake, the static pressure measured at Hydrant A would be:

$$P_S = (0.433\,\text{psi/ft})\,(160\,\text{ft}) = 69\,\text{psi}$$

Or in S.I. Units:

$$\text{P}_\text{S} = (0.098\,\text{bar/m})(49\,\text{m}) = 4.8\,\text{bar}$$

In graphing the water supply available, 69 psi (4.8 bar) would be plotted above 0 gpm (0 L/min). If Hydrant B is opened and water begins to flow, the pressure at A is reduced, reflecting the magnitude of the friction loss between the lake and Hydrant A. Any flow may be chosen to compute a loss. If 1500 gpm (5678 L/min) is used, the Hazen–Williams formula yields a loss of about 24 psi (1.6 bar). The static pressure of 69 psi (4.8 bar) is reduced by 24 psi (1.6 bar), giving a residual pressure of 45 psi (3.2 bar), which is plotted above 1500 gpm (5678 L/min). Figure 11.2a and b show the graphical representation of the water supply at Hydrant A.

If the lake from the first example is replaced by an elevated tank as depicted in Fig. 11.3, the technique for analysis is essentially the same. One important difference does exist, however. Tanks have a limited capacity compared to a lake; if water flows for any length of time, the height of water in the tank will begin to diminish. We are faced with a continually changing static pressure point. Graphically depicting the water supply is not possible without the use of calculus and will not result in a straight line curve. The problem may be simplified by taking the practical approach that the water height in the tank will only be considered to be as high as the bottom of the tank. Although the results from such an assumption will be incorrect, they will always be on the safe side. Until the tank is empty, there will always be at least a little more pressure available than the supply curve indicates.

In Fig. 11.3, a static pressure of 43 psi (3.0 bar) rather than 52 psi (3.6 bar) would be used. If 1500 gpm (5678 L/min) were again used to calculate a friction loss, the resulting water supply graph is shown in Figs. 11.5 and 11.6. The dotted lines in Fig. 11.4a and b represent the supply available at the moment water begins to flow from a full tank. The lower solid lines represent the supply available at the instant the tank becomes empty, and thus represents a safe approximation.

Graphically Combining Elevated Tanks with Existing Water Supplies

It is important to understand that as long as the pressure within a system remains high enough to keep the tank full, there will be no contribution from the elevated

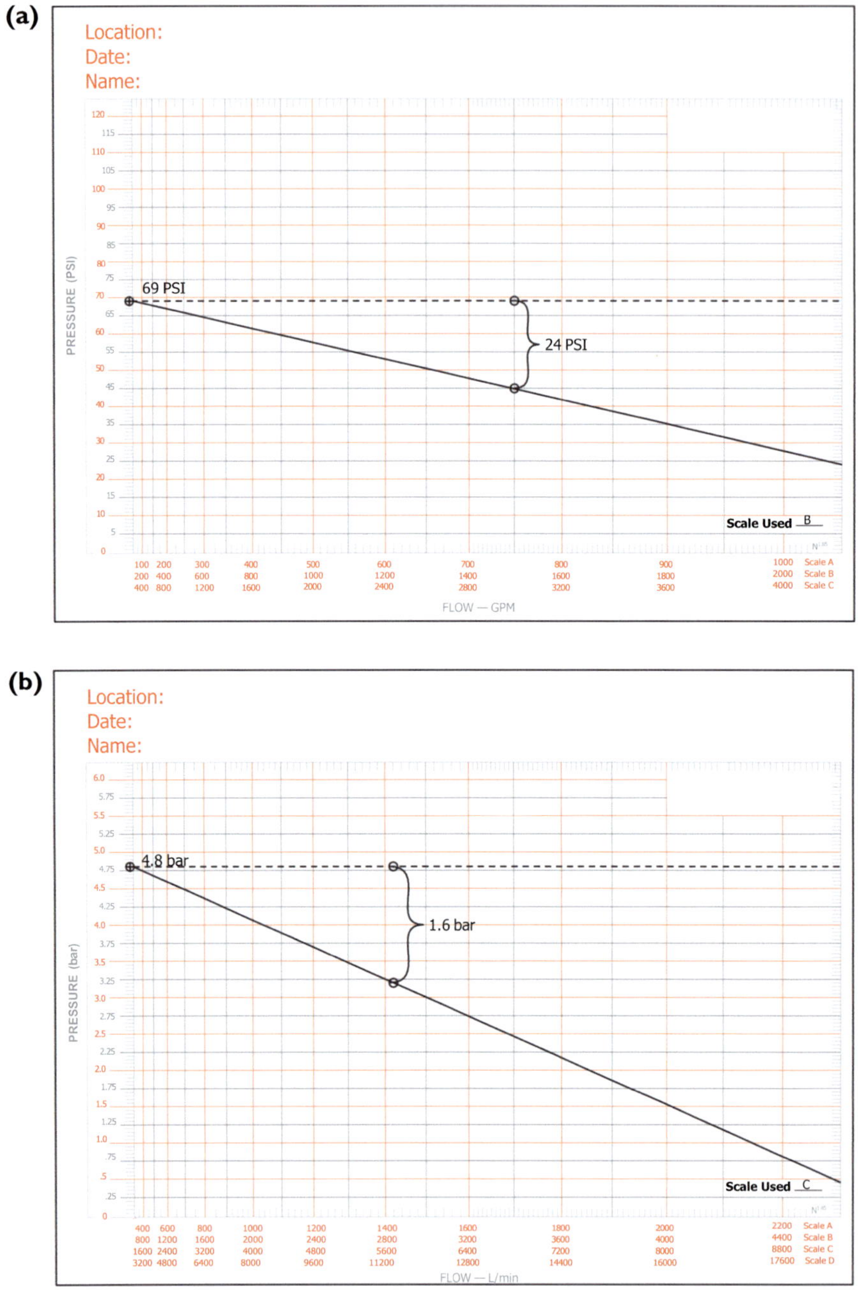

Fig. 11.2 **a** The water supply available at Hydrant A shown in Fig. 11.1. **b** The water supply available at Hydrant A shown in Fig. 11.1 (S.I. units)

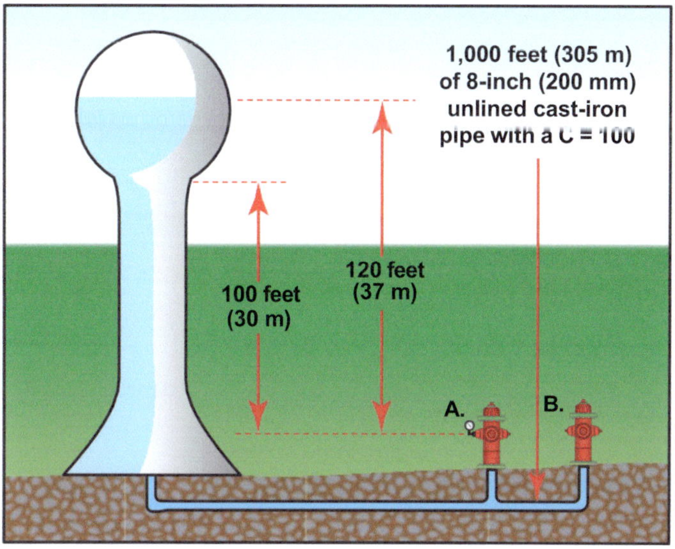

Fig. 11.3 An example of an elevated storage tank

tank. Elevated tanks are typically equipped with an elevation-controlled valve which permits water to enter the tank from the supply system until water has built up to the desired level. It will then close off automatically, preventing the tank from being overfilled. If the pressure on the system diminishes to a point below the full tank magnitude, the elevation valve then opens allowing water to come out of the tank.

Figure 11.5 will be used as an example for graphically combining the water supply contributed by a tank to that supplied by an existing system. Figure 11.5 depicts an industrial loop taking its primary supply from a public water supply system but supported by a private elevated tank.

The most practical approach to evaluating the available water supply involves conducting two water flow tests, one with the elevated tank out of service and one with the municipal supply out of service. With the elevated tank shut off, suppose a test at Hydrant A yielded a static pressure of 66 psi (4.5 bar) and a residual pressure of 34 psi (2.3 bar) with 1260 gpm (4770 L/min) flowing.

Next, with the municipal supply shut off and the tank returned to service, test results adjusted to the bottom of the tank gave a static pressure of 39 psi (2.7 bar) and a residual pressure of 29 psi (2.0 bar) with 1080 gpm (4090 L/min) flowing.

Curves from both tests must first be plotted to combine the two. Remember that until the municipal water pressure drops below the plotted static pressure of the tank, the tank will not contribute. Below the static pressure the flows are additive at each pressure point. Therefore, if the flows represented by each curve at each pressure point are added together, a combined curve can be constructed (Fig. 11.6a and b). The combined curve in Figs. 11.8 and 11.9 represent the total water supply available from both sources, the public supply and the tank.

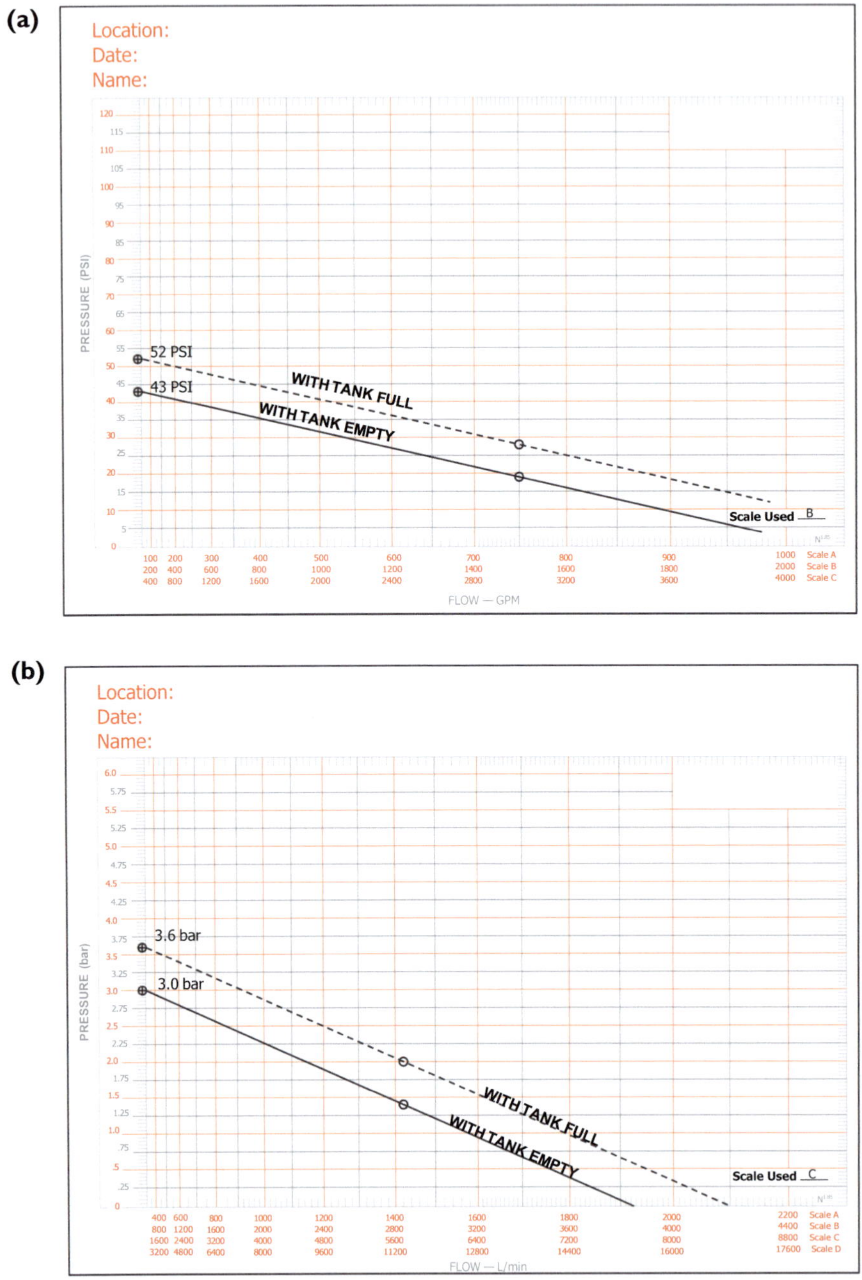

Fig. 11.4 **a** The water supply available at Hydrant A shown in Fig. 11.3. **b** The water supply available at Hydrant A shown in Fig. 11.3 (S.I. units)

Fig. 11.5 A water supply system supported by an elevated tank

Since it is tedious to add the flows at various pressure points, it is possible to use a short-cut approximation. In many cases, the two curves intersect at some point. If this occurs, a convenient technique is to use the pressure at which the two curves intersect and simply doubling the flow. This doubled flow and the pressure at the intersection is used as one of two points to identify the combined curve. The other point is where the tank first begins to contribute water (Fig. 11.7a and b).

This short-cut method is more expedient but is an approximation. However, comparison of Figs. 11.8 and 11.9 to Figs. 11.10 and 11.11 will show that the approximation is on the safe side. This method coupled with using the bottom of the tank as the elevation reference point should provide a truly conservative estimate of the contribution of an elevated tank.

Modern Considerations

In times past it was very common to see industrial plants with a private, elevated tank on site as a backup for the public water supply. One that always comes to mind is a famous pool table manufacturer that had the tank painted like the eight ball. However, there are several problems associated with providing fire protection water supplies from elevated tanks. The most important of those is the limitation of pressure available from an elevated tank. A tank has to be 100 ft (30.5 m) high to generate a pressure of only 43 psi (3.0 bar). This pressure is not adequate to meet many modern fire system requirements. For example, Early Suppression Fast Response sprinklers used for warehouse protection can require as much as 50 psi (3.4 bar) or 75 psi (5.2 bar) at the most hydraulically demanding sprinkler. System demands for other high-challenge occupancies also require tanks to be of impractical heights to provide adequate pressure. For this reason, it makes more sense to use ground-level storage and a fire pump for the water supply redundancy required at many highly protected facilities.

Elevated tanks provide a very reliable water supply. It is only necessary to keep the tanks full and the valves open to be assured that water will flow when needed. However, tanks do require considerable maintenance and often require protection

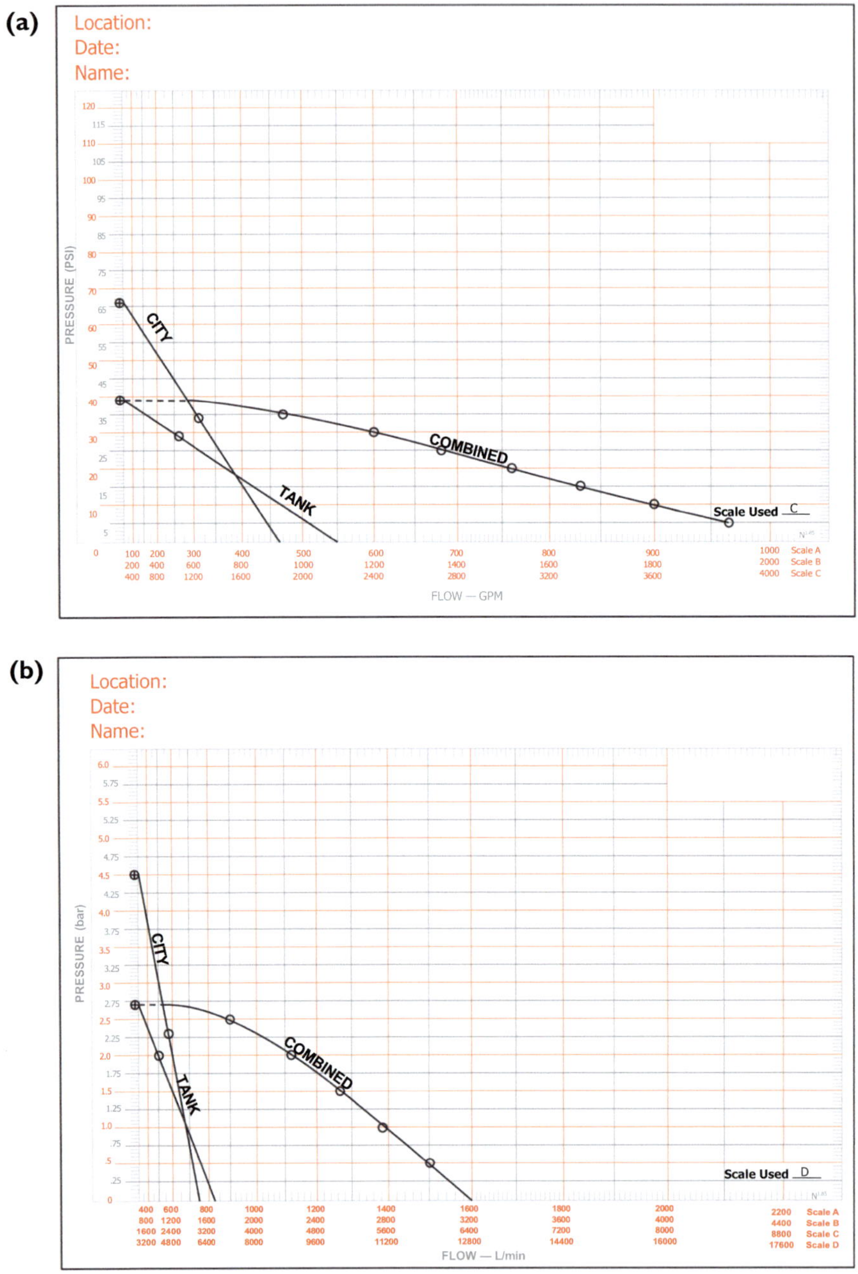

Fig. 11.6 **a** The combination of a municipal supply and an elevated tank. **b** The combination of a municipal supply and an elevated tank (S.I. units)

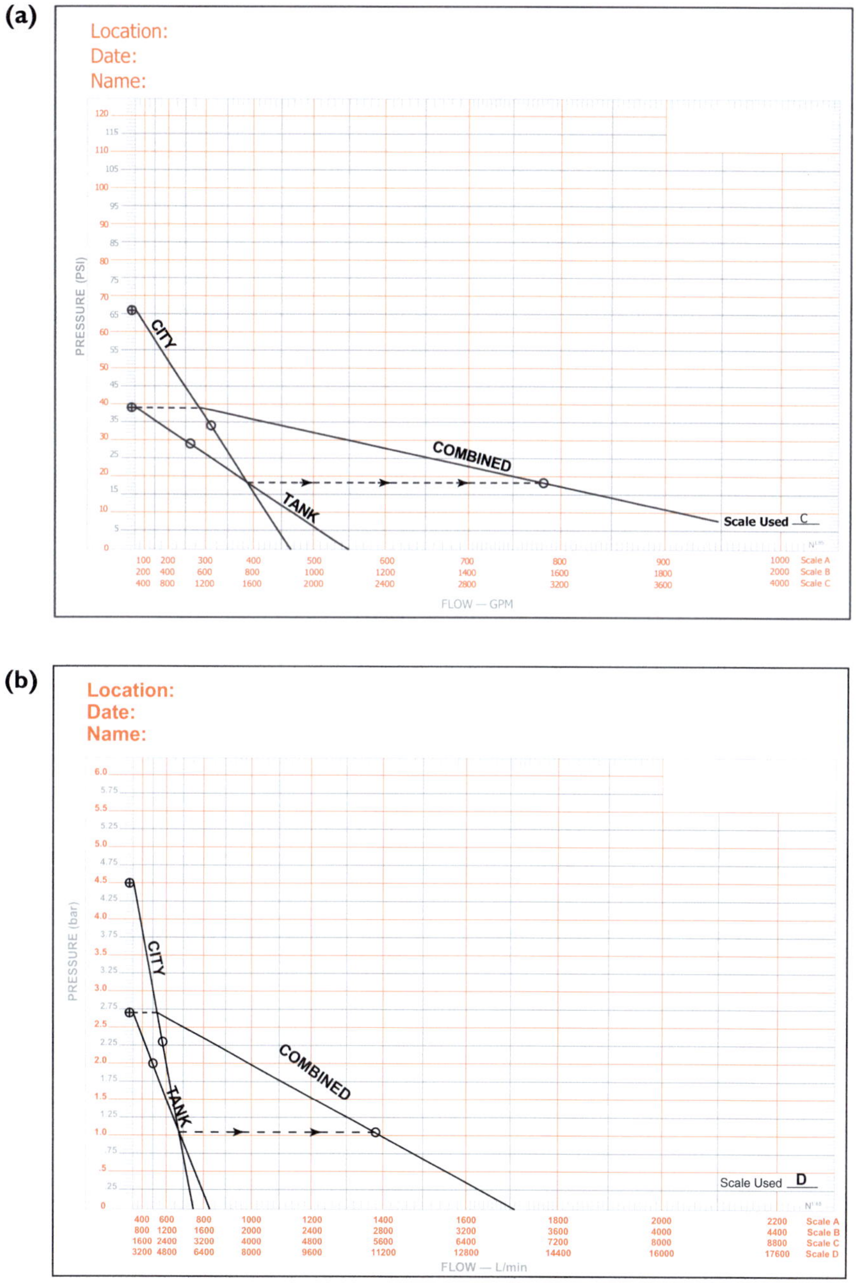

Fig. 11.7 **a** Short- cut combination of a municipal supply and an elevated tank. **b** Short- cut combination of a municipal supply and an elevated tank (S.I. units)

against freezing. They are an attractive nuisance and are commonly subject to vandalism and malicious mischief. Combining all of the reasons above, it can be easily understood why private elevated storage tanks are rarely constructed today.

Summary of Important Concepts

1. The difference between static and residual pressures is a result of friction loss.
2. Water supplies coming from purely static sources can be graphically illustrated through simple friction loss calculations.
3. Municipal supplies are typically pressurized by pumps, not tanks. Tanks are filled by the pressure existing in the system.
4. The combined supply of a municipal system and a tank can be graphically illustrated.
5. A conservative approximation will result from using the bottom of the tank as the elevation reference point and the intersection of the tank and municipal curves.

Exercises

11.1. How high would a tank have to be to generate a pressure of 75 psi (5.2 bar) when full?

11.2. How high would a tank have to be to generate a pressure of 125 psi (8.6 bar) when full?

11.3. If a tank is 85 ft (25.9 m) high, what city water pressure would have to be maintained to keep the tank full?

11.4. If a tank is 108 ft (32.9 m) high, what city water pressure would have to be maintained to keep the tank full?

11.5. If a tank is 92 ft (28 m) high, to what pressure would the city supply have to drop before the tank would begin to feed water into the system?

11.6. If a tank is 59 ft (18 m) high, to what pressure would the city supply have to drop before the tank would begin to feed water into the system?

11.7. In Fig. 11.8, plot the expected water supply from the elevated tank. Make your graph valid at Point A. Assume piping is cement-lined cast iron, $C = 140$.

11.8. In Fig. 11.9, graphically represent a conservative prediction of the water supply that would be available at Point A.

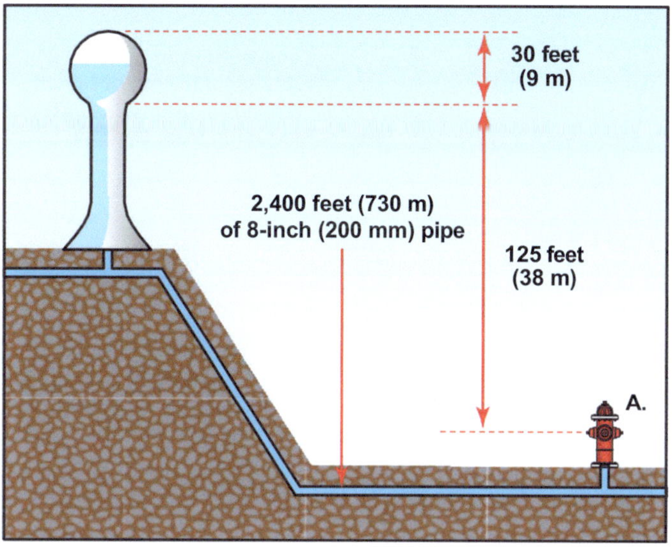

Fig. 11.8 Illustration for Exercise 11.7

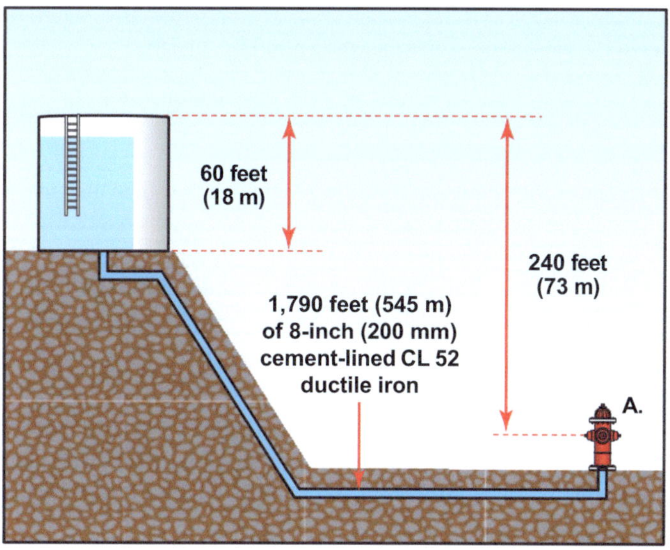

Fig. 11.9 Illustration for Exercise 11.8

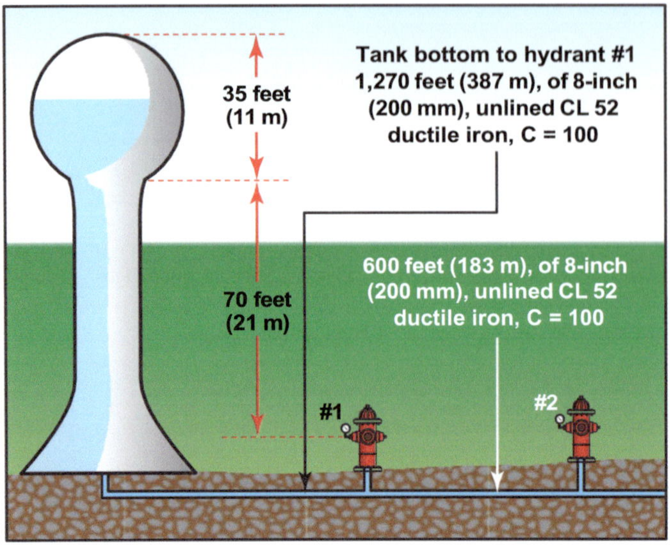

Fig. 11.10 Illustration for Exercise 11.9

11.9. Given the following situations shown in Fig. 11.10:

a. How many gallons (liters) will the elevated tank hold?
b. If the tank is full, what is the static pressure at Hydrant 1? What static pressure would exist at Hydrant 2?
c. If one of the outlets on Hydrant 1 were opened when the tank is full, what would be the approximated velocity of the water stream discharging from the outlet? Give your answer in ft/sec (m/sec).
d. With the tank shut off, the city water supply can provide the following at Hydrant 2: static: 80 psi (5.5 bar) residual: 8 psi (0.6 bar) flow: 625 gpm (2366 L/min) Predict a water supply curve for the tank alone and combine it with the city water supply at Hydrant 2.

11.10. For the situation shown in Fig. 11.11, establish a theoretical water supply curve for the elevated tank and combine the supply from the tank with a city supply that gave the following test results: static: 80 psi (5.5 bar) residual: 20 psi (1.4 bar) flow: 620 gpm (2347 L/min)

a. How many gallons (liters) will the elevated tank hold?
b. If the tank is full, what static pressure would exist at the test hydrant?
c. If one of the outlets on the test hydrant is opened when the tank is full, what will be the approximate velocity of the water stream discharging from the outlet? Give your answer in ft/sec (m/sec).

Selected answers are found in Appendix K.

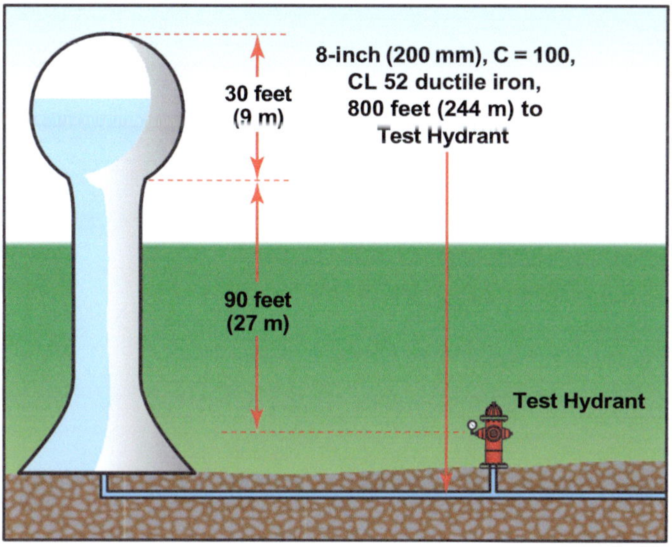

Fig. 11.11 Illustration for Exercise 11.10

Supplemental Reading

National Fire Protection Association. The Fire Protection Handbook. National Fire Protection Association: Quincy, MA, 21st edition, Sec. 15, Ch. 2.

Chapter 12
Establishing Water Supply Requirements

Abstract This chapter introduces three techniques to calculate exterior hose stream requirements for exterior fire fighting, including the Iowa State, National Fire Academy, and ISO methods. Additionally, the International Fire Code method is also introduced. The concepts of standpipes and hoses in accordance with NFPA 14 are introduced including the typical requirements. A method of estimating water flow requirements for sprinkler systems is introduced.

Water supply analysis establishes how much water is available and determines how much pressure and flow are needed for adequate fire protection. In some cases, industry standards and codes give specific requirements. In others, requirements are left substantially to the judgment of design engineers or authorities having jurisdiction.

This chapter addresses how to determine the requirements for exterior and interior manual fire suppression and water-based fire extinguishing systems.

Water Supply Requirements for Exterior Manual Firefighting Operations

Determining water supply requirements for manual firefighting operations has always been more of an art than a science. The more commonly used formulas have never really addressed differences in the fuel load of contents and are really nothing more than gross approximations. An historical formula that is no longer used based the water supply requirements on population. However, it became apparent that sparsely populated industrial communities may have a greater fire protection water supply demand than densely populated residential areas.

Four techniques for calculating fire flow requirements that are considered unscientific approximations will be presented. However, they have proven to provide conservative results that most fire department personnel consider adequate.

P. Brock and V. Charter, *Fire Protection Hydraulics and Water Supply Analysis*, The Society of Fire Protection Engineers Series,
https://doi.org/10.1007/978-3-032-09877-1_12

A formula traditionally used in fire service circles was developed by the Fire Service Extension Division of Iowa State University and is typically referred to as the Iowa State Formula [1]. The formula is as follows:

$$Q = (0.01)(V) \tag{12.1}$$

where:

$\boldsymbol{Q}$ = the required fire flow in gpm.

$\boldsymbol{V}$ = the building volume in cubic feet.

And, in S.I. Units:

$$Q = (1.337)(V) \tag{12.1a}$$

where:

$\boldsymbol{Q}$ = the required fire flow in liters per minute.

$\boldsymbol{V}$ = the building volume in cubic meters.

Note that no consideration is given to combustibility of construction or contents. More recently, the National Fire Academy has begun promoting the following formula [1].

$$Q = (1/3)(A) \tag{12.2}$$

where:

$\boldsymbol{Q}$ = the required fire flow in gpm.

$\boldsymbol{A}$ = the building area of all floors in square feet.

And, in S.I. Units:

$$Q = (13.6)(A) \tag{12.2a}$$

where:

$\boldsymbol{Q}$ = the required fire flow in liters per minute.

$\boldsymbol{A}$ = the building area of all floors in square meters.

Again, note that compensation is not made for variations in construction type or hazard of occupancy. However, both formulas contemplate full building involvement and could be revised appropriately for fireground operations if only a portion of a building were involved in fire.

The following formula was developed for insurance grading purposes and is published by the Insurance Services Office [1]. The formula in its basic form is as follows:

$$Q = (18)(\mathrm{C})(A)^{0.5} \tag{12.3}$$

where:

$\boldsymbol{Q}$ = needed fire flow in gpm.

$\boldsymbol{A}$ = building area of all floors in square feet.

C = a factor based upon construction as follows:

C = 1.5 for wood frame,
C = 1.0 for wood-joisted masonry,
C = 0.8 for unprotected noncombustible, and
C = 0.6 for fire resistive.

And, in S.I. Units:

$$Q = (224)(C)(A)^{0.5} \tag{12.3a}$$

where

$\boldsymbol{Q}$ = needed fire flow in liters per minute.

$\boldsymbol{A}$ = building area of all floors in square meters.

C = a factor based upon construction as follows:

C = 1.5 for wood frame,
C = 1.0 for wood-joisted masonry,
C = 0.8 for unprotected noncombustible, and
C = 0.6 for fire resistive.

Use of the formula usually requires rounding the results to the nearest 250 gpm (1000 L/min). Also, a minimum value of 500 gpm (2000 L/min) is imposed, with a maximum of 8000 gpm (30,000 L/min) for wood frame and wood- joisted masonry construction. A 6000 gpm (23,000 L/min) maximum value is specified for other construction types.

This formula may be adjusted for differences in occupancy combustibility. For noncombustible contents, the results may be reduced by as much as 25%. For rapid burning contents, the results could be increased by up to 25%. This formula also permits adjustments to compensate for exposures to other buildings. This insurance-related formula is more versatile, but is still an approximate method. The objective of this formula is to determine the fire flow needed to contain a fully developed fire to the building of fire origin and thus prevent conflagration.

Some fire codes also provide water supply requirements. The International Fire Code [2] (IFC) provides a table listing required minimum fire flows and required flow duration. The values in the table are based upon construction type and building area. No adjustments are indicated for either contents or exposure. See Table 12.1 for the IFC values.

Table 12.1 Minimum required fire-flow and flow duration for buildings

Fire-flow calculation area (square feet)					Fire flow (gallons per minute)[b]	Flow duration
Type IA and IB[a]	Type IIA and IIIA[a]	Type IV and V-A[a]	Type IIB and IIIB[a]	Type V-B[a]		
0–22,700	0–12,700	0–8200	0–5900	0–3600	1500	2
22,701–30,200	12,701–17,000	8201–10,900	5901–7900	3601–4800	1750	
30,201–38,700	17,001–21,800	10,901–12,900	7901–9800	4801–6200	2000	
38,701–48,300	21,801–24,200	12,901–17,400	9801–12,600	6201–7700	2250	
48,301–59,000	24,201–33,200	17,401–21,300	12,601–15,400	7701–9400	2500	
59,001–70,900	33,201–39,700	21,301–25,500	15,401–18,400	9401–11,300	2750	
70,901–83,700	39,701–47,100	25,501–30,100	18,401–21,800	11,301–13,400	3000	3
83,701–97,700	47,101–54,900	30,101–35,200	21,801–25,900	13,401–15,600	3250	
97,701–112,700	54,901–63,400	35,201–40,600	25,901–29,300	15,601–18,000	3500	
112,701–128,700	63,401–72,400	40,601–46,400	29,301–33,500	18,001–20,600	3750	
128,701–145,900	72,401–82,100	46,401–52,500	33,501–37,900	20,601–23,300	4000	4
145,901–164,200	82,101–92,400	52,501–59,100	37,901–42,700	23,301–26,300	4250	
164,201–183,400	92,401–103,100	59,101–66,000	42,701–47,700	26,301–29,300	4500	
183,401–203,700	103,101–114,600	66,001–73,300	47,701–53,000	29,301–32,600	4750	
203,701–225,200	114,601–126,700	73,301–81,100	53,001–58,600	32,601–36,000	5000	
225,201–247,700	126,701–139,400	81,101–89,200	58,601–65,400	36,001–39,600	5250	
247,701–271,200	139,401–152,600	89,201–97,700	65,401–70,600	39,601–43,400	5500	
271,201–295,900	152,601–166,500	97,701–106,500	70,601–77,000	43,401–47,400	5750	
295,901–Greater	166,501–Greater	106,501–115,800	77,001–83,700	47,401–51,500	6000	
–	–	115,801–125,500	83,701–90,600	51,501–55,700	6250	
–	–	125,501–135,500	90,601–97,900	55,701–60,200	6500	

(continued)

Table 12.1 (continued)

Fire-flow calculation area (square feet)					Fire flow (gallons per minute)[b]	Flow duration
Type IA and IB[a]	Type IIA and IIIA[a]	Type IV and V-A[a]	Type IIB and IIIB[a]	Type V-B[a]		
–	–	135,501–145,800	97,901–106,800	60,201–64,800	6750	
–	–	145,801–156,700	106,801–113,200	64,801–69,600	7000	
–	–	156,701–167,900	113,201–121,300	69,601–74,600	7250	
–	–	167,901–179,400	121,301–129,600	74,601–79,800	7500	
–	–	179,401–191,400	129,601–138,300	79,801–85,100	7750	
–	–	191,401–Greater	138,301–Greater	85,101–Greater	8000	

As detailed in IFC Appendix B, Table B105.1(2) [2]

For SI: 1 square foot = 0.0929 m^2, 1 gallon per minute = 3785 L/min, 1 pound per square inch = 6.895 kPa

[a]Types of construction are based on the *International Building Code*

[b]Measured at 20 psi residual pressure

For the purpose of illustration and comparison, fire flows are calculated using each of the four methods in the following examples.

Example 12.1

Determine the required fire flow for a single-story, unprotected, noncombustible building (International Building Code Type IIB) housing a moderate hazard industrial occupancy. The building has dimensions of 200 ft (60 m) by 100 ft (30 m) and is one story high with ceiling height of 16 ft (5 m).

Iowa State Formula:

$$Q = (0.01)(200\,\text{ft} \times 100\,\text{ft} \times 16\,\text{ft}) = 3200\,\text{gpm}$$

And:

$$Q = (1.337)(60\,\text{m} \times 30\,\text{m} \times 5\,\text{m}) = 12,030\,\text{L/ min}$$

National Fire Academy Formula:

$$Q = (1/3)(200\,\text{ft} \times 100\,\text{ft}) = 6666\,\text{gpm}$$

And:

$$Q = (13.6)(60\,\text{m} \times 30\,\text{m}) = 24,480\,\text{L/ min}$$

ISO Formula:

$$Q = (18)(0.8)(200\,\text{ft} \times 100\,\text{ft})^{0.5} = 2036 \text{ or } 2000\,\text{gpm}$$

And:

$$Q = (224)(0.8)(60\,\text{m} \times 30\,\text{m})^{0.5} = 7603 \text{ or } 8000\text{ L/ min}$$

International Fire Code

The construction type is indicated to be Type IIB. With a building floor area of 20,000 ft^2 (1800 m^2), Table 12.1 indicates that the required fire flow should be 3000 gpm (11,355 L/min).

Example 12.2

Determine the required fire flow for a three-story building with brick exterior walls and unprotected wood floors and roof if the building is 60 ft (18.3 m) wide and 120 ft (36.6 m) long with ceilings 10 ft (3 m) high on each floor.

Iowa State Formula:

$$Q = (0.01)(60\,\text{ft} \times 120\,\text{ft} \times 30\,\text{ft}) = 2160\,\text{gpm}$$

And:

$$Q = (1.337)(18.3\,\text{m} \times 36.6\,\text{m} \times 9\,\text{m}) = 8060\,\text{L/min}$$

National Fire Academy Formula:

$$Q = (1/3)(3 \times 60\,\text{ft} \times 120\,\text{ft}) = 7200\,\text{gpm}$$

And:

$$Q = (13.6)(3 \times 18.3\,\text{m} \times 36.6\,\text{m}) = 27,330\,\text{L/min}$$

ISO Formula:

$$Q = (18)(1.0)(3 \times 60\,\text{ft} \times 120\,\text{ft})^{0.5} = 2645 \text{ or } 2750\,\text{gpm}$$

And:

$$Q = (224)(1.0)(3 \times 18.3\,\text{m} \times 36.6\,\text{m})^{0.5} = 10,040 \text{ or } 10,000\,\text{L/min}$$

International Fire Code

The construction type is indicated to be Type IIIB. With a total building floor area, including all floors, of 21,600 ft^2 (2000 m^2), Table 12.1 indicates that the required fire flow should be 3000 gpm (11,355 L/min).

It is clear from the two examples that the National Fire Academy Formula is the most conservative of the four. Remember that all of these methods contemplate manual firefighting operations involving fully developed fires in unsprinklered buildings. The methods for establishing the fire protection water supplies for sprinklered buildings will be addressed later in this chapter and in Chap. 13, Hydraulic Calculation of Automatic Sprinkler Systems.

To be meaningful, fire flow requirements must have both a flow and pressure component. For all of the techniques discussed, the flow requirement is typically projected to be at a residual pressure of 20 psi (1.4 bar). It is considered that pressures for firefighting operations will be provided through fire department pumping apparatus. Therefore, the water system needs only to provide enough pressure to supply the pumpers. Traditionally, 20 psi (1.4 bar) has been considered the minimum acceptable pressure. However, this is not a hard and fast rule. A higher minimum pressure might be imposed on communities with severe elevation changes. Other systems might permit pressures as low as 10 psi (0.7 bar) with no adverse effects.

All of the formulas give unrealistic results for high-rise buildings. This is the reason that the ISO approach specifies maximum values. The International Fire Code requirements also top out at 6000 gpm (22,710 L/min) for the types of construction normally used with tall buildings. Reasonable maximums should also be established if using the Iowa State or National Fire Academy formulas.

Water Supply Requirements for Standpipe and Hose Systems

For standpipe and hose systems, the water supply requirements are more straightforward. NFPA 14, *Standard for the Installation of Standpipe and Hose Systems provides specific requirements* [3]. NFPA 14 defines three classes of standpipe service with different requirements for each class.

Class I service essentially consists of a system intended to be used by the fire department. On older systems, 2 1/2-in. (65 mm) hose connections were most often located in stairwells at every floor level. More recent editions of the standard also have an exception that where required by the authority having jurisdiction the hose connections can be located on the highest intermediate landings between the floors. Class I systems are required to have 500 gpm (1893 L/min) available for the first standpipe plus 250 gpm (946 L/min) for each additional standpipe. A 1250 gpm (4731 L/min) maximum is specified. A residual pressure of 100 psi (6.9 bar) is required at the most hydraulically demanding hose connection. The most hydraulically demanding hose connection is the one with the greatest amount of pressure lost to friction and elevation when delivering 500 gpm (1893 L/min).

For Class I service in buildings other than high rises, NFPA 14 permits the flow and pressure to be supplied to the standpipe by means of fire department pumping apparatus. This makes sense because Class I service is intended exclusively for Fire Department use. However, note that a standard 1000 gpm (3785 L/min) pumper can be expected to develop only 150 psi (10.3 bar) while delivering 1000 gpm (3785 L/min). This means that buildings over approximately 100 ft (30 m) tall need the support of a pump installation inside the building. This is why an automatic water supply is required for high-rise buildings.

Class II service is intended to be used primarily by building occupants as first aid fire appliances. This consists of hose reels or cabinets with 1 1/2-in. (38 mm) hose. Older systems may contain 75 ft (23 m) of unlined cotton or linen hose. Modern systems should have 100 ft (30 m) of lined or synthetic hose. One-inch (25 mm) listed, non-collapsible hose may be found in some light hazard installations.

Class II service requires only 100 gpm (378 L/min), with no increase for multiple standpipes. The pressure requirement is 65 psi (4.5 bar) at the most remote connection with 100 gpm (378 L/min) flowing. Since the Class II service is for building occupants, the pressure and flow must be available independent of fire department pumping apparatus. This means that a pump will be needed for most buildings with the Class II service.

Class III service is a combination of Classes I and II. Class III water supply requirements are identical to Class I requirements. NFPA 14 requires automatic water supplies for Class III systems. The support of the systems through fire department connections is considered an auxiliary supply. Even though this requirement has debatable merits for the Class I service, alternate design approaches should be undertaken only after obtaining permission from governing authorities.

Water supplies for standpipe systems also provide water for sprinkler systems. In these combined systems, if the buildings are fully sprinklered, the sprinkler demand does not have to be added to the standpipe demand. The standpipe demand alone is used. However, if the building is only partially sprinklered, the sum of the standpipe demand and the sprinkler demand must be used.

This section is not intended to address the design of standpipe and hose systems, only the water supply requirements. The detailed design specifications may be found in NFPA 14. However, it might be noted that 1 1/2-in. (38 mm) hose stations are rarely installed in modern buildings and those that exist are rapidly being removed. Fire departments will not use these small hoselines because building owners historically have not maintained the hose. To use poorly maintained hose could jeopardize the safety of firefighters. Also, the existence of hose stations is evidence that building owners expect building occupants to use the hose. Under federal law (OSHA) this means building occupants must be trained in the use of small hoselines. To avoid the complications and expense of hose maintenance and employee training, most hoselines have been removed. Also, the International Building Code only calls for Class I systems except in unsprinklered multiple story buildings where Class III systems are required [4].

This is also a good time to point out that the small hoseline systems required by the NFPA 13 for protecting warehouses containing high-piled storage are not intended to be systems complying with NFPA 14. Because of that, you will find that the pressure and flow requirements for warehouse hose stations are entirely different than those dictated by NFPA 14.

Water Supply Requirements for Automatic Sprinkler Systems

There are two recognized design techniques for automatic sprinkler systems. The older, rarely used approach is the pipe schedule design technique. The more modern method involves hydraulic calculations. The pipe schedule method will be discussed first.

Pipe schedule systems are those in which the pipe sizing is based upon the number of sprinklers supplied, as dictated by tables in NFPA 13, *Standard for the Installation of Sprinkler Systems* [5]. Systems designed in this manner have been around for more than a century, and their performance record has been outstanding. From a fire control standpoint, there is nothing wrong with pipe schedule design. However, certain severe hazards require more water than the typical pipe schedule system can deliver. From an economical standpoint, a hydraulic analysis might reveal that smaller pipe than required by the pipe schedules can provide adequate protection. Under proper occupancy considerations, pipe schedule design provides excellent protection, but perhaps not the most economical.

Unfortunately, the water supply requirements for pipe schedule systems are substantially left up to the judgment of designers or authorities having jurisdiction. At best, NFPA 13 provides only a range of suggested values for a few occupancy groups. For light hazard occupancies, a range of 500–750 gpm (1893–2893 L/min) is suggested for the acceptable flows at the base of the riser. For ordinary hazard occupancies, a range of 850–1500 gpm (3218–5678 L/min) is suggested at the base of the riser. It should be noted that these ranges include the hose stream allowance. The required duration of the water supply requirement is also a range, where light hazard is 30–60 min and ordinary hazard is 60–90 min [5].

It is easy to understand why specific requirements have not been established with so many possible variations in occupancy, contents, and construction. This understanding does not reduce the burden upon the authorities when determining specific requirements. Judgment must be exercised and can be based upon building size, ceiling height, construction type, and other factors affecting the quantity of fuel available.

NFPA 13 also provides pressure requirements for pipe schedule systems. They are the required residual pressures at the location of the highest sprinkler in the building. Light hazard occupancies require 15 psi (1 bar) and ordinary hazard occupancies require 20 psi (1.4 bar) at the highest sprinkler in the building. To establish ground level pressure requirements, pressure must be added to compensate for elevation losses. However, with the pipe schedule design method, friction losses need not be evaluated. It should also be noted that the pressure demands listed in NFPA 13 are for small buildings. Buildings over 5000 ft^2 (465 m^2) in size are required to have a residual pressure of 50 psi (3.4 bar) available at the top line of sprinklers for both occupancy classes [5].

Even with only rough guidelines provided by the codes and standards on determining water supply requirements, pipe schedule designs have provided excellent protection for a long time. This is not a lucky accident, but rather a reflection of the conservative nature of establishing sprinkler pipe sizing using the pipe schedule method.

Hydraulically Designed Water-Based Extinguishing Systems

For the hydraulically designed automatic sprinkler systems, mathematical calculations are used to arrive at specific flow and pressure requirements. Chapter 13 addresses this more complicated design technique in detail. The water supply demands for other water-based systems such as foam systems, water spray systems, or foam-water sprinkler systems are determined through minor variations in the hydraulic calculation technique for sprinkler systems.

For example, with the deluge water spray systems, the primary variation is that all nozzles are open and will deliver water. With the conventional sprinkler system design, only a few sprinklers are considered to be open.

With the foam system, differences include the use of the Darcy-Weisbach technique for calculating friction loss before the foam concentrate is mixed with water. The number of gpm required is also based on square feet of surface to be covered. Other minor differences exist. However, design or evaluation of these systems requires knowledge of hydraulic calculations. Instruction on this technique will be given in Chap. 13.

Estimating Sprinkler System Flow Demands

For both the pipe schedule design method and the hydraulic calculation design methods, an estimate for the flow demand for a sprinkler system can be easily calculated. For the hydraulically designed systems, NFPA 13 specifies required sprinkler water delivery rates [5]. These delivery rates are referred to as *design densities*, and they have a corresponding area of sprinkler operation. Chapter 13 discusses these concepts in detail.

In summary, light and ordinary hazard occupancies require a minimum of 1500 ft^2 (140 m^2) area of sprinkler operation while extra hazard occupancies require a minimum of 2500 ft^2 (230 m^2) area of sprinkler operation. The corresponding design densities will differ based upon the specific occupancy. For example, at the given areas of operation above, light hazard requires 0.10 gpm/ft^2 (4.1 mm/min), ordinary hazard group 1 requires 0.15 gpm/ft^2 (6.1 mm/min), ordinary hazard group 2 requires 0.20 gpm/ft^2 (8.1 mm/min), extra hazard group 1 requires 0.30 gpm/ft^2 (12.2 mm/min), and extra hazard requires 0.40 gpm/ft^2 (16.3 mm/min). Furthermore, there are other design densities and corresponding areas that can be selected within NFPA 13 based upon the circumstances of the design [5].

Additionally, NFPA 13 has a hose stream requirement to provide additional water for final manual fire extinguishment by the fire department, ranging from 100 gpm (379 L/min) for light hazard to 250 gpm for ordinary hazard (946 L/min) to 500 gpm for extra hazard (1893 L/min) [5].

The flow for a sprinkler system can be estimated by multiplying the density by the area of sprinkler operation, adding a 20% safety factor and then adding in the hose stream demand. An equation for this calculation is as follows:

$$Q = (1.2)(d)(A) + \mathrm{HS} \tag{12.4}$$

where

Q = the estimated sprinkler system demand in gpm (L/min).

d = required sprinkler density in gpm/ft^2 (mm/min).

A = area of sprinkler operation in ft^2 (m^2) and

HS = the hose stream requirement in gpm (L/min).

Example 12.3

What is an estimated sprinkler water flow demand for an Ordinary Hazard Group 1 occupancy?

Solution:

$$Q = (1.2)(d)(A) + \mathrm{HS}$$

$$Q = (1.2)(0.15)(1500) + 250$$

$$Q = 520\,\mathrm{gpm}$$

The solution in S.I. units is:

$$Q = (1.2)(d)(A) + \mathrm{HS}$$

$$Q = (1.2)(6.1)(139) + 946$$

$$Q = 1963\,\mathrm{L/\,min}$$

The actual water flow and pressure requirements for modern fire sprinkler systems must be determined through the mathematical process referred to as hydraulic calculations. This mathematical process is the topic of Chap. 13.

Duration of Water Supplies

The required duration of fire protection water supplies is essential information when designing water storage facilities. The required duration of sprinkler system water supplies is given for pipe schedule designs or in various locations in NFPA 13, depending upon the occupancy or hazard protected.

Where a duration range is given, the higher number is typically recommended unless the system alarms signal directly to the fire department. For example, if a fire pump taking water from a tank supplies water for sprinklers in a light hazard building where the pipe schedule demand is considered to be 500 gpm (1893 L/min), the required duration of the water supply would be 60 min with a local alarm only. The tank volume would have to be at least 30,000 gallons (60 min × 500 gpm) or 113,580 L (60 min × 1893 L/min).

The duration of water supplies for exterior manual firefighting is recommended to be at least 2 h. The International Fire Code (see Table 12.1) requires a 2-h duration for required fire flows of up to 2750 gpm (10,400 L/min), a 3-h duration for required fire flows from 3000 gpm (11,355 L/min) to 3750 gpm (14,190 L/min), and a 4-h duration for required fire flows exceeding 3750 gpm (14,190 L/min). The required duration for standpipe and hose systems is 30 min for all Class I, II, and III service.

Summary of Important Concepts

1. For exterior fire fighting, judgment must be exercised in determining water supply requirements.
2. Three techniques to calculate exterior hose stream requirements are the following:

   Iowa State

   $$Q = (0.01)(\text{V})\text{and in S.I. Units } Q = (1.337)(\text{V})$$

   National Fire Academy

   $$Q = (1/3)(\text{A})\text{and in S.I. Units } Q = (13.6)(\text{A})$$

   ISO

   $$Q = (18)(\text{C})(\text{A})^{0.5}\text{and in S.I. Units } Q = (224)(\text{C})(\text{A})^{0.5}$$

3. Three classifications of Standpipe and Hose Service are defined by NFPA 14.
4. Class I and III require 500 gpm (1893 L/min) for the first standpipe and 250 gpm (946 L/min) for each additional standpipe up to 1250 gpm (4732 L/min).
5. Class II service requires 100 gpm (378 L/min).

6. Class I and III standpipe systems require a 100 psi (6.9 bar) residual pressure at the most remote outlet.
7. Judgment is the key to establishing water supply requirements for pipe schedule sprinkler systems.
8. The hydraulically designed sprinkler system plus other water-based extinguishing systems require knowledge of a mathematical calculation technique which is described in Chap. 13.
9. The water flow requirements for sprinkler systems may be estimated from:

$$Q = (1.2)(d)(A) + \text{HS}$$

Exercises

12.1. Use all four methods given in the text to establish the water supply requirements for manual firefighting for the following buildings:

a. A 24 foot × 66 foot (7.3 m × 20 m), 2-story wood frame, 22 ft (6.7 m) to the top of the roof (IBC Type VB)
b. A 160 foot × 500 foot (49 m × 152 m), 1-story unprotected noncombustible building, with ceiling height of 22 ft (6.7 m) (IBC Type IIB)
c. A 60 foot × 80 foot (18 m × 24 m), 5 -story fire-resistive building, 55 ft (16.8 m) high (IBC Type IA)

12.2. Use all four methods given in the text to establish the water supply requirements for manual firefighting for the following buildings:

a. A 30 foot × 60 foot (9 m × 18 m), 1-story wood frame building, 10 ft (3 m) high (IBC Type VB)
b. A 40 foot × 80 foot (12.2 m × 24 m), 3-story building with brick walls, wood floors, and roof, 36 ft (11 m) high (IBC Type IIIB)
c. A 10 -story 120 foot × 80 foot (36.6 m × 24 m) fire-resistive office building120 ft (36.6 m)) high (IBC Type IA)

12.3. Neglecting friction loss, what would be the ground level pressure and flow requirement for a Class I standpipe system consisting of three standpipe risers with the highest connection 85 ft (26 m) above the ground?

12.4. Neglecting friction loss, what would be the ground level pressure and flow requirement for a Class I standpipe system consisting of two standpipe risers with the highest connection 140 ft (42.7 m) above the ground?

12.5. Neglecting friction loss, what would be the ground level pressure and flow requirement for a Class II standpipe system with five hose cabinets per floor and with the highest cabinet 36 ft (11 m) above the ground?

12.6. Neglecting friction loss, what would be the ground level pressure and flow requirement for a Class II standpipe system with four hose cabinets per floor and with the highest cabinet 72 ft (22 m) above the ground?

12.7. Estimate the flow requirement for sprinkler systems protecting the following:

a. An office building (Light Hazard)
b. A parking garage (Ordinary Hazard Group 1)
c. An automobile paint shop (Extra Hazard Group 2)

12.8. Estimate the flow requirement for sprinkler systems protecting the following:

a. A department store (Ordinary Hazard Group 2)
b. A hotel (Light Hazard)
c. A saw mill (Extra Hazard Group 1)

12.9. What size of storage tank would be required to hold sufficient water for each demand from Exercise 12.1? Use the duration requirements from Table 12.1.

12.10. What size of storage tank would be required to hold sufficient water for each demand from Exercise 12.2? Use the duration requirements from Table 12.1.

12.11. If a tank has a continuous available supply of 100 gpm (378 L/min), how large must a tank be to hold sufficient water to meet each demand from Exercise 12.7? Assume the duration requirements are 30 min for Light Hazard, 60 min for Ordinary Hazard and 90 min for Extra Hazard.

12.12. If a tank has a continuous available supply of 70 gpm (265 L/min), how large must it be to hold sufficient water to meet each demand from exercise 12.8? Assume the duration requirements are 30 min for Light Hazard, 60 min for Ordinary Hazard and 90 min for Extra Hazard.

Selected Answers Are Found in Appendix K.

References

1. Wieder, Michael A., *Fire Service Hydraulics and Water Supply*, Oklahoma State University Fire Protection Publications, Stillwater, OK. Second Edition.
2. *International Fire Code*, International Code Council, Inc., Country Club Hills, IL, 2024 Edition, Appendix B.
3. NFPA 14, *Standard for the Installation of Standpipe and Hose Systems*, 2024 Edition (Quincy: National Fire Protection Association, 2023)
4. *International Building Code*, International Code Council, Inc., Country Club Hills, IL, 2024 Edition.
5. NFPA 13, *Standard for the Installation of Sprinkler Systems,* 2025 Edition (Quincy: National Fire Protection Association, 2024)

Supplemental Readings

National Fire Protection Association. The Fire Protection Handbook. National Fire Protection Association: Quincy, MA, 21st edition, Section 15, Ch. 2.

Chapter 13
Hydraulic Calculations of Automatic Sprinkler Systems

Abstract This chapter introduces the most widely used technique for modern sprinkler system design using hydraulic calculations. The approach can be broken down into 16 basic steps. NFPA 13, Chapters 19 and 28 provide important information in aiding in these calculations. Velocity pressures are also introduced. Computer aided systems are commonly used today, but understanding the background is a key aspect of understanding the computer programs.

The most popular method for designing modern automatic sprinkler systems involves using a mathematical calculation technique referred to as hydraulic calculations. Hydraulic calculations is the only acceptable design technique used for protection of many special hazards, such as high-piled and rack storage of commodities. Also, hydraulic calculations can be used to demonstrate that systems with piping smaller than required by the pipe schedules can provide good protection. This reduces materials and labor costs on light and ordinary hazard systems. Another benefit is that use of hydraulic calculations permits the designer to take advantage of the superior friction loss characteristics of thin wall piping, copper tubing, and plastic piping.

This chapter provides an overview of the calculation steps in the hydraulic design technique. The presentation is based upon the requirements of NFPA 13, *Standard for the Installation of Sprinkler Systems*; Chapter 19, Design Approaches and Chapter 28, Plans and Calculations. Additionally, information on the differences and similarities of the design requirements between the 2022 and 2025 editions are presented. However, the presentation may be generally applied to any hydraulic design situation. A more exhaustive coverage is given to the subject by Wass and Fleming in their text [1].

P. Brock and V. Charter, *Fire Protection Hydraulics and Water Supply Analysis*, The Society of Fire Protection Engineers Series,
https://doi.org/10.1007/978-3-032-09877-1_13

Basic Steps for Hydraulic Calculations

The hydraulic design technique can be summarized in 16 basic steps. These steps can be used as a guide to system design or as a checklist for plan review by governing authorities. (The steps will be described in detail in the example in the following section.) The 16 steps are as follows:

Step 1: Identify occupancy or hazard to be protected.

Step 2:Determine the size of the area of sprinkler operation.

Step 3: Determine the required design density.

Step 4: Establish the number of sprinklers contained in the area of sprinkler operation.

Step 5: Determine the shape and location of the area of sprinkler operation.

Step 6: Calculate the required minimum flow from the first sprinkler.

Step 7: Calculate the required minimum pressure for the first sprinkler.

Step 8: Calculate friction loss between the first and second sprinklers.

Step 9: Calculate the flow from the second sprinkler.

Step 10: Repeat Steps 8 and 9 for successive sprinklers until all operating sprinklers on the first branch line are calculated.

Step 11: If the design area extends to the opposite side of the cross main, Steps 6 through 10 are repeated for the opposite side. The intersecting branch lines must be balanced to the highest pressure demand.

Step 12: Calculate a k-factor for the first riser nipple with additional factors calculated for non-identical lines.

Step 13: Repeat Steps 8 and 9 for riser nipples (instead of sprinklers) until all riser nipples within the design area are added in.

Step 14: Compute friction loss to the point of supply with compensation for elevation changes, valves and fittings, and differences in materials of underground piping.

Step 15: Add the hose stream requirement.

Step 16: Compare calculated demand with water supply available.

Calculation Example

Step 1

Identify the occupancy or hazard to be protected. Choosing the correct occupancy or hazard classification is one of the most important and difficult steps. This step is important because a mistake here can leave the system inadequate. It is difficult because there are many potential choices, which include the special hazard possibilities. For example, with rack storage of materials, the designer must know what commodity is being stored, storage height, ceiling height, aisle width, temperature rating of sprinklers, etc. NFPA 13 contains definitions of the various occupancy and commodity classifications in Chap. 4. Helpful lists of examples are also contained in Annex A of NFPA 13. However, no list is ever complete and good judgment may often have to be used.

An error in any consideration can leave the system design inadequate. The problem is complicated by the knowledge that whatever the occupancy classification is today, it will probably change next year or even next week.

For this example, consider the occupancy Ordinary Hazard, Group 2 as defined by NFPA 13.

Step 2

Determine the size of the area of sprinkler operation. The area of sprinkler operation is often referred to as the "design area." This is the area in the building where all of the sprinklers will be considered to be open and discharging water. All other sprinklers are considered closed. More than 100 years of successful sprinkler system operations has revealed that in most fires only a few sprinklers open. Therefore, it is a reasonable design assumption that only a portion of the sprinklers will operate. Of course, with deluge-type systems, the entire system must be considered the design area.

Sometimes the authority having jurisdiction or insurance industry will dictate the size of the design area. If requirements are not imposed, the designer has the option of using any size of area that is permitted by the pertinent design standard. Within the 2022 edition of NFPA 13, two approaches are provided. One for new occupancies that has a simple table with two choices for each occupancy. The second approach is for existing occupancies, providing curves that give a range of design areas. Design areas may be so small as to involve only two or three sprinklers or as large as 5000 or 6000 ft^2. For the same occupancy classification, smaller design areas will always result in smaller gpm requirements and usually more economical designs. Larger design areas generally provide a greater factor of safety.

It should be noted that in the 2025 edition of NFPA 13 eliminated the existing system criteria figure and a designer may need to review prior editions to understand existing designs. This example is for existing occupancy criteria to ensure understanding of the more complicated figures compared to the table criteria of new occupancies.

Furthermore, there are additional rules pertaining to the design area beyond those mentioned above. NFPA 13 permits the use of what is called the Room Design Method for establishing the size of the design area. In this method, the most hazardous or largest room may be used as the design area. This can be of benefit to the designer if the largest room is less than 1500 ft^2 in size. If this method is used, the walls enclosing the rooms and doors to the rooms throughout the building must meet specific requirements dictated by the standard.

In other cases, such as in warehouses protected by Control Mode Special Application (CMSA) or Early Suppression Fast Response (ESFR) sprinklers, the design area will be based upon a specified number of sprinklers. The design areas in residential properties up to four stories high are also based upon a specified number of sprinklers rather than floor area.

It is known that dry systems and double interlock preaction systems operate more slowly than wet systems and typically experience a greater number of sprinklers opening during a fire. For this reason, the design area is required to be increased by 30% after establishing the density for their systems. A 30% increase is also required for buildings with steeply pitched ceilings above the sprinklers.

Finally, for light and ordinary hazard occupancies protected by wet systems using quick response sprinklers, a reduction in design area of up to 40% is permitted where ceiling heights do not exceed 20 ft.

For this example, a design area of 2000 ft^2*(186* m^2*) will be used for an existing occupancy.*

Step 3

Determine the required design density. The design density is the minimum amount of water that must be delivered from every sprinkler and is expressed in units of gpm per square foot of floor area or millimeters per minute in S.I. units.. Most design standards contain graphs or tables which allow determination of the required density. The density will be based upon the occupancy or hazard classification and the size of the design area. There may also be adjustments to the density that will be required or permitted due to variables such as storage height. Remember that when protection is from dry pipe systems or double interlocked preaction systems, the design area must be increased by 30% after the density has been established.

For this example, NFPA 13 requires 0.19 gpm/ft^2*(7.7 mm/min) for an existingOrdinary Hazard, Group 2 and a 2000* ft^2*(186* m^2*) sprinkler operating area.*

Step 4

Establish the number of sprinklers contained in the design area. Not only must the size of the design area be determined, but from a practical standpoint, the designer must know how many sprinklers compose the design area. This is done by application of the following formula:

$$N = \frac{A_d}{A_s} \tag{13.1}$$

where

N = the number of sprinklers in the design area.

$\mathbf{A_d}$ = the size of the design area in square feet (square meters).

$\mathbf{A_s}$ = the actual area protected per individual sprinklers in square feet (square meters).

If A_s varies from sprinkler to sprinkler, a conservative design will result from using the smallest value of A_s. Another method is to calculate the areas protected by each individual sprinkler and use enough sprinklers to give a cumulative area equal to or greater than the design area.

As indicated in the sprinkler system plan shown in Fig. 13.1, each sprinkler protects 120 ft² (10 ft × 12 ft) or 11.1 m² (3.7 m × 3.0 m). Therefore, the number of sprinklers in the design area must be as follows:

$$N = \frac{A_d}{A_s}$$

$$N = \frac{2000\,\text{ft}^2}{120\,\text{ft}^2/\text{sprinkler}} = 16.7 \text{ or } 17 \text{ sprinklers}$$

(Fractions are typically rounded up to the next whole number.)

And, in S.I. Units:

$$N = \frac{186\,\text{m}^2}{11.1\,\text{m}^2/\text{sprinkler}} = 16.8 \text{ or } 17 \text{ sprinklers}$$

Step 5

Determine the shape and location of the design area. At this step it is necessary to establish the shape and location of the design area. Chapter 28 of NFPA 13 requires the design area to be rectangular with a width parallel to the branch lines of at least 1.2 times the square root of the design area. However, other authorities may require a factor of 1.4. The larger the factor, the more demanding the system will be. Thus, the minimum width of the design area is calculated as follows:

$$W = (1.2)(\text{Design Area})^{0.5} \tag{13.2}$$

The design area is required to be the most hydraulically demanding one in the building. For systems that are single occupancy, symmetrical, and non-gridded, *hydraulically most demanding* equates to *physically farthest away* from the supply. However, for buildings that are multiple occupancy and multiple hazard or where gridded or non-uniform piping schemes are used, the location of the most demanding area may not be obvious. In such a case, multiple calculations must be completed to determine the most demanding.

In the example problem, the width of the design area is computed as follows:

$$W = (1.2)\left(2000\,\text{ft}^2\right)^{0.5} = 53.7\,\text{ft}$$

And, in S.I. Units:

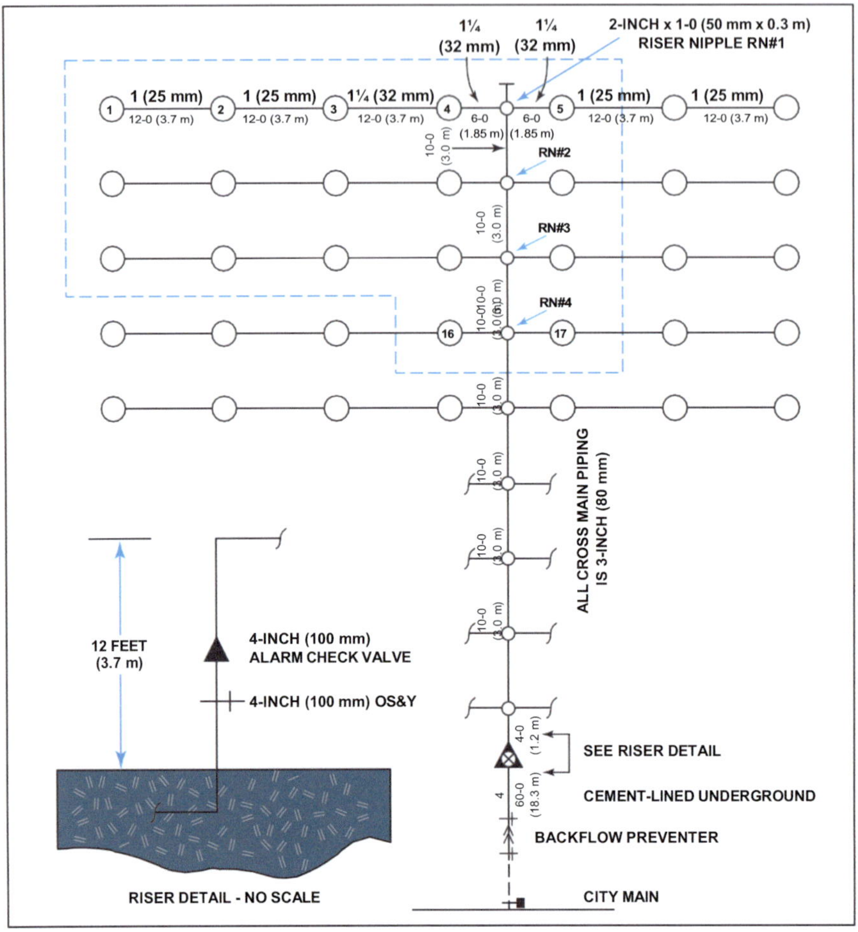

Fig. 13.1 Sprinkler system for hydraulic calculation example

$$W = (1.2)(184\,\mathrm{m}^2)0.5 = 16.4\,\mathrm{m}$$

Therefore, the design area must be at least 53.7 ft (16.4 m) wide. This width can be equated to some number of sprinklers when the distance between sprinklers are the same by dividing it by the distance between sprinklers on the branch lines. Thus,

$$N = \frac{53.7\,\mathrm{ft}}{12\,\mathrm{ft}/\,\mathrm{sprinkler}} = 4.475\text{ or }5\text{ sprinklers}$$

And, in S.I. Units:

$$N = \frac{16.4\,\text{m}}{3.7\,\text{m}/\text{sprinkler}} = 4.4 \text{ or } 5 \text{ sprinklers}$$

Again, the result has been rounded up to the next whole number. The design area has now been determined to be 5 sprinklers wide and consist of a total of 17 sprinklers. The design area should be outlined as shown in Fig. 13.1. The technique for identifying the most demanding 17 sprinklers is to begin counting with the end sprinkler on the longest of the end branch lines.

If there are an equal number of sprinklers on each side of the cross main, counting may start on either side. In the example, there are four sprinklers on the left and only three on the right. Therefore, the first sprinkler is the one on the end of the left branch line. Beginning there, the design area is five sprinklers wide, as shown, including one sprinkler from the right side of the cross main.

Two more sets of five sprinklers each are included from the next two riser nipples, as indicated in the drawing. This gives a total of 15 sprinklers. From the fourth riser nipple, two additional sprinklers must be chosen to give a total of 17. Choose the two sprinklers nearest the cross main to give the most demanding design area. The reason for this will be demonstrated later.

The complete design area is as outlined by the dotted line in Fig. 13.1. All sprinklers within the dotted line are considered to be open and discharging water. All other sprinklers are considered closed and have no impact upon the calculations.

Step 6

Calculate the required minimum flow from the first sprinkler. The required minimum flow from the first sprinkler (and actually any sprinkler on the system) is determined as follows:

$$q = (\text{design density})(\text{area per sprinkler}) \tag{13.3}$$

In the example, this gives

$$q = (0.19\,\text{gpm/sq ft})(120\,\text{sq ft}) = 22.8\,\text{gpm}$$

And, in S.I. Units:

$$q = (7.7\,\text{mm/min})\left(11.1\,\text{m}^2\right)\left(10^6\,\text{mm}^2/\text{m}^2\right)\left(1\,\text{L}/10^6\,\text{mm}^3\right)$$
$$q = 85.5\,\text{L/min}$$

This value for the flow is entered onto the worksheet (Fig. 13.2a, b, p. 248).

Step 7

Calculate the required minimum pressure for the first sprinkler. The pressure required to deliver the minimum flow from the first sprinkler is computed as follows:

HYDRAULIC SPRINKLER CALCULATIONS

SHEET 1 OF 3

STEP NO.	NOZZLE IDENT. AND LOCATION	FLOW IN GPM	PIPE SIZE	PIPE FITTINGS AND DEVICES	EQUIV. PIPE LENGTH	FRICTION LOSS PSI/FOOT	PRESSURE SUMMARY	NORMAL PRESSURE	NOTES $K=5.6$	REF. STEP
	SPKLR #1	q Q 22.8	1 (1.049)		L 12-0 F - T 12-0	C=120 .166	Pt 16.6 Pe - Pf 2.0	Pt Pv Pn	$Q=.19\times120$ $P=\left(\frac{22.8}{5.6}\right)^2$	
	SPKLR #2	q 24.2 Q 47.0	1		L 12-0 F - T 12-0	.632	Pt 18.6 Pe - Pf 7.6	Pt Pv Pn	$q=5.6\sqrt{18.6}$	
	SPKLR #3	q 28.7 Q 75.7	1 1/4 (1.38)		L 12-0 F - T 12-0	.402	Pt 26.2 Pe - Pf 4.8	Pt Pv Pn	$q=5.6\sqrt{26.2}$	
	SPKLR #4	q 31.2 Q 106.9	1 1/4	(1) Tee	L 6-0 F 6 T 12-0	.760	Pt 31.0 Pe - Pf 9.1	Pt Pv Pn	$q=5.6\sqrt{31}$	
		q Q			L F T		Pt 40.1 Pe Pf	Pt Pv Pn		
		q Q			L F T		Pt Pe Pf	Pt Pv Pn		
	SPKLR #5	q Q 22.8	1 1/4	(1) Tee	L 6-0 F 6 T 12-0	.044	Pt 16.6 Pe - Pf 0.5	Pt Pv Pn	RIGHT SIDE OF CROSS MAIN	
		q Q			L F T		Pt 17.1 Pe Pf	Pt Pv Pn	$Q_{ADJ}=22.8\sqrt{\frac{40.1}{17.1}}$ $=34.9$	
		q Q			L F T		Pt Pe Pf	Pt Pv Pn		
	RISER NIPPLE	q 106.9 +34.9 Q 141.8	2 (2.067)	(1) Tee	L 1-0 F 10 T 11-0	.179	Pt 40.1 Pe - Pf 2.0	Pt Pv Pn		
		q Q			L F T		Pt 42.1 Pe Pf	Pt Pv Pn	$K=\frac{141.8}{\sqrt{42.1}}$ $K=21.85$	
		q Q			L F T		Pt Pe Pf	Pt Pv Pn		

Fig. 13.2 Worksheets 1, 2, and 3 for hydraulic calculations

HYDRAULIC SPRINKLER CALCULATIONS

SHEET 2 OF 3

STEP NO.	NOZZLE IDENT. AND LOCATION	FLOW IN GPM	PIPE SIZE	PIPE FITTINGS AND DEVICES	EQUIV. PIPE LENGTH	FRICTION LOSS PSI/FOOT	PRESSURE SUMMARY	NORMAL PRESSURE	NOTES	REF STEP
	RN #1	q Q 141.8	3 (3.068)		L 10-0 F – T 10-0	C=120 .026	Pt 42.1 Pe – Pf .3	Pt Pv Pn	K=21.85	
	RN #2	q 142.3 Q 284.1	3		L 10-0 F – T 10-0	.095	Pt 42.4 Pe – Pf .9	Pt Pv Pn	$q=21.85\sqrt{42.4}$	
	RN #3	q 143.8 Q 427.9	3		L 10-0 F – T 10-0	.202	Pt 43.3 Pe – Pf 2.0	Pt Pv Pn	$q=21.85\sqrt{43.3}$	
	RN #4	q 73.8 Q 501.7	3	(1) LTE	L 54-0 F 5 T 59-0	.271	Pt 45.3 Pe – Pf 16.0	Pt Pv Pn	$q=10.96\sqrt{45.3}$	
	TO BOTTOM OF RISER	q – Q 501.7	4 (4.026)	(1) AL. VA. (1) GA. VA.	L 12-0 F 24-0 T 36-0	.072	Pt 61.3 Pe 5.6 Pf 2.6	Pt Pv Pn	→ .433 x 13 ft.	
	TO CITY MAIN	q – Q 501.7	4 (4.1)	(1) LTE=6 (1) TEE=20 (1) GA.VA.=2	L 60 F 41 T 101	C=140 .049	Pt 69.5 Pe 3.8 Pf 4.9	Pt Pv Pn	← Backflow Preventer TOTAL DEMAND	
	HOSE	q 250 Q 751.7		28X1.33X1.1	L F T		Pt 78.2 Pe Pf	Pt Pv Pn	752gpm @ 78psi	
		q Q			L F T	DIAMETER FACTOR= $\left(\frac{4.1}{4.026}\right)^{4.87}$	Pt Pe Pf	Pt Pv Pn		
		q Q			L F T		Pt Pe Pf	Pt Pv Pn		
		q Q			L F T		Pt Pe Pf	Pt Pv Pn		
		q Q			L F T		Pt Pe Pf	Pt Pv Pn		
		q Q			L F T		Pt Pe Pf	Pt Pv Pn		

Fig. 13.2 (continued)

HYDRAULIC SPRINKLER CALCULATIONS

SHEET 3 OF 3

STEP NO.	NOZZLE IDENT. AND LOCATION	FLOW IN GPM	PIPE SIZE	PIPE FITTINGS AND DEVICES	EQUIV. PIPE LENGTH	FRICTION LOSS PSI/FOOT	PRESSURE SUMMARY	NORMAL PRESSURE	NOTES	REF. STEP
	SPKLR #16	q Q 22.8	1 1/4	(1) TEE	L 6-0 F 6 T 12-0	C=120 .044	Pt 16.6 Pe – Pf .5	Pt Pv Pn		
		q Q			L F T		Pt 17.1 Pe Pf	Pt Pv Pn		
		q Q			L F T		Pt Pe Pf	Pt Pv Pn	DEMAND FOR SPKLR #17 WILL BE SAME AS #16	
	TO BOTTOM OF RN #4	22.8 q +22.8 Q 45.6	2	(1) TEE	L 1-0 F 10 T 11-0	.022	Pt 17.1 Pe – Pf .2	Pt Pv Pn		
		q Q			L F T		Pt 17.3 Pe Pf	Pt Pv Pn	$K=\frac{45.6}{\sqrt{17.3}}$ K=10.96	
		q Q			L F T		Pt Pe Pf	Pt Pv Pn		
		q Q			L F T		Pt Pe Pf	Pt Pv Pn		
		q Q			L F T		Pt Pe Pf	Pt Pv Pn		
		q Q			L F T		Pt Pe Pf	Pt Pv Pn		
		q Q			L F T		Pt Pe Pf	Pt Pv Pn		
		q Q			L F T		Pt Pe Pf	Pt Pv Pn		
		q Q			L F T		Pt Pe Pf	Pt Pv Pn		

Fig. 13.2 (continued)

$$P = (q/k)^2 \tag{13.4}$$

where

P = the required pressure in psi (bar).

q = the required flow at the first sprinkler in gpm (L/min).

k = the nominal discharge coefficient of the sprinklers being used.

The nominal discharge coefficient, **k**, is dependent upon the orifice diameter of the sprinkler being used. The nominal discharge coefficients for the common orifice diameters may be found in Appendix I of this text.

In the current example, sprinklers with a ½-in. (12.7 mm) orifice diameter are to be used having a **k** of 5.6 (80). Therefore, the required minimum pressure is as follows:

$$P = (22.8)^2/(5.6)^2 = 16.6\,\text{psi}$$

And, in S.I. Units:

$$P = (85.5)^2/(80)^2 = 1.14\,\text{bar}$$

Note that Chapter 28 of NFPA 13 prescribes a minimum allowable pressure of 7 psi (0.5 bar). If the above calculation yields a pressure less than 7 psi (0.5 bar), 7 psi (0.5 bar) must be used. The flow must then also be adjusted to what will be delivered by 7 psi (0.5 bar) according to the equation given in Step 9. Special sprinklers or special situations may call for different minimum pressures. Sprinklers with special listings or special discharge patterns may require specific minimum pressures different from the 7 psi (0.5 bar) required by NFPA 13. For example, ESFR sprinklers have minimum pressure requirements in some cases as high as 75 psi (5.2 bar).

Step 8

Calculate friction loss between the first and second sprinklers. Both the beginning pressure and the beginning flow should now have been written onto the worksheet (see Figs. 13.2 sheet 1 and 13.3 sheet 1). The next step is to calculate the pressure lost to friction as the water flows from the second sprinkler to the first sprinkler. The Hazen-Williams formula (see Chap. 5 of this text) is used for this computation using the flow from the first sprinkler, the pipe size which feeds the first sprinkler, the length of pipe which separates the first two sprinklers, and the appropriate coefficient of roughness. The appropriate coefficient of roughness is dictated by NFPA 13 and depends upon the piping material and the type of system. In this example, Schedule 40 black steel pipe is being used, and the system is a wet system. The required C-factor is 120. The friction loss between the first two sprinklers is therefore calculated to be the following:

HYDRAULIC SPRINKLER CALCULATIONS

SHEET 1 OF 3

STEP NO.	NOZZLE IDENT. AND LOCATION	FLOW IN LPM	PIPE SIZE	PIPE FITTINGS AND DEVICES	EQUIV. PIPE LENGTH	FRICTION LOSS BAR/M	PRESSURE SUMMARY	NORMAL PRESSURE	NOTES K=80	REF. STEP
	SPKLR	q	25		L 3.7	C=120	Pt 1.14	Pt	$q=(7.7)\times(11.1)$	
			(26.6)		F –	.037	Pe –	Pv		
	#1	Q 85.5			T 3.7		Pf .14	Pn		
	SPKLR	q 90.5	25		L 3.7		Pt 1.28	Pt	$q=80\sqrt{1.28}$	
					F –	.141	Pe –	Pv		
	#2	Q 176			T 3.7		Pf .52	Pn		
	SPKLR	q 107.3	32		L 3.7		Pt 1.80	Pt	$q=80\sqrt{1.8}$	
			(35.1)		F –	.088	Pe –	Pv		
	#3	Q 283.3			T 3.7		Pf .33	Pn		
	SPKLR	q 116.8	32		L 1.85		Pt 2.13	Pt	$q=80\sqrt{2.13}$	
				(1) Tee	F 1.8	.167	Pe –	Pv		
	#4	Q 400.1			T 3.65		Pf .61	Pn		
		q			L		Pt 2.74	Pt		
					F		Pe	Pv		
		Q			T		Pf	Pn		
		q			L		Pt	Pt		
					F		Pe	Pv		
		Q			T		Pf	Pn		
	SPKLR	q	32		L 1.85		Pt 1.14	Pt	RIGHT SIDE	
				(1) Tee	F 1.8	.010	Pe –	Pv	OF	
	#5	Q 85.5			T 3.65		Pf .04	Pn	CROSS MAIN	
		q			L		Pt 1.18	Pt	$Q_{ADJ}=85.5\sqrt{\frac{2.74}{1.18}}$	
					F		Pe	Pv	$=130.3$	
		Q			T		Pf	Pn		
		q			L		Pt	Pt		
					F		Pe	Pv		
		Q			T		Pf	Pn		
	RISER	q 400.1 +130.3	50		L 0.3		Pt 2.74	Pt		
			(52.5)	(1) Tee	F 3.0	.040	Pe –	Pv		
	NIPPLE	Q 530.4			T 3.3		Pf .13	Pn		
		q			L		Pt 2.87	Pt	$K=\frac{530.4}{\sqrt{2.87}}$	
					F		Pe	Pv		
		Q			T		Pf	Pn	$K=312.85$	
		q			L		Pt	Pt		
					F		Pe	Pv		
		Q			T		Pf	Pn		

Fig. 13.3 Worksheets 1, 2, and 3 for hydraulic calculations (S.I. units)

HYDRAULIC SPRINKLER CALCULATIONS

SHEET 2 OF 3

STEP NO	NOZZLE IDENT. AND LOCATION	FLOW IN LPM	PIPE SIZE	PIPE FITTINGS AND DEVICES	EQUIV. PIPE LENGTH	FRICTION LOSS BAR/M	PRESSURE SUMMARY	NORMAL PRESSURE	NOTES	REF STEP
	RN #1	q Q 530.4	80 (77.9)		L 3.0 F – T 3.0	C=120 .006	Pt 2.87 Pe – Pf .02	Pt Pv Pn		
	RN #2	q 531.8 Q 1062.2	80		L 3.0 F – T 3.0	.021	Pt 2.89 Pe – Pf .06	Pt Pv Pn	q=312.85√2.89	
	RN #3	q 537.3 Q 1599.5	80		L 3.0 F – T 3.0	.045	Pt 2.95 Pe – Pf .13	Pt Pv Pn	q=312.85√2.95	
	RN #4	q 274 Q 1873.5	80	(1) LTE	L 16.2 F 1.5 T 17.7	.06	Pt 3.08 Pe – Pf 1.06	Pt Pv Pn	q=156.1√3.08	
	TO BOTTOM OF RISER	q – Q 1873.5	100 (102.3)	(1) AL. VA. (1) GA. VA.	L 3.7 F 7.3 T 11.0	.016	Pt 4.14 Pe .39 Pf .18	Pt Pv Pn	.098 x 4m	
	TO CITY MAIN	q – Q 1873.5	100 (104)	(1) LTE=1.5 (1) TEE=6.1 (1) GA.VA.=.6 8.2x1.33x1.1	L 18.3 F 12.0 T 30.3	C=140 .011	Pt 4.71 Pe .26 Pf .33	Pt Pv Pn	Backflow Preventer TOTAL DEMAND 2820lpm @ 5.3bar	
	HOSE	q 946 Q 2819.5			L F T		Pt 5.3 Pe Pf	Pt Pv Pn		
		q Q			L F T	DIAMETER FACTOR= $\left(\frac{104}{102.3}\right)^{4.87}$	Pt Pe Pf	Pt Pv Pn		
		q Q			L F T		Pt Pe Pf	Pt Pv Pn		
		q Q			L F T		Pt Pe Pf	Pt Pv Pn		
		q Q			L F T		Pt Pe Pf	Pt Pv Pn		
		q Q			L F T		Pt Pe Pf	Pt Pv Pn		

Fig. 13.3 (continued)

HYDRAULIC SPRINKLER CALCULATIONS

SHEET 3 OF 3

STEP NO.	NOZZLE IDENT. AND LOCATION	FLOW IN LPM	PIPE SIZE	PIPE FITTINGS AND DEVICES	EQUIV. PIPE LENGTH	FRICTION LOSS BAR/M	PRESSURE SUMMARY	NORMAL PRESSURE	NOTES	REF. STEP
	SPKLR #16	q	32	(1) TEE	L 1.85	.010	Pt 1.14	Pt		
					F 1.8		Pe -	Pv		
		Q 85.5			T 3.65		Pf .04	Pn		
		q			L		Pt 1.18	Pt		
					F		Pe	Pv		
		Q			T		Pf	Pn		
		q			L		Pt	Pt	DEMAND FOR SPKLR #17 WILL BE SAME AS #16	
					F		Pe	Pv		
		Q			T		Pf	Pn		
	TO BOTTOM OF RN #4	85.5	50	(1) TEE	L 0.3	.005	Pt 1.18	Pt		
		q +85.5			F 3.0		Pe -	Pv		
		Q 171			T 3.3		Pf .02	Pn		
		q			L		Pt 1.2	Pt	$K=\frac{171}{\sqrt{1.2}}$	
					F		Pe	Pv	$K=156.1$	
		Q			T		Pf	Pn		
		q			L		Pt	Pt		
					F		Pe	Pv		
		Q			T		Pf	Pn		
		q			L		Pt	Pt		
					F		Pe	Pv		
		Q			T		Pf	Pn		
		q			L		Pt	Pt		
					F		Pe	Pv		
		Q			T		Pf	Pn		
		q			L		Pt	Pt		
					F		Pe	Pv		
		Q			T		Pf	Pn		
		q			L		Pt	Pt		
					F		Pe	Pv		
		Q			T		Pf	Pn		
		q			L		Pt	Pt		
					F		Pe	Pv		
		Q			T		Pf	Pn		
		q			L		Pt	Pt		
					F		Pe	Pv		
		Q			T		Pf	Pn		

Fig. 13.3 (continued)

$$P_{\mathrm{f}} = \frac{(4.52)(22.8\,\mathrm{gpm})^{1.85}(12\,\mathrm{ft})}{(120)^{1.85}(1.049)^{4.87}} = 2.0\,\mathrm{psi}$$

And, in S.I. Units:

$$P_{\mathrm{f}} = \frac{(6.05)(85.5\,\mathrm{L/min})^{1.85}(10^{5})(3.7\,\mathrm{m})}{(120)^{1.85}(26.6)^{4.87}} = 0.14\,\mathrm{bar}$$

This friction loss must then be added to the pressure required at the first sprinkler (see Figs. 13.2 sheet 1 and 13.3 sheet 1). The sum of the two pressures gives the pressure required at the second sprinkler. If this pressure is not available at the second sprinkler, the required flow will not be delivered to the first sprinkler.

Nominal pipe sizes may be used. However, most designers use the actual internal diameter, which is usually larger, and gives less friction loss. Note that compensation is not made for the fitting into which the sprinkler is screwed. This is in accordance with the design conventions dictated by NFPA 13.

Step 9

Calculate the flow from the second sprinkler. Flow from the second sprinkler is calculated from the following equation:

$$q = (k)\left(\sqrt{P}\right) \tag{13.5}$$

where

P = the pressure at the second sprinkler.

In the example, the flow from the second sprinkler will be the following:

$$q = (5.6)\left(\sqrt{18.6}\right) = 24.2\,\mathrm{gpm}$$

And, in S.I. Units:

$$q = (80)\left(\sqrt{1.28}\right) = 90.5\,\mathrm{L/min}$$

This flow is recorded on the worksheet as shown and added to the flow from the first sprinkler. The sum of these two flows is the total flow that will be traveling through the pipe between the second and third sprinklers.

Step 10

Repeat Steps 8 and 9 for successive sprinklers until all operating sprinklers on the first branch line are calculated. The total flow from Step 9 is used to calculate the pressure lost to friction between the second and third sprinklers. The friction loss is again added to the pressure required at the second sprinkler to get the pressure demand at the third sprinkler. This pressure demand is then used to calculate the flow from the third sprinkler in the manner described in Step 9. This process is repetitive until all flowing sprinklers on the first branch line are calculated.

Note that in the friction loss calculation between the final sprinkler on the branch line and the top of the riser nipple, the fitting at the top of the riser nipple must be included. The size of the fitting is determined by the size of the branch line piping attached to the fitting. The equivalent length is then established from the *Equivalent Pipe Length Chart*, found in Chapter 28 of NFPA 13. It is important to note that the table for equivalent pipe lengths is specifically for Schedule 40 black steel pipe with a C of 120. If other C values are being used, the equivalent length value from the table must be multiplied by the factor given in the C Value Multiplier Table in NFPA 13. Additionally, the equivalent length must be multiplied by the factor calculated from the formula shown in Chapter 28 for piping materials having different diameters than Schedule 40 steel.

As can be seen from the worksheet, the total demand for the left-hand branch line has been calculated to be 106.9 gpm at 40.1 psi (400.1 L/min at 2.74 bar).

Step 11

Most sprinkler systems will probably have longer branch lines than exist in this example. If that is the case and the design area does not extend to the other side of the cross main, the next calculation procedure is described in Step 12.

If the design area extends to the opposite side of the cross main, Steps 6 through 10 are repeated for the opposite side. The intersecting branch lines must then be balanced to the highest-pressure demand. The branch line calculation process must simply be repeated to establish the demand for those sprinklers considered to be operating on the opposite side of the cross main. In this example, only a single sprinkler is considered operating on the right-hand side. Note from the worksheet that the calculated demand for the right-hand side is 22.8 gpm at 17.1 psi (85.5 L/min at 1.18 bar).

At this point in the calculations, two different pressure and flow demands have been established at the top of the riser nipple. Since it is not possible that two different pressures can exist at the same point at the same time, the higher of the two pressures must be used and the flow into the low-pressure line "balanced" or adjusted up to this pressure. In general, if the pressure imbalance is less than 0.5 psi (0.03 bar) it may be ignored and the two previously calculated flow demands simply added together.

For pressure imbalance greater than 0.5 psi (0.03 bar), the flow into the lower pressure line must be adjusted to the higher pressure by the following equation:

$$Q_{\text{adj}} = (Q_{\text{L}})\left(\sqrt{\frac{P_{\text{H}}}{P_{\text{L}}}}\right) \tag{13.6}$$

where:

$\mathbf{Q_L}$ = the calculated flow in the low pressure line.

$\mathbf{P_L}$ = the lower of the two pressures.

$\mathbf{P_H}$ = the higher of the two pressures.

$\mathbf{Q_{adj}}$ = the adjusted flow into the low pressure line.

Equation 13.6 can be used in any hydraulic calculation process where a pressure imbalance exists at a junction point.

In the example, the adjusted flow demand for the single sprinkler on the right-hand side becomes the following:

$$Q_{\text{adj}} = (22.8)\left(\sqrt{\frac{40.1}{17.1}}\right) = 34.9\,\text{gpm}$$

And, in S.I. Units:

$$Q_{\text{adj}} = (85.5)\left(\sqrt{\frac{2.74}{1.18}}\right) = 130.3\,\text{L/min}$$

This adjusted flow is then added to the previously calculated flow from the high pressure line to get the total flow demand for the first riser nipple. In the example, the total flow demand at the top of the first riser nipple becomes 141.8 gpm (530.4 L/min) and this must be available at a residual pressure of 40.1 psi (2.74 bar).

Step 12

Calculate a k-factor for the first riser nipple with additional factors calculated for non-identical lines. This total flow and pressure demand is next used to calculate friction loss to the bottom of the riser nipple. Standard convention requires that the fitting at the bottom of the riser nipple be included in the riser nipple calculation and assigned the same diameter as the riser nipple. An easy way to remember to which pipe a fitting is added is that a fitting value is always added to the downstream section of pipe.

Note that there is a change of elevation here. However, the calculation procedure may be correctly simplified by delaying consideration of elevation to the end of the calculation. Delaying this consideration is legitimate because the k-factor is a mathematical reflection of the sprinkler and piping hardware in place and is not affected by elevation.

Once the total pressure and flow calculation has been established at the bottom of the riser nipple, a k-factor may be established. This k-factor is established as follows:

$$K = \frac{Q}{\sqrt{P}} \tag{13.7}$$

Once the pressure demands at the other riser nipples have been established, this k-factor may be used to calculate the flow from identical riser nipples. As shown in Figs. 13.2 sheet 1 and 13.3 sheet 1, the k-factors for the first riser nipple are 21.85 and 312.85 in S.I. units. These k-factors will apply to the second and third riser nipples shown in Fig. 13.1. Calculation of the flows from the second and third riser nipples are shown on the worksheets in Figs. 13.2 sheet 2 and 13.3 sheet 2. Since the fourth riser nipple has only two operating sprinklers, a new k-factor using the previous procedure must be established for this final riser nipple.

At this point, the use of the two sprinklers closest to the cross main on the fourth riser nipple must be justified. The k-factor for this riser nipple is 10.96 (156.1) as shown in Fig. 13.2 sheet 3 and Fig. 13.3 sheet 3. A larger friction loss would result if the sprinklers on the end of the line had been used. However, since pressure is found in the denominator of Eq. 13.7, a larger pressure demand would result in the computation of a smaller k-factor. When the flow for the fourth riser nipple is to be calculated, the pressure demand has already been established and must be large enough to supply the three previous riser nipples. Therefore, the only thing to be established is how many additional gallons to add for the final, two-sprinkler riser nipple. The greater flow demand will come from the largest k-factor. The largest k-factor comes from using the two sprinklers closest to the cross main.

Step 13

Repeat Steps 8 and 9 for riser nipples (instead of sprinklers) until all riser nipples within the design area are added in. The flow from the first riser nipple is used to calculate the friction loss to the second riser nipple (see Figs. 13.2 sheet 2 and 13.3 sheet 2). The friction loss is added to the pressure required at the bottom of the first riser nipple to get the pressure demand at the second riser nipple. This pressure is used with the computed k-factor for the first riser nipple to get the flow required for the second riser nipple using Eq. 13.5. This process is identical to the one used for calculating the branch lines. Repeat the process until all of the riser nipples in the design area have been added in; do not forget to use the different k-factor for nonidentical riser nipples such as the fourth one.

Step 14

Compute friction loss to the point of supply with compensation for elevation changes, valves, and fittings and differences in materials of underground piping. No more flow is added after all of the riser nipples in the design area have been included. The remainder of the calculations involves calculating friction loss to the point of supply. Compensation must be made for all valves and fittings, including backflow preventers, and for the change of piping materials that often occurs with the underground piping.

Note that the table in Chapter 28 of NFPA 13 does not contain equivalent lengths for alarm check valves, dry pipe valves, deluge valves, and other devices of nonuniform design. The best approach is to obtain the information from manufacturers. However, a reasonable approximation for most of these devices will result from using the value for a swing check valve of the appropriate size. Values to add for backflow preventers are dependent on the flow rate and must be obtained from the manufacturer.

At this point, compensation needs to be made for elevation. The height to the highest sprinkler must be multiplied by 0.433 psi/ft (0.098 bar/m) and added to the total pressure demand.

Step 15

Add the hose stream requirement. This step involves the addition of water for hose stream requirements. The amount of water to be added depends upon the hazard being protected and the pertinent standard. In our example, Chapter 19 of NFPA 13 requires 250 gpm (946 L/min) for Ordinary Hazard Occupancies. If inside hose stations exist, up to 100 gpm (379 L/min) of the total hose stream demand must be added to the calculations at the point of the attachment of the hoselines inside the building. This will have some impact on friction loss and the total pressure demand.

If yard hydrants exist, the total outside hose demand must be added at the point of hydrant branch attachment to the underground piping supplying the sprinklers. If yard hydrants or inside hose stations do not exist (which is the most common case), the total hose requirement is simply added on as a last step without any impact upon pressure.

Even though the hose requirement is primarily for the benefit of the fire department, the standards make no provision to recognize the arrival of firefighters with fire department pumping equipment. The hose stream demand plus the sprinkler demand must be available at the residual pressure required by the sprinklers. This approach does not take into account the ability of the fire department to add pressure to the water supply system by use of its pumper trucks. This provides an additional and substantial safety factor in the design.

The worksheet in Figs. 13.2 (sheet 2) and 13.3 (sheet 2) indicate that the final system demand is 752 gpm (2820 L/min) at a residual pressure of 78 psi (5.3 bar).

Step 16

Compare calculated demand with water supply available. The final step in the calculation process is the comparison of the final demand with the available water supply. The demand point must fall on or below the supply curve to provide standard protection. Considering the probability of eventual water supply and piping deterioration, and the possibility of occupancy and hazard changes, the best designs will provide system demands that are well below the supply curves. In fact, many authorities having jurisdiction impose arbitrary safety margins between the demand point and the supply curve. Although the intent of these safety margins is commendable, the specification of an arbitrary pressure or percent factor does not consider the variability of the slope of water supply curves. As illustrated in Fig. 13.4a, b, a 5 psi (0.34 bar) cushion can provide a good safety margin in Supply B and a meaningless few gpm (L/min) in Supply A.

The best approach for establishing safety margins is to evaluate each case and exercise good engineering judgment. Unfortunately, this approach is difficult to legislate and difficult to enforce with fairness and consistency. If authorities having jurisdiction insist that a safety factor be imposed, a gpm (L/min) addition is probably better

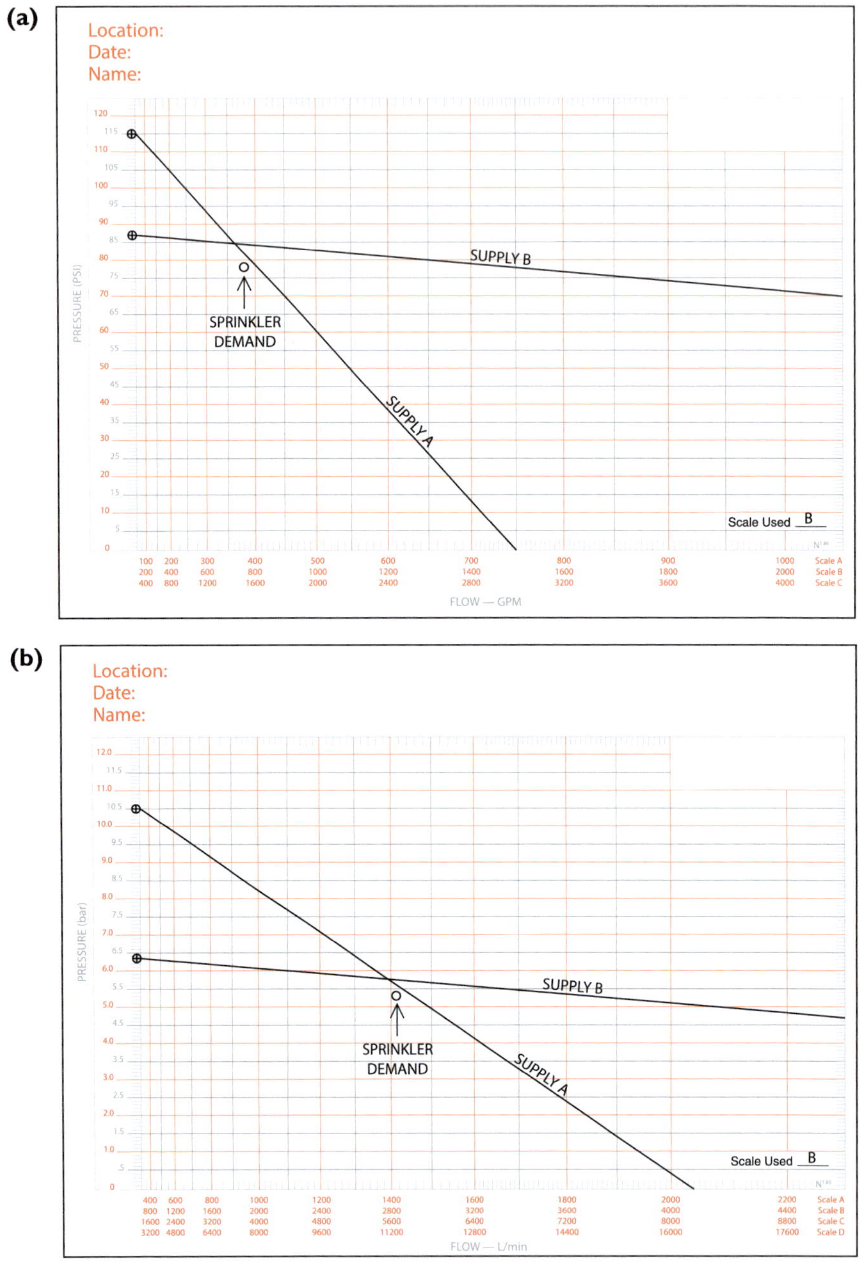

Fig. 13.4 Impact of the variation in curve on arbitrary safety factors

than a factor based upon pressure. It also should be kept in mind that the hose stream addition already provides a very significant safety factor.

If the demand point falls well below the curve, the designer has the option of reducing pipe sizes to develop a more economical design. If the demand point falls above the curve, pipe sizes must be increased or the water supply improved (by use of a fire pump).

Some insurance carriers impose additional design safety factors; for example, they may impose maximum water velocities. Historically, maximum permitted water velocities have ranged from 20 to 32 fps (6–10 m/s). Regulation on velocity encourages shorter branch lines and larger pipe sizes and discourages designers from taking full advantage of high pressure water supply systems. However, the NFPA standards do not impose limitations on velocity.

Calculations Considering the Impact of Velocity Pressures

When water travels through a piping system, it possesses both kinetic and potential energy (Chap. 3, Hydrokinetics). The total energy may therefore be expressed as follows:

$$P_t = P_n + P_v$$

where:

$\mathbf{P_t}$ = the total energy.

$\mathbf{P_n}$ = the potential energy component.

$\mathbf{P_v}$ = the kinetic energy component.

The potential energy component ($\mathbf{P_n}$) is also called the "normal" pressure because it acts in a direction perpendicular to the walls of the piping. It is actually the normal pressure which causes water to flow out of openings in the pipe. For a more theoretically accurate calculation of sprinkler systems, the normal pressure must be determined in order to calculate the flow from any sprinkler or riser nipple, and Eq. 13.5 becomes the following:

$$q = (k)(\sqrt{P_n}) \tag{13.8}$$

The kinetic energy component ($\mathbf{P_v}$) is also called the velocity pressure. Unfortunately, since the actual velocity pressure can be calculated only after knowing the flow that is being solved for, this technique involves a trial and error procedure.

The calculation procedure where velocity pressure is addressed is identical to the technique previously described, except at Steps 9 and 13. The following procedure is used whether dealing with the flow from the second and subsequent sprinklers or the second and subsequent riser nipples.

Returning to the previous calculation example to illustrate this technique, the procedure is the same up to the point of determining the flow from the second sprinkler (Step 9). The flow from the second sprinkler must be estimated. Note that the estimate must be at least as large as the flow from the first sprinkler. This estimate is added to the flow from the first sprinkler and the sum of the two is used to calculate the velocity pressure by using the following equation:

$$P = \frac{(0.001123)(Q)^2}{(D)^4} \tag{13.9a}$$

where

Q = the sum of the flow from the previous sprinkler(s) and the estimated flow from the next sprinkler in gpm.

D = the actual internal diameter of the pipe upstream of the sprinkler, for which the flow was estimated, in inches.

0.001123 = a conversion factor which yields psi.

In S.I. Units the equation becomes:

$$P = \frac{(2.185)(Q)^2}{(D)^4} \tag{13.9b}$$

where:

Q = the sum of the flow from the previous sprinkler(s) and the estimated flow from the next sprinkler in L/min.

D = the actual inside diameter of the pipe upstream of the sprinkler, for which the flow was estimated, in millimeters.

2.185 = a conversion factor which yields bar.

In the example, a flow identical to the first sprinkler is used as the estimate. The velocity pressure thus becomes the following:

$$P_v = \frac{(0.001123)(45.6)^2}{(1.049)^4} = 1.9\,\text{psi}$$

And, in S.I. Units:

$$P_v = \frac{(2.185)(85.5)^2}{(26.6)^4} = 0.13\,\text{bar}$$

This calculated velocity pressure is subtracted from the total pressure existing at the second sprinkler to give the normal pressure. The normal pressure is then used with Eq. 13.8 to find out if the estimate is acceptable. If the result of this calculation is within $\pm$ 0.1 gpm ($\pm$ 0.4 L/min), the estimate is acceptable. When calculating the flow into riser nipples if the calculation is within $\pm$1.0 gpm (3.8 L/min), the estimate may be considered acceptable. If the calculation results in a flow larger than the estimate, then the estimate was too small and must be adjusted upward and the trial repeated. If the calculated flow is smaller than the estimated value, then the estimate must be reduced and tried again. If your first estimate is not close enough, your second estimate will work out if you use as this second try the result of the checking calculation from Eq. 13.8. This is demonstrated in Figs. 13.5 sheet 1 and 13.6 sheet 1 in establishing the flow for Sprinkler #3.

Figures 13.5(sheet 1) and 13.6 (sheet 1) show that the initial estimate was within 0.1 gpm (0.4 L/min) and considered correct. Therefore, the flow of 22.8 gpm (85.5 L/min) is used as the flow from the second sprinkler. Figures 13.5 (sheet 1) and 13.6 (sheet 1) show the entire branch line calculated using this technique. If these results are compared with the calculations shown in Figs. 13.2 (sheet 1) and 13.3(sheet 1), where velocity pressure is not subtracted, note that by subtracting velocity pressures, the final flow and pressure is lower. Therefore, the first technique is the more conservative of the two and provides (but usually not a great amount) additional safety factors. NFPA 13 recognizes both techniques as acceptable.

In cases of large branch lines or cross-main piping where the friction loss leading to a sprinkler or riser nipple is small relative to the velocity pressure that will be created, the flow from the second sprinkler or riser nipple may actually be less than the flow from the first. This is referred to as the "second head phenomenon". Be careful that the final flow established from the second sprinkler or riser nipple is large enough to meet minimum system requirements. It may be necessary to arbitrarily increase end-sprinkler or riser-nipple flows to get the magnitude of flow needed at all subsequent points. This is illustrated in Fig. 13.5(sheets 2 and 3) Fig. 13.6(sheets 2 and 3) where the flow from the second riser nipple is being established.

Gridded Designs and Computer Assistance

Up to this point, the discussion has been limited to nongridded systems without the mention of computers. From a practical standpoint, there is no time-efficient technique for the computation of gridded designs by hand. Most people consider life too short for such a time-consuming process. However, there are a number of excellent computer programs which expediently calculate gridded designs as well as the more traditional non-gridded designs. Some of these programs have the ability to work in both English and Standard International units. Some will also make calculations using velocity pressures. Most all calculations today are performed using computers, and several programs even incorporate drafting capabilities. If so equipped, the program automatically retrieves the pipe diameters, lengths and configuration once the drawing is completed and makes the data entry process more efficient.

HYDRAULIC SPRINKLER CALCULATIONS

SHEET 1 OF 3

STEP NO.	NOZZLE IDENT. AND LOCATION	FLOW IN GPM	PIPE SIZE	PIPE FITTINGS AND DEVICES	EQUIV. PIPE LENGTH	FRICTION LOSS PSI/FOOT	PRESSURE SUMMARY	NORMAL PRESSURE	NOTES K=5.6	REF. STEP
	SPKLR	q	1		L 12-0	C=120	Pt 16.6	Pt	Q=.19x120	
	#1		(1.049)		F –	.166	Pe –	Pv	$P=\left(\frac{22.8}{5.6}\right)^2$	
		Q 22.8			T 12-0		Pf 2.0	Pn		
		q			L		Pt 18.6 →	Pt 18.6 →	Estimate 22.8 gpm	
					F		Pe	Pv -1.9	from 2nd Sprinkler.	
		Q			T		Pf	Pn 16.7	@ 45.6 gpm	
		q			L		Pt	Pt	P=1.9 CHECK:	
					F		Pe	Pv	$q=5.6\sqrt{16.7}$	
		Q			T		Pf	Pn	q=22.9∴ OK	
		q			L		Pt	Pt		
					F		Pe	Pv		
		Q			T		Pf ↓	Pn		
	SPKLR	q 22.8	1		L 12-0		Pt 18.6	Pt		
	#2				F –	.598	Pe –	Pv		
		Q 45.6			T 12-0		Pf 7.2	Pn		
		q			L		Pt 25.8	Pt 25.8	Estimate 24.4 gpm	
					F		Pe	Pv -1.5	from 3rd Sprinkler.	
		Q			T		Pf	Pn 24.3	@ 70 gpm & 1 1/4"	
		q			L		Pt	Pt	~~Pv=1.5~~ CHECK:	
					F		Pe	Pv	$q=5.6\sqrt{24.3}$	
		Q			T		Pf	Pn	q=27.6	
		q			L		Pt	Pt 25.8	Estimate too low!	
					F		Pe	Pv -1.7	TRY 27.6	
		Q			T		Pf	Pn 24.1	@ 73.2 gpm, Pv=1.7	
	SPKLR	q 27.6	1 1/4		L 12-0		Pt 25.8	Pt	CHECK:	
	#3		(1.38)		F –	.377	Pe –	Pv	$q=5.6\sqrt{24.1}$	
		Q 73.2			T 12-0		Pf 4.5	Pn	q=27.5∴ OK	
		q			L		Pt 30.3	Pt 30.3	Estimate 29.1 gpm	
					F		Pe	Pv -3.2	from 4th Sprinkler.	
		Q			T		Pf	Pn 27.1	@ 102.3 gpm, Pv=3.2	
	SPKLR	q 29.1	1 1/4	(1) Tee	L 6-0		Pt 30.3	Pt	CHECK:	
	#4				F 6	.701	Pe	Pv	$q=5.6\sqrt{27.1}$	
		Q 102.3			T 12		Pf 8.4	Pn	q=29.2∴ OK	
		q			L		Pt 38.7	Pt		
					F		Pe	Pv		
		Q			T		Pf	Pn		

Fig. 13.5 Branch line calculations using velocity pressures

HYDRAULIC SPRINKLER CALCULATIONS

SHEET 2 OF 3

STEP NO.	NOZZLE IDENT. AND LOCATION	FLOW IN GPM	PIPE SIZE	PIPE FITTINGS AND DEVICES	EQUIV. PIPE LENGTH	FRICTION LOSS PSI/FOOT	PRESSURE SUMMARY	NORMAL PRESSURE	NOTES	REF. STEP
	SPKLR #5	q Q 22.8	1 1/4	(1) TEE	L 6-0 F 6 T 12-0	.044	Pt 16.6 Pe – Pf 0.5	Pt Pv Pn	Right Side of Cross Main	
		q Q			L F T		Pt 17.1 Pe Pf	Pt Pv Pn	$Q_{ADJ}=22.8\sqrt{\frac{38.7}{17.1}}$ $Q_{ADJ}=34.3$	
	TO BOTTOM OF RN #1	102.3 q +34.3 Q 136.6	2 (2.067)	(1) TEE	L 1-0 F 10 T 11-0	.167	Pt 38.7 Pe – Pf 1.8	Pt Pv Pn		
		q Q			L F T		Pt 40.5 Pe Pf	Pt Pv Pn	$K=\frac{136.6}{\sqrt{40.5}}$ $K=21.46$	
		q Q			L F T		Pt Pe Pf	Pt Pv Pn		
	RN #1	q Q 136.6	3 (3.068)		L 10-0 F – T 10-0	.024	Pt 40.5 Pe – Pf .2	Pt Pv Pn		
		q Q			L F T		Pt 40.7 Pe Pf	Pt 40.7 Pv -.9 Pn 39.8	Estimate 136.6 gpm from RN #2 @ 273.2 gpm, Pv=.9	
		q Q			L F T		Pt Pe Pf	Pt Pv Pn	CHECK: $q=21.46\sqrt{39.8}$ $q=135.4$	
		q Q			L F T		Pt Pe Pf	Pt Pv Pn	Estimate is too high, plus 2nd Head Phenomenon Must Increase Flow @	
		q Q			L F T		Pt Pe Pf	Pt Pv Pn	RN #1 to get enough @ RN #2	
	REVISE RN #1	q Q 140	3		L 10-0 F – T 10-0	.026	Pt 42.6 Pe Pf .3	Pt Pv Pn	Revise RN #1 to 140 gpm: and $Pt=\left(\frac{140}{21.46}\right)^2=42.6$	
		q Q			L F T		Pt 42.9 Pe Pf	Pt Pv Pn		

Fig. 13.5 (continued)

HYDRAULIC SPRINKLER CALCULATIONS

SHEET 3 OF 3

STEP NO.	NOZZLE IDENT. AND LOCATION	FLOW IN GPM	PIPE SIZE	PIPE FITTINGS AND DEVICES	EQUIV. PIPE LENGTH	FRICTION LOSS PSI/FOOT	PRESSURE SUMMARY	NORMAL PRESSURE	NOTES	REF. STEP
	NEW RN #1	q Q 140	3		L 10-0 F – T 10-0	.026	Pt 42.6 Pe – Pf .3	Pt Pv Pn		
		q Q			L F T		Pt 42.9 Pe Pf	Pt 42.9 Pv -1.0 Pn 41.9	Estimate 138 gpm from RN #2 @ 278 gpm, Pv=1.0	
		q Q			L F T		Pt Pe Pf	Pt Pv Pn	CHECK: $q=21.46\sqrt{41.9}$ $q=138.9$ ∴ OK and $q \geq 136.6$	
	RN #2	q 138 Q 278	3		L 10-0 F – T 10-0	.091	Pt 42.9 Pe – Pf .9	Pt Pv Pn		
		q Q			L F T		Pt 43.8 Pe Pf	Pt 43.8 Pv -2.2 Pn 41.6	Estimate 138 gpm from RN #3 @ 416 gpm, Pv=2.2	
		q Q			L F T		Pt Pe Pf	Pt Pv Pn	CHECK: $q=21.46\sqrt{41.6}$ $q=138.4$ ∴ OK	
	RN #3	q 138 Q 416	3		L 10-0 F – T 10-0	.192	Pt 43.8 Pe – Pf 1.9	Pt Pv Pn		
		q Q			L F T		Pt 45.7 Pe Pf	Pt 45.7 Pv -3.0 Pn 42.7	Estimate 72 gpm from RN #4 @ 488 gpm, Pv=3.0	
		q Q			L F T		Pt Pe Pf	Pt Pv Pn	CHECK: $q=10.96\sqrt{42.7}$ $q=71.6$ ∴ OK	
	RN #4	q 72 Q 488	3	(1) LTE	L 54-0 F 5 T 59-0	.258	Pt 45.7 Pe – Pf 15.2	Pt Pv Pn		
		q Q			L F T		Pt 60.9 Pe Pf	Pt Pv Pn	Continue to City Main in Normal Fashion.	
		q Q			L F T		Pt Pe Pf	Pt Pv Pn		

Fig. 13.5 (continued)

HYDRAULIC SPRINKLER CALCULATIONS

SHEET 1 OF 3

STEP No.	NOZZLE IDENT. AND LOCATION	FLOW IN LPM	PIPE SIZE	PIPE FITTINGS AND DEVICES	EQUIV. PIPE LENGTH	FRICTION LOSS BAR/M	PRESSURE SUMMARY	NORMAL PRESSURE	NOTES K=80	REF. STEP
	SPKLR	q	25		L 3.7	C=120	Pt 1.14	Pt	q=(7.7) x (11.1)	
			(26.6)		F –	.037	Pe –	Pv		
	#1	Q 85.5			T 3.7		Pf .14	Pn		
		q			L		Pt 1.28 →	Pt 1.28 →	Estimate 85.5 lpm	
					F		Pe	Pv -.13	from 2nd Sprinkler.	
		Q			T		Pf	Pn 1.15	@ 171 lpm, Pv=.13	
		q			L		Pt	Pt	CHECK:	
					F		Pe	Pv	$q=80\sqrt{1.15}$	
		Q			T		Pf ↓	Pn	q=85.8∴ OK	
	SPKLR	q 85.5	25		L 3.7		Pt 1.28	Pt		
					F –	.134	Pe –	Pv		
	#2	Q 171			T 3.7		Pf .5	Pn		
		q			L		Pt 1.78	Pt 1.78	Estimate 90 lpm	
					F		Pe	Pv -.1	from 3rd Sprinkler.	
		Q			T		Pf	Pn 1.68	@ 26 lpm, 32mm Pv=.1	
		q			L		Pt	Pt	CHECK:	
					F		Pe	Pv	$q=80\sqrt{1.66}$	
		Q			T		Pf	Pn	q=103.7 lpm	
		q			L		Pt	Pt 1.78	Estimate too low!	
					F		Pe	Pv -.11	TRY 103.7	
		Q			T		Pf	Pn 1.67	@ 274.7 lpm, Pv=.11	
	SPKLR	q 103.7	32		L 3.7		Pt 1.78	Pt	CHECK:	
			(35.1)		F –	.083	Pe –	Pv	$q=80\sqrt{1.67}$	
	#3	Q 274.7			T 3.7		Pf .31	Pn	q=103.4∴ OK	
		q			L		Pt 2.09	Pt 2.09	Estimate 109.3 lpm	
					F		Pe	Pv -.21	@ 384 lpm, Pv=.21	
		Q			T		Pf	Pn 1.88		
		q			L		Pt	Pt	CHECK:	
					F		Pe	Pv	$q=80\sqrt{1.88}$	
		Q			T		Pf	Pn	q=109.7∴ OK	
	SPKLR	q 109.3	32		L 1.85		Pt 2.09	Pt		
				(1) Tee	F 1.8	.155	Pe –	Pv		
	#4	Q 384			T 3.65		Pf .57	Pn		
		q			L		Pt 2.66	Pt		
					F		Pe	Pv		
		Q			T		Pf	Pn		

Fig. 13.6 Branch line calculations using velocity pressures (S.I. Units)

HYDRAULIC SPRINKLER CALCULATIONS

SHEET 2 OF 3

STEP NO.	NOZZLE IDENT. AND LOCATION	FLOW IN LPM	PIPE SIZE	PIPE FITTINGS AND DEVICES	EQUIV. PIPE LENGTH	FRICTION LOSS BAR/M	PRESSURE SUMMARY	NORMAL PRESSURE	NOTES	REF. STEP
	SPKLR	q	32	(1) TEE	L 1.85	.010	Pt 1.14	Pt	Right Side of	
					F 1.8		Pe -	Pv	Cross Main	
	#5	Q 85.5			T 3.65		Pf 0.4	Pn		
		q			L		Pt 1.18	Pt	$Q_{ADJ}=85.5\sqrt{\frac{2.66}{1.18}}$	
					F		Pe	Pv		
		Q			T		Pf	Pn	$Q_{ADJ}=128.4$	
	TO	384	50		L 0.3		Pt 2.66	Pt		
	BOTTOM OF	q +128.4	(52.5)	(1) TEE	F 3.0	.037	Pe -	Pv		
	RN #1	Q 512.4			T 3.3		Pf .12	Pn		
		q			L		Pt 2.78	Pt	$K=\frac{512.4}{\sqrt{2.78}}$	
					F		Pe	Pv		
		Q			T		Pf	Pn	K=307.3	
		q			L		Pt	Pt		
					F		Pe	Pv		
		Q			T		Pf	Pn		
	RN	q	80		L 3.0		Pt 2.78	Pt		
			(77.9)		F -	.005	Pe -	Pv		
	#1	Q 512.4			T 3.0		Pf .01	Pn		
		q			L		Pt 2.79	Pt 2.79	Estimate 512.4	
					F		Pe	Pv -.06	from RN #2	
		Q			T		Pf	Pn 2.73	@ 1024.8 lpm, Pv=.06	
		q			L		Pt	Pt	CHECK:	
					F		Pe	Pv	$q=307.3\sqrt{2.73}$	
		Q			T		Pf	Pn	q=507.7	
		q			L		Pt	Pt	Estimate is too high,	
					F		Pe	Pv	plus 2nd Head Phenomenon	
		Q			T		Pf	Pn	Must Increase Flow @	
		q			L		Pt	Pt	RN #1 to get enough @	
					F		Pe	Pv	RN #2	
		Q			T		Pf	Pn		
	REVISE	q	80		L 3.0		Pt 2.97	Pt	Revise RN #1 to	
	RN				F -	.006	Pe -	Pv	530 lpm; and	
	#1	Q 530			T 3.0		Pf .02	Pn	$Pt=\left(\frac{530}{307.3}\right)^2=2.97$	
		q			L		Pt 2.99	Pt		
					F		Pe	Pv		
		Q			T		Pf	Pn		

Fig. 13.6 (continued)

HYDRAULIC SPRINKLER CALCULATIONS

SHEET 3 OF 3

STEP NO.	NOZZLE IDENT. AND LOCATION	FLOW IN LPM	PIPE SIZE	PIPE FITTINGS AND DEVICES	EQUIV. PIPE LENGTH	FRICTION LOSS BAR/M	PRESSURE SUMMARY	NORMAL PRESSURE	NOTES	REF. STEP
	NEW	q	80		L 3.0	.006	Pt 2.97	Pt		
	RN				F –		Pe –	Pv		
	#1	Q 530			T 3.0		Pf .02	Pn		
		q			L		Pt 2.99	Pt 2.99	Estimate 525	
					F		Pe	Pv -.07	from RN #2	
		Q			T		Pf	Pn 2.92	@ 1055 lpm, Pv=.07	
		q			L		Pt	Pt	CHECK:	
					F		Pe	Pv	q=307.3$\sqrt{2.92}$	
		Q			T		Pf	Pn	q=525.1∴ OK	
		q 525	80		L 3.0	.021	Pt 2.99	Pt	and q≥512.4	
	RN				F –		Pe –	Pv		
	#2	Q 1055			T 3.0		Pf .06	Pn		
		q			L		Pt 3.05	Pt 3.05	Estimate 525	
					F		Pe	Pv -.15	from RN #3	
		Q			T		Pf	Pn 2.90	@ 1580 lpm, Pv=.15	
		q			L		Pt	Pt	CHECK:	
					F		Pe	Pv	q=307.3$\sqrt{2.9}$	
		Q			T		Pf	Pn	q=523.3∴ OK	
		q 525	80		L 3.0	.043	Pt 3.05	Pt		
	RN				F –		Pe –	Pv		
	#3	Q 1580			T 3.0		Pf .13	Pn		
		q			L		Pt 3.18	Pt 3.18	Estimate 270	
					F		Pe	Pv -.20	from RN #4	
		Q			T		Pf	Pn 2.98	@ 1850 lpm, Pv=.20	
		q			L		Pt	Pt	CHECK:	
					F		Pe	Pv	q=156.1$\sqrt{2.98}$	
		Q			T		Pf	Pn	q=269.5∴ OK	
		q 270	80		L 16.2	.059	Pt 3.18	Pt		
	RN			(1) LTE	F 1.5		Pe –	Pv		
	#4	Q 1850			T 17.7		Pf 1.04	Pn		
		q			L		Pt 4.22	Pt	Continue to City Main	
					F		Pe	Pv	in Normal Fashion.	
		Q			T		Pf	Pn		
		q			L		Pt	Pt		
					F		Pe	Pv		
		Q			T		Pf	Pn		

Fig. 13.6 (continued)

It is possible to use most of these programs without any in-depth knowledge of hydraulic principles. However, individuals who meet this description are not design professionals, but computer operators.

Summary of Important Concepts

1. The most widely used technique for modern sprinkler system design involves hydraulic calculations.
2. The hydraulic design process can be broken down into 16 basic steps.
3. Chapters 19 and 28 from NFPA 13 are the most important in performing hydraulic calculations.
4. It is acceptable design technique either to ignore or to consider velocity pressure.
5. Computer programs exist for design of gridded sprinkler systems.

Exercises

13.1. Establish the sprinkler system water supply demand for the system shown in Fig. 13.7. Ignore velocity pressures.

13.2. Establish the sprinkler system water supply demand for the system shown in Fig. 13.8. Ignore velocity pressures.
13.3. Recalculate the system in Fig. 13.7 considering velocity pressures.
13.4. Recalculate the system in Fig. 13.8 considering velocity pressures.

Selected answers are found in Appendix K.

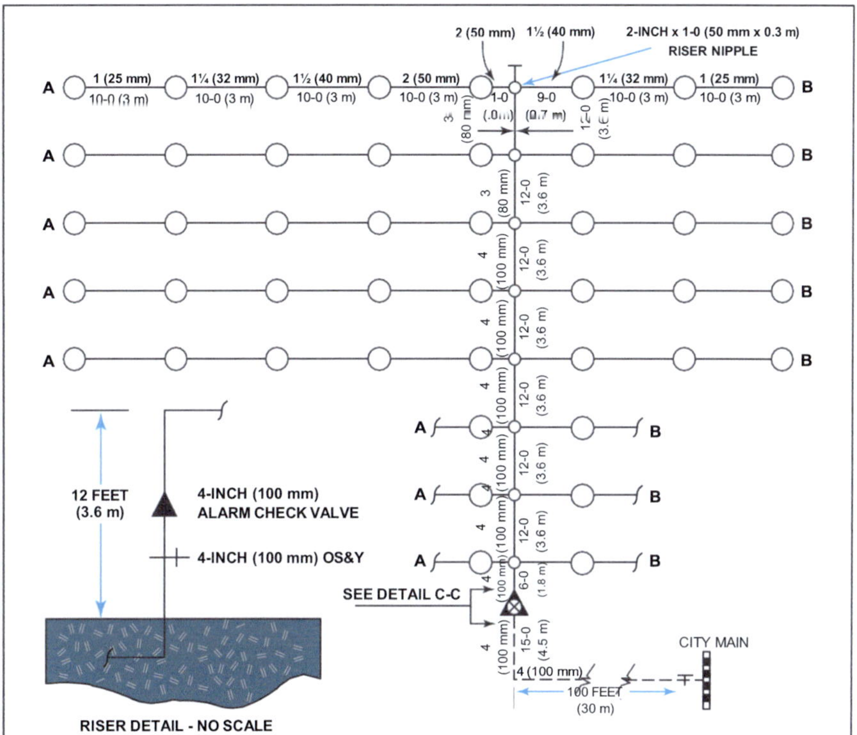

Hydraulically calculate this system on the basis of a new Ordinary Hazard, Group 2 Occupancy—a 1,500 square foot (139m2) design area, and sprinkler discharge coefficients of 5.6 (80). Branch line piping is Schedule 40 but cross main and riser is Schedule 10.

Underground is lined, Class 52 ductile iron.

Fig. 13.7 Sprinkler system for Exercises 13.1 and 13.3

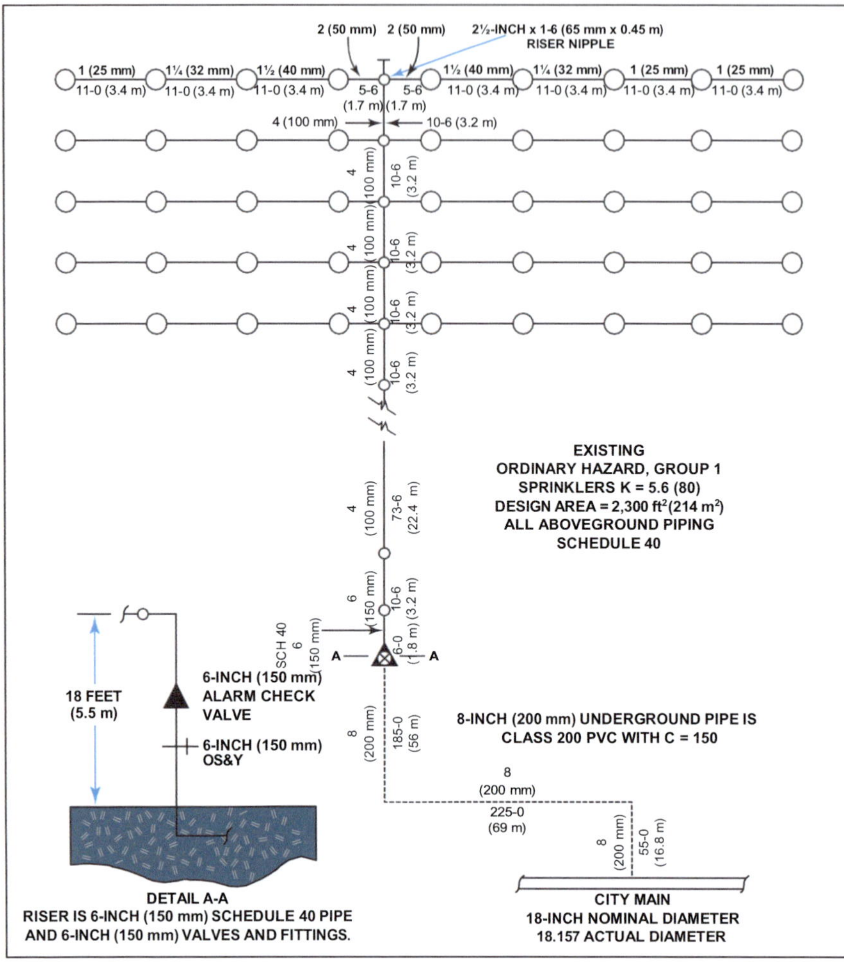

Fig. 13.8 Sprinkler system for Exercises 13.2 and 13.4

Reference

1. Wass, H. S., & Fleming, R. P. *Sprinkler hydraulics : a guide to fire system hydraulic calculations* (Third edition.). Springer International Publishing AG, 2020.

Supplemental Readings

Hickey, Harry E. *Hydraulics for Fire Protection*. National Fire Protection Association: Quincy, MA, 1980, Ch.8.

National Fire Protection Association. NFPA 13, *The Standard for the Installation of Sprinkler Systems*. Quincy, MA, 2022 Edition.

National Fire Protection Association. NFPA 13, *The Standard for the Installation of Sprinkler Systems*. Quincy, MA, 2025 Edition

Appendix A
Actual Internal Pipe Diameters

Table A.1 Aboveground piping nominal and actual inside diameters

Nominal diameters										
Inches	0.75	1	1.25	1.5	2	2.5	3	4	6	8
(Millimeters)	(20)	(25)	(32)	(40)	(50)	(65)	(80)	(100)	(150)	(200)
Actual inside diameters										
Sch. 10 steel	0.884	1.097	1.442	1.682	2.157	2.635	3.26	4.26	6.357	8.249
	(22.4)	(27.9)	(36.6)	(42.7)	(54.8)	(66.9)	(82.8)	(108.2)	(161.5)	(209.5)
Sch. 40 steel	0.824	1.049	1.38	1.61	2.067	2.469	3.068	4.026	6.065	7.981
	(21.0)	(26.6)	(35.1)	(40.9)	(52.5)	(62.7)	(77.9)	(102.3)	(154.1)	(202.7)
XL steel		1.104	1.452	1.687	2.154					
		(28.0)	(36.9)	(42.8)	(54.7)					
Type K copper	0.745	0.995	1.245	1.481	1.959	2.435	2.907	3.857	5.741	7.583
	(18.9)	(25.3)	(31.6)	(37.6)	(49.8)	(61.8)	(73.8)	(98.0)	(145.8)	(192.6)
Type L copper	0.785	1.025	1.265	1.505	1.985	2.465	2.945	3.905	5.845	7.725
	(19.9)	(26.0)	(32.1)	(38.2)	(50.4)	(62.6)	(74.8)	(99.2)	(148.5)	(196.2)
Type M copper	0.811	1.055	1.291	1.527	2.009	2.495	2.981	3.935	5.881	7.785
	(20.6)	(26.8)	(32.8)	(38.8)	(51.0)	(63.4)	(75.7)	(99.9)	(149.4)	(197.7)
CPVC plastic	0.884	1.109	1.400	1.602	2.003	2.423	2.951			
	(22.5)	(28.2)	(35.6)	(40.7)	(50.9)	(61.5)	(75.0)			

P. Brock and V. Charter, *Fire Protection Hydraulics and Water Supply Analysis*, The Society of Fire Protection Engineers Series, https://doi.org/10.1007/978-3-032-09877-1

Table A.2 Underground piping nominal and actual inside diameters

Nominal diameters							
Inches	4	6	8	10	12	14	16
(Millimeters)	(100)	(150	(200)	(250)	(300)	(350)	(400)
Actual inside diameters							
Unlined ductile iron							
Class 50		6.4	8.51	10.52	12.58	14.64	16.72
		(162.6)	(216.2)	(267.2)	(319.5)	(371.9)	(424.7)
Class 51	4.28	6.34	8.45	10.46	12.52	14.58	16.66
	(108.7)	(161.0)	(214.6)	(265.7)	(318.0)	(370.3)	(423.2)
Class 52	4.155	6.275	8.385	10.4	12.46	14.52	16.4
	(105.5)	(159.4)	(213.0)	(264.2)	(316.5)	(368.8)	(416.6)
Cement-lined ductile iron							
Class 50		6.28	8.39	10.4	12.46	14.45	16.53
		(159.5)	(213.1)	(264.2)	(316.5)	(367.0)	(419.9)
Class 51	4.16	6.22	8.33	10.34	12.4	14.39	16.47
	(105.7)	(158.0)	(211.6)	(262.6)	(315.0)	(365.5)	(418.3)
Class 52	4.1	6.16	8.27	10.28	12.34	14.33	16.41
	(104.1)	(156.5)	(210.1)	(261.1)	(313.4)	(364.0)	(416.8)
Unlined cast iron	4.1	6.14	8.23	10.22	12.24	14.28	16.32
	(104.1)	(156.0)	(209.0)	(259.6)	(310.9)	(362.7)	(414.5)
Enamel-lined cast iron	3.98	6.02	8.11	10.1	12.12	14.09	16.13
	(101.1)	(152.9)	(206.0)	(256.5)	(307.8)	(357.9)	(409.7)
Cement-lined cast iron	3.85	5.89	7.98	9.97	11.99	13.9	15.94
	(97.8)	(149.6)	(202.7)	(253.2)	(304.5)	(353.1)	(404.9)
Class 150 PVC	4.24	6.09	7.98	9.79	11.65		
	(107.7)	(154.7)	(202.7)	(248.7)	(295.9)		
Class 200 PVC	4.08	5.86	7.68	9.42	11.2		
	(103.6)	(148.8)	(195.1)	(239.3)	(284.5)		
Permastran	4.28	6.32	8.23	10.23	12.12		
	(108.7)	(160.5)	(209.0)	(259.8)	(307.8)		
Class 150 blue brute	4..27	6.13	8.04	9.87	11.73		
	(108.5)	(155.7)	(204.2)	(250.7)	(297.9)		
Class 200 blue brute	4.11	5.91	7.76				
	(104.4)	(150.1)	(197.1)				
Class 100 asbestos cement	4.0	6.0	8.0	10.0	12.0	13.59	15.5
	(101.6)	(152.4)	(203.2)	(254.0)	(304.8)	(345.2)	(393.7)
Class 150 asbestos cement	4.0	5.85	7.85	10.1	12.0	14.0	16.0
	(101.6)	(148.6)	(199.4)	(254.0)	(304.8)	(355.6)	(406.4)
Class 200 asbestos cement	4.0	5.7	7.6	9.63	11.56	13.59	15.5
	(101.6)	(144.8)	(193.0)	(244.6)	(293.6)	(345.2)	(393.7)

Appendix B
Friction Loss Tables for Schedule 40 Steel Pipe

P. Brock and V. Charter, *Fire Protection Hydraulics and Water Supply Analysis*, The Society of Fire Protection Engineers Series, https://doi.org/10.1007/978-3-032-09877-1

Table B.1 Water flowing in 1-in. (25 mm) schedule 40 steel pipe I.D. = 1.049 in. (26.6 mm)

Q		P_f		P_f		Velocity	
		$C = 100$		$C = 120$			
GPM	L/min	psi/ft	bar/m	psi/ft	bar/m	fps	m/s
10	37.9	0.051	0.012	0.036	0.008	3.7	1.1
11	41.6	0.060	0.014	0.043	0.010	4.1	1.2
12	45.4	0.071	0.016	0.051	0.012	4.5	1.4
13	49.2	0.082	0.019	0.059	0.013	4.8	1.5
14	53.0	0.094	0.021	0.067	0.015	5.2	1.6
15	56.8	0.107	0.024	0.076	0.017	5.6	1.7
16	60.6	0.121	0.028	0.086	0.020	5.9	1.8
17	64.4	0.135	0.031	0.096	0.022	6.3	1.9
18	68.1	0.150	0.034	0.107	0.024	6.7	2.0
19	71.9	0.166	0.038	0.118	0.027	7.0	2.1
20	75.7	0.182	0.042	0.130	0.030	7.4	2.3
21	79.5	0.200	0.046	0.142	0.032	7.8	2.4
22	83.3	0.217	0.050	0.155	0.035	8.2	2.5
23	87.1	0.236	0.054	0.169	0.038	8.5	2.6
24	90.8	0.255	0.058	0.182	0.042	8.9	2.7
25	94.6	0.276	0.063	0.197	0.045	9.3	2.8
26	98.4	0.296	0.068	0.211	0.048	9.6	2.9
27	102.2	0.318	0.072	0.227	0.052	10.0	3.1
28	106.0	0.340	0.078	0.243	0.055	10.4	3.2
29	109.8	0.363	0.083	0.259	0.059	10.8	3.3
30	113.6	0.386	0.088	0.276	0.063	11.1	3.4
31	117.3	0.410	0.094	0.293	0.067	11.5	3.5
32	121.1	0.435	0.099	0.310	0.071	11.9	3.6
33	124.9	0.460	0.105	0.329	0.075	12.2	3.7
34	128.7	0.487	0.111	0.347	0.079	12.6	3.8
35	132.5	0.513	0.117	0.366	0.084	13.0	4.0
36	136.3	0.541	0.123	0.386	0.088	13.4	4.1
37	140.1	0.569	0.130	0.406	0.093	13.7	4.2
38	143.8	0.598	0.136	0.427	0.097	14.1	4.3
39	147.6	0.627	0.143	0.448	0.102	14.5	4.4
40	151.4	0.657	0.150	0.469	0.107	14.8	4.5
41	155.2	0.688	0.157	0.491	0.112	15.2	4.6
42	159.0	0.719	0.164	0.513	0.117	15.6	4.7
43	162.8	0.751	0.171	0.536	0.122	16.0	4.9
44	166.6	0.784	0.179	0.560	0.128	16.3	5.0

(continued)

Table B.1 (continued)

Q		P_f		P_f		Velocity	
		$C = 100$		$C = 120$			
GPM	L/min	psi/ft	bar/m	psi/ft	bar/m	fps	m/s
45	170.3	0.817	0.186	0.583	0.133	16.7	5.1
46	174.1	0.851	0.194	0.608	0.139	17.1	5.2
47	177.9	0.886	0.202	0.632	0.144	17.4	5.3
48	181.7	0.921	0.210	0.657	0.150	17.8	5.4
49	185.5	0.957	0.218	0.683	0.156	18.2	5.5
50	189.3	0.993	0.227	0.709	0.162	18.6	5.7
51	193.1	1.030	0.235	0.735	0.168	18.9	5.8
52	196.8	1.068	0.244	0.762	0.174	19.3	5.9
53	200.6	1.106	0.252	0.790	0.180	19.7	6.0
54	204.4	1.145	0.261	0.817	0.186	20.0	6.1
55	208.2	1.185	0.270	0.846	0.193	20.4	6.2
56	212.0	1.225	0.279	0.874	0.199	20.8	6.3
57	215.8	1.266	0.289	0.903	0.206	21.1	6.4
58	219.6	1.307	0.298	0.933	0.213	21.5	6.6
59	223.3	1.349	0.308	0.963	0.220	21.9	6.7
60	227.1	1.392	0.317	0.993	0.227	22.3	6.8
61	230.9	1.435	0.327	1.024	0.234	22.6	6.9
62	234.7	1.479	0.337	1.055	0.241	23.0	7.0
63	238.5	1.523	0.347	1.087	0.248	23.4	7.1
64	242.3	1.568	0.358	1.119	0.255	23.7	7.2
65	246.1	1.614	0.368	1.152	0.263	24.1	7.4
66	249.8	1.660	0.379	1.185	0.270	24.5	7.5
67	253.6	1.707	0.389	1.218	0.278	24.9	7.6
68	257.4	1.754	0.400	1.252	0.286	25.2	7.7
69	261.2	1.802	0.411	1.286	0.293	25.6	7.8
70	265.0	1.851	0.422	1.321	0.301	26.0	7.9
71	268.8	1.900	0.433	1.356	0.309	26.3	8.0
72	272.5	1.950	0.445	1.392	0.317	26.7	8.1
73	276.3	2.000	0.456	1.428	0.326	27.1	8.3
74	280.1	2.051	0.468	1.464	0.334	27.5	8.4
75	283.9	2.103	0.480	1.501	0.342	27.8	8.5
76	287.7	2.155	0.492	1.538	0.351	28.2	8.6
77	291.5	2.208	0.504	1.576	0.359	28.6	8.7

(continued)

Table B.1 (continued)

Q		P_f		P_f		Velocity	
		$C = 100$		$C = 120$			
GPM	L/min	psi/ft	bar/m	psi/ft	bar/m	fps	m/s
78	295.3	2.261	0.516	1.614	0.368	28.9	8.8
79	299.0	2.315	0.528	1.652	0.377	29.3	8.9
80	302.8	2.370	0.541	1.691	0.386	29.7	9.0
81	306.6	2.425	0.553	1.731	0.395	30.1	9.2
82	310.4	2.480	0.566	1.770	0.404	30.4	9.3
83	314.2	2.537	0.579	1.810	0.413	30.8	9.4
84	318.0	2.593	0.592	1.851	0.422	31.2	9.5
85	321.8	2.651	0.605	1.892	0.432	31.5	9.6
86	325.5	2.709	0.618	1.933	0.441	31.9	9.7
87	329.3	2.767	0.631	1.975	0.451	32.3	9.8
88	333.1	2.827	0.645	2.017	0.460	32.6	10.0
89	336.9	2.886	0.658	2.060	0.470	33.0	10.1
90	340.7	2.947	0.672	2.103	0.480	33.4	10.2
91	344.5	3.007	0.686	2.146	0.490	33.8	10.3
92	348.3	3.069	0.700	2.190	0.500	34.1	10.4
93	352.0	3.131	0.714	2.234	0.510	34.5	10.5
94	355.8	3.193	0.728	2.279	0.520	34.9	10.6
95	359.6	3.256	0.743	2.324	0.530	35.2	10.7
96	363.4	3.320	0.757	2.370	0.541	35.6	10.9
97	367.2	3.384	0.772	2.415	0.551	36.0	11.0
98	371.0	3.449	0.787	2.462	0.562	36.4	11.1
99	374.8	3.515	0.802	2.508	0.572	36.7	11.2
100	378.5	3.581	0.817	2.556	0.583	37.1	11.3

Table B.2 Water flowing in 1¼-in. (32 mm) schedule 40 steel pipe I.D. = 1.38 in. (35.1 mm)

Q		P_f		P_f		Velocity	
		$C = 100$		$C = 120$			
GPM	L/min	psi/ft	bar/m	psi/ft	bar/m	fps	m/s
10	37.9	0.013	0.003	0.009	0.002	2.1	0.7
11	41.6	0.016	0.004	0.011	0.003	2.4	0.7
12	45.4	0.019	0.004	0.013	0.003	2.6	0.8
13	49.2	0.022	0.005	0.015	0.003	2.8	0.8
14	53.0	0.025	0.006	0.018	0.004	3.0	0.9
15	56.8	0.028	0.006	0.020	0.005	3.2	1.0
16	60.6	0.032	0.007	0.023	0.005	3.4	1.0
17	64.4	0.036	0.008	0.025	0.006	3.6	1.1
18	68.1	0.039	0.009	0.028	0.006	3.9	1.2
19	71.9	0.044	0.010	0.031	0.007	4.1	1.2
20	75.7	0.048	0.011	0.034	0.008	4.3	1.3
21	79.5	0.052	0.012	0.037	0.008	4.5	1.4
22	83.3	0.057	0.013	0.041	0.009	4.7	1.4
23	87.1	0.062	0.014	0.044	0.010	4.9	1.5
24	90.8	0.067	0.015	0.048	0.011	5.1	1.6
25	94.6	0.072	0.016	0.052	0.012	5.4	1.6
26	98.4	0.078	0.018	0.056	0.012	5.6	1.7
27	102.2	0.084	0.019	0.060	0.013	5.8	1.8
28	106.0	0.089	0.020	0.064	0.014	6.0	1.8
29	109.8	0.095	0.021	0.068	0.015	6.2	1.9
30	113.6	0.102	0.023	0.072	0.016	6.4	2.0
31	117.3	0.108	0.024	0.077	0.017	6.6	2.0
32	121.1	0.114	0.026	0.082	0.018	6.8	2.1
33	124.9	0.121	0.027	0.086	0.019	7.1	2.2
34	128.7	0.128	0.029	0.091	0.021	7.3	2.2
35	132.5	0.135	0.030	0.096	0.022	7.5	2.3
36	136.3	0.142	0.032	0.102	0.023	7.7	2.3
37	140.1	0.150	0.034	0.107	0.024	7.9	2.4
38	143.8	0.157	0.035	0.112	0.025	8.1	2.5
39	147.6	0.165	0.037	0.118	0.026	8.3	2.5
40	151.4	0.173	0.039	0.123	0.028	8.6	2.6
41	155.2	0.181	0.041	0.129	0.029	8.8	2.7
42	159.0	0.189	0.043	0.135	0.030	9.0	2.7
43	162.8	0.198	0.044	0.141	0.032	9.2	2.8
44	166.6	0.206	0.046	0.147	0.033	9.4	2.9

(continued)

Table B.2 (continued)

Q		P_f		P_f		Velocity	
		$C = 100$		$C = 120$			
GPM	L/min	psi/ft	bar/m	psi/ft	bar/m	fps	m/s
45	170.3	0.215	0.048	0.153	0.034	9.6	2.9
46	174.1	0.224	0.050	0.160	0.036	9.8	3.0
47	177.9	0.233	0.052	0.166	0.037	10.1	3.1
48	181.7	0.242	0.054	0.173	0.039	10.3	3.1
49	185.5	0.252	0.057	0.180	0.040	10.5	3.2
50	189.3	0.261	0.059	0.186	0.042	10.7	3.3
51	193.1	0.271	0.061	0.193	0.043	10.9	3.3
52	196.8	0.281	0.063	0.200	0.045	11.1	3.4
53	200.6	0.291	0.065	0.208	0.047	11.3	3.5
54	204.4	0.301	0.068	0.215	0.048	11.6	3.5
55	208.2	0.312	0.070	0.222	0.050	11.8	3.6
56	212.0	0.322	0.072	0.230	0.052	12.0	3.7
57	215.8	0.333	0.075	0.238	0.053	12.2	3.7
58	219.6	0.344	0.077	0.245	0.055	12.4	3.8
59	223.3	0.355	0.080	0.253	0.057	12.6	3.8
60	227.1	0.366	0.082	0.261	0.059	12.8	3.9
61	230.9	0.377	0.085	0.269	0.061	13.1	4.0
62	234.7	0.389	0.087	0.278	0.062	13.3	4.0
63	238.5	0.401	0.090	0.286	0.064	13.5	4.1
64	242.3	0.412	0.093	0.294	0.066	13.7	4.2
65	246.1	0.424	0.095	0.303	0.068	13.9	4.2
66	249.8	0.437	0.098	0.312	0.070	14.1	4.3
67	253.6	0.449	0.101	0.320	0.072	14.3	4.4
68	257.4	0.461	0.104	0.329	0.074	14.6	4.4
69	261.2	0.474	0.107	0.338	0.076	14.8	4.5
70	265.0	0.487	0.109	0.347	0.078	15.0	4.6
71	268.8	0.500	0.112	0.357	0.080	15.2	4.6
72	272.5	0.513	0.115	0.366	0.082	15.4	4.7
73	276.3	0.526	0.118	0.375	0.084	15.6	4.8
74	280.1	0.540	0.121	0.385	0.087	15.8	4.8
75	283.9	0.553	0.124	0.395	0.089	16.1	4.9
76	287.7	0.567	0.127	0.405	0.091	16.3	5.0
77	291.5	0.581	0.131	0.414	0.093	16.5	5.0

(continued)

Table B.2 (continued)

Q		P_f		P_f		Velocity	
		$C = 100$		$C = 120$			
GPM	L/min	psi/ft	bar/m	psi/ft	bar/m	fps	m/s
78	295.3	0.595	0.134	0.424	0.095	16.7	5.1
79	299.0	0.609	0.137	0.435	0.098	16.9	5.2
80	302.8	0.623	0.140	0.445	0.100	17.1	5.2
81	306.6	0.638	0.143	0.455	0.102	17.3	5.3
82	310.4	0.652	0.147	0.466	0.105	17.5	5.3
83	314.2	0.667	0.150	0.476	0.107	17.8	5.4
84	318.0	0.682	0.153	0.487	0.109	18.0	5.5
85	321.8	0.697	0.157	0.498	0.112	18.2	5.5
86	325.5	0.712	0.160	0.508	0.114	18.4	5.6
87	329.3	0.728	0.164	0.519	0.117	18.6	5.7
88	333.1	0.743	0.167	0.531	0.119	18.8	5.7
89	336.9	0.759	0.171	0.542	0.122	19.0	5.8
90	340.7	0.775	0.174	0.553	0.124	19.3	5.9
91	344.5	0.791	0.178	0.565	0.127	19.5	5.9
92	348.3	0.807	0.181	0.576	0.129	19.7	6.0
93	352.0	0.823	0.185	0.588	0.132	19.9	6.1
94	355.8	0.840	0.189	0.599	0.135	20.1	6.1
95	359.6	0.856	0.192	0.611	0.137	20.3	6.2
96	363.4	0.873	0.196	0.623	0.140	20.5	6.3
97	367.2	0.890	0.200	0.635	0.143	20.8	6.3
98	371.0	0.907	0.204	0.647	0.146	21.0	6.4
99	374.8	0.924	0.208	0.660	0.148	21.2	6.5
100	378.5	0.942	0.212	0.672	0.151	21.4	6.5

Table B.3 Water flowing in 1½-in. (40 mm) schedule 40 steel pipe I.D. = 1.61 in. (40.9 mm)

Q		P_f $C = 100$		P_f $C = 120$		Velocity	
GPM	L/min	psi/ft	bar/m	psi/ft	bar/m	fps	m/s
20	75.7	0.023	0.005	0.016	0.004	3.2	1.0
22	83.3	0.027	0.006	0.019	0.004	3.5	1.1
24	90.8	0.032	0.007	0.023	0.005	3.8	1.2
26	98.4	0.037	0.008	0.026	0.006	4.1	1.3
28	106.0	0.042	0.010	0.030	0.007	4.4	1.3
30	113.6	0.048	0.011	0.034	0.008	4.7	1.4
32	121.1	0.054	0.012	0.039	0.009	5.1	1.5
34	128.7	0.060	0.014	0.043	0.010	5.4	1.6
36	136.3	0.067	0.015	0.048	0.011	5.7	1.7
38	143.8	0.074	0.017	0.053	0.012	6.0	1.8
40	151.4	0.082	0.018	0.058	0.013	6.3	1.9
42	159.0	0.089	0.020	0.064	0.014	6.6	2.0
44	166.6	0.097	0.022	0.069	0.016	7.0	2.1
46	174.1	0.106	0.024	0.075	0.017	7.3	2.2
48	181.7	0.114	0.026	0.082	0.018	7.6	2.3
50	189.3	0.123	0.028	0.088	0.020	7.9	2.4
52	196.8	0.133	0.030	0.095	0.021	8.2	2.5
54	204.4	0.142	0.032	0.101	0.023	8.5	2.6
56	212.0	0.152	0.034	0.109	0.025	8.8	2.7
58	219.6	0.162	0.037	0.116	0.026	9.2	2.8
60	227.1	0.173	0.039	0.123	0.028	9.5	2.9
62	234.7	0.184	0.042	0.131	0.030	9.8	3.0
64	242.3	0.195	0.044	0.139	0.031	10.1	3.1
66	249.8	0.206	0.047	0.147	0.033	10.4	3.2
68	257.4	0.218	0.049	0.155	0.035	10.7	3.3
70	265.0	0.230	0.052	0.164	0.037	11.1	3.4
72	272.5	0.242	0.055	0.173	0.039	11.4	3.5
74	280.1	0.255	0.058	0.182	0.041	11.7	3.6
76	287.7	0.268	0.060	0.191	0.043	12.0	3.7
78	295.3	0.281	0.063	0.200	0.045	12.3	3.8
80	302.8	0.294	0.067	0.210	0.047	12.6	3.9
82	310.4	0.308	0.070	0.220	0.050	13.0	3.9
84	318.0	0.322	0.073	0.230	0.052	13.3	4.0
86	325.5	0.336	0.076	0.240	0.054	13.6	4.1
88	333.1	0.351	0.079	0.250	0.057	13.9	4.2

(continued)

Table B.3 (continued)

Q		P_f $C = 100$		P_f $C = 120$		Velocity	
GPM	L/min	psi/ft	bar/m	psi/ft	bar/m	fps	m/s
90	340.7	0.366	0.083	0.261	0.059	14.2	4.3
92	348.3	0.381	0.086	0.272	0.061	14.5	4.4
94	355.8	0.396	0.090	0.283	0.064	14.9	4.5
96	363.4	0.412	0.093	0.294	0.067	15.2	4.6
98	371.0	0.428	0.097	0.306	0.069	15.5	4.7
100	378.5	0.445	0.101	0.317	0.072	15.8	4.8
102	386.1	0.461	0.104	0.329	0.074	16.1	4.9
104	393.7	0.478	0.108	0.341	0.077	16.4	5.0
106	401.3	0.495	0.112	0.353	0.080	16.7	5.1
108	408.8	0.513	0.116	0.366	0.083	17.1	5.2
110	416.4	0.530	0.120	0.378	0.086	17.4	5.3
112	424.0	0.548	0.124	0.391	0.088	17.7	5.4
114	431.5	0.566	0.128	0.404	0.091	18.0	5.5
116	439.1	0.585	0.132	0.418	0.094	18.3	5.6
118	446.7	0.604	0.137	0.431	0.097	18.6	5.7
120	454.2	0.623	0.141	0.445	0.101	19.0	5.8
122	461.8	0.642	0.145	0.458	0.104	19.3	5.9
124	469.4	0.662	0.150	0.472	0.107	19.6	6.0
126	477.0	0.682	0.154	0.487	0.110	19.9	6.1
128	484.5	0.702	0.159	0.501	0.113	20.2	6.2
130	492.1	0.722	0.163	0.515	0.117	20.5	6.3
132	499.7	0.743	0.168	0.530	0.120	20.9	6.4
134	507.2	0.764	0.173	0.545	0.123	21.2	6.5
136	514.8	0.785	0.178	0.560	0.127	21.5	6.5
138	522.4	0.807	0.182	0.576	0.130	21.8	6.6
140	530.0	0.828	0.187	0.591	0.134	22.1	6.7
142	537.5	0.850	0.192	0.607	0.137	22.4	6.8
144	545.1	0.873	0.197	0.623	0.141	22.8	6.9
146	552.7	0.895	0.202	0.639	0.144	23.1	7.0
148	560.2	0.918	0.208	0.655	0.148	23.4	7.1
150	567.8	0.941	0.213	0.672	0.152	23.7	7.2
152	575.4	0.965	0.218	0.688	0.156	24.0	7.3
154	583.0	0.988	0.223	0.705	0.159	24.3	7.4
156	590.5	1.012	0.229	0.722	0.163	24.6	7.5

(continued)

Table B.3 (continued)

Q		P_f $C = 100$		P_f $C = 120$		Velocity	
GPM	L/min	psi/ft	bar/m	psi/ft	bar/m	fps	m/s
158	598.1	1.036	0.234	0.739	0.167	25.0	7.6
160	605.7	1.061	0.240	0.757	0.171	25.3	7.7
162	613.2	1.085	0.245	0.774	0.175	25.6	7.8
164	620.8	1.110	0.251	0.792	0.179	25.9	7.9
166	628.4	1.135	0.257	0.810	0.183	26.2	8.0
168	635.9	1.161	0.262	0.828	0.187	26.5	8.1
170	643.5	1.186	0.268	0.847	0.191	26.9	8.2
172	651.1	1.212	0.274	0.865	0.196	27.2	8.3
174	658.7	1.239	0.280	0.884	0.200	27.5	8.4
176	666.2	1.265	0.286	0.903	0.204	27.8	8.5
178	673.8	1.292	0.292	0.922	0.208	28.1	8.6
180	681.4	1.319	0.298	0.941	0.213	28.4	8.7
182	688.9	1.346	0.304	0.961	0.217	28.8	8.8
184	696.5	1.373	0.311	0.980	0.222	29.1	8.9
186	704.1	1.401	0.317	1.000	0.226	29.4	9.0
188	711.7	1.429	0.323	1.020	0.231	29.7	9.1
190	719.2	1.457	0.330	1.040	0.235	30.0	9.2
192	726.8	1.486	0.336	1.061	0.240	30.3	9.2
194	734.4	1.515	0.342	1.081	0.244	30.7	9.3
196	741.9	1.544	0.349	1.102	0.249	31.0	9.4
198	749.5	1.573	0.356	1.123	0.254	31.3	9.5
200	757.1	1.603	0.362	1.144	0.259	31.6	9.6

Table B.4 Water flowing in 2-in. (50 mm) schedule 40 steel pipe I.D. = 2.067 in. (52.5 mm)

Q		P_f $C = 100$		P_f $C = 120$		Velocity	
GPM	L/min	psi/ft	bar/m	psi/ft	bar/m	fps	m/s
40	151.4	0.024	0.005	0.017	0.004	3.8	1.2
42	159.0	0.026	0.006	0.019	0.004	4.0	1.2
44	166.6	0.029	0.007	0.021	0.005	4.2	1.3
46	174.1	0.031	0.007	0.022	0.005	4.4	1.3
48	181.7	0.034	0.008	0.024	0.005	4.6	1.4
50	189.3	0.037	0.008	0.026	0.006	4.8	1.5
52	196.8	0.039	0.009	0.028	0.006	5.0	1.5
54	204.4	0.042	0.010	0.030	0.007	5.2	1.6
56	212.0	0.045	0.010	0.032	0.007	5.4	1.6
58	219.6	0.048	0.011	0.034	0.008	5.6	1.7
60	227.1	0.051	0.012	0.037	0.008	5.8	1.8
62	234.7	0.054	0.012	0.039	0.009	6.0	1.8
64	242.3	0.058	0.013	0.041	0.009	6.1	1.9
66	249.8	0.061	0.014	0.044	0.010	6.3	1.9
68	257.4	0.065	0.015	0.046	0.010	6.5	2.0
70	265.0	0.068	0.015	0.049	0.011	6.7	2.0
72	272.5	0.072	0.016	0.051	0.012	6.9	2.1
74	280.1	0.075	0.017	0.054	0.012	7.1	2.2
76	287.7	0.079	0.018	0.057	0.013	7.3	2.2
78	295.3	0.083	0.019	0.059	0.013	7.5	2.3
80	302.8	0.087	0.020	0.062	0.014	7.7	2.3
82	310.4	0.091	0.021	0.065	0.015	7.9	2.4
84	318.0	0.095	0.022	0.068	0.015	8.1	2.5
86	325.5	0.100	0.023	0.071	0.016	8.3	2.5
88	333.1	0.104	0.024	0.074	0.017	8.4	2.6
90	340.7	0.108	0.025	0.077	0.017	8.6	2.6
92	348.3	0.113	0.026	0.081	0.018	8.8	2.7
94	355.8	0.117	0.027	0.084	0.019	9.0	2.8
96	363.4	0.122	0.028	0.087	0.020	9.2	2.8
98	371.0	0.127	0.029	0.091	0.020	9.4	2.9
100	378.5	0.132	0.030	0.094	0.021	9.6	2.9
102	386.1	0.137	0.031	0.097	0.022	9.8	3.0
104	393.7	0.142	0.032	0.101	0.023	10.0	3.0
106	401.3	0.147	0.033	0.105	0.024	10.2	3.1
108	408.8	0.152	0.034	0.108	0.025	10.4	3.2

(continued)

Table B.4 (continued)

Q		P_f $C = 100$		P_f $C = 120$		Velocity	
GPM	L/min	psi/ft	bar/m	psi/ft	bar/m	fps	m/s
110	416.4	0.157	0.036	0.112	0.025	10.6	3.2
112	424.0	0.162	0.037	0.116	0.026	10.8	3.3
114	431.5	0.168	0.038	0.120	0.027	10.9	3.3
116	439.1	0.173	0.039	0.124	0.028	11.1	3.4
118	446.7	0.179	0.040	0.128	0.029	11.3	3.5
120	454.2	0.184	0.042	0.132	0.030	11.5	3.5
122	461.8	0.190	0.043	0.136	0.031	11.7	3.6
124	469.4	0.196	0.044	0.140	0.032	11.9	3.6
126	477.0	0.202	0.046	0.144	0.033	12.1	3.7
128	484.5	0.208	0.047	0.148	0.034	12.3	3.7
130	492.1	0.214	0.048	0.153	0.035	12.5	3.8
132	499.7	0.220	0.050	0.157	0.036	12.7	3.9
134	507.2	0.226	0.051	0.161	0.037	12.9	3.9
136	514.8	0.233	0.053	0.166	0.038	13.1	4.0
138	522.4	0.239	0.054	0.170	0.039	13.2	4.0
140	530.0	0.245	0.056	0.175	0.040	13.4	4.1
142	537.5	0.252	0.057	0.180	0.041	13.6	4.2
144	545.1	0.258	0.058	0.184	0.042	13.8	4.2
146	552.7	0.265	0.060	0.189	0.043	14.0	4.3
148	560.2	0.272	0.062	0.194	0.044	14.2	4.3
150	567.8	0.279	0.063	0.199	0.045	14.4	4.4
152	575.4	0.286	0.065	0.204	0.046	14.6	4.4
154	583.0	0.293	0.066	0.209	0.047	14.8	4.5
156	590.5	0.300	0.068	0.214	0.048	15.0	4.6
158	598.1	0.307	0.069	0.219	0.050	15.2	4.6
160	605.7	0.314	0.071	0.224	0.051	15.4	4.7
162	613.2	0.321	0.073	0.229	0.052	15.6	4.7
164	620.8	0.329	0.074	0.235	0.053	15.7	4.8
166	628.4	0.336	0.076	0.240	0.054	15.9	4.9
168	635.9	0.344	0.078	0.245	0.056	16.1	4.9
170	643.5	0.351	0.080	0.251	0.057	16.3	5.0
172	651.1	0.359	0.081	0.256	0.058	16.5	5.0
174	658.7	0.367	0.083	0.262	0.059	16.7	5.1
176	666.2	0.375	0.085	0.267	0.061	16.9	5.1

(continued)

Table B.4 (continued)

Q		P_f $C = 100$		P_f $C = 120$		Velocity	
GPM	L/min	psi/ft	bar/m	psi/ft	bar/m	fps	m/s
178	673.8	0.383	0.087	0.273	0.062	17.1	5.2
180	681.4	0.391	0.088	0.279	0.063	17.3	5.3
182	688.9	0.399	0.090	0.284	0.064	17.5	5.3
184	696.5	0.407	0.092	0.290	0.066	17.7	5.4
186	704.1	0.415	0.094	0.296	0.067	17.9	5.4
188	711.7	0.423	0.096	0.302	0.068	18.0	5.5
190	719.2	0.432	0.098	0.308	0.070	18.2	5.6
192	726.8	0.440	0.100	0.314	0.071	18.4	5.6
194	734.4	0.449	0.102	0.320	0.072	18.6	5.7
196	741.9	0.457	0.103	0.326	0.074	18.8	5.7
198	749.5	0.466	0.105	0.332	0.075	19.0	5.8
200	757.1	0.475	0.107	0.339	0.077	19.2	5.9
202	764.7	0.483	0.109	0.345	0.078	19.4	5.9
204	772.2	0.492	0.111	0.351	0.080	19.6	6.0
206	779.8	0.501	0.113	0.358	0.081	19.8	6.0
208	787.4	0.510	0.115	0.364	0.082	20.0	6.1
210	794.9	0.519	0.118	0.371	0.084	20.2	6.1
212	802.5	0.529	0.120	0.377	0.085	20.4	6.2
214	810.1	0.538	0.122	0.384	0.087	20.5	6.3
216	817.6	0.547	0.124	0.391	0.088	20.7	6.3
218	825.2	0.557	0.126	0.397	0.090	20.9	6.4
220	832.8	0.566	0.128	0.404	0.091	21.1	6.4

Table B.5 Water flowing in 2½-in. (65 mm) schedule 40 steel pipe I.D. = 2.469 in. (62.7 mm)

Q		P_f $C = 100$		P_f $C = 120$		Velocity	
GPM	L/min	psi/ft	bar/m	psi/ft	bar/m	fps	m/s
60	227.1	0.022	0.0049	0.015	0.0035	4.0	1.2
62	234.7	0.023	0.0052	0.016	0.0037	4.2	1.3
64	242.3	0.024	0.0055	0.017	0.0039	4.3	1.3
66	249.8	0.026	0.0058	0.018	0.0042	4.4	1.3
68	257.4	0.027	0.0061	0.019	0.0044	4.6	1.4
70	265.0	0.029	0.0065	0.020	0.0046	4.7	1.4
72	272.5	0.030	0.0068	0.022	0.0049	4.8	1.5
74	280.1	0.032	0.0072	0.023	0.0051	5.0	1.5
76	287.7	0.033	0.0076	0.024	0.0054	5.1	1.6
78	295.3	0.035	0.0079	0.025	0.0057	5.2	1.6
80	302.8	0.037	0.0083	0.026	0.0059	5.4	1.6
82	310.4	0.038	0.0087	0.027	0.0062	5.5	1.7
84	318.0	0.040	0.0091	0.029	0.0065	5.6	1.7
86	325.5	0.042	0.0095	0.030	0.0068	5.8	1.8
88	333.1	0.044	0.0099	0.031	0.0071	5.9	1.8
90	340.7	0.046	0.0103	0.033	0.0074	6.0	1.8
92	348.3	0.047	0.0108	0.034	0.0077	6.2	1.9
94	355.8	0.049	0.0112	0.035	0.0080	6.3	1.9
96	363.4	0.051	0.0116	0.037	0.0083	6.4	2.0
98	371.0	0.053	0.0121	0.038	0.0086	6.6	2.0
100	378.5	0.055	0.0125	0.040	0.0090	6.7	2.0
102	386.1	0.057	0.0130	0.041	0.0093	6.8	2.1
104	393.7	0.060	0.0135	0.043	0.0096	7.0	2.1
106	401.3	0.062	0.0140	0.044	0.0100	7.1	2.2
108	408.8	0.064	0.0145	0.046	0.0103	7.2	2.2
110	416.4	0.066	0.0150	0.047	0.0107	7.4	2.2
112	424.0	0.068	0.0155	0.049	0.0110	7.5	2.3
114	431.5	0.071	0.0160	0.050	0.0114	7.6	2.3
116	439.1	0.073	0.0165	0.052	0.0118	7.8	2.4
118	446.7	0.075	0.0170	0.054	0.0122	7.9	2.4
120	454.2	0.078	0.0176	0.055	0.0125	8.0	2.5
122	461.8	0.080	0.0181	0.057	0.0129	8.2	2.5
124	469.4	0.082	0.0187	0.059	0.0133	8.3	2.5
126	477.0	0.085	0.0192	0.061	0.0137	8.4	2.6
128	484.5	0.087	0.0198	0.062	0.0141	8.6	2.6

(continued)

Table B.5 (continued)

Q		P_f $C = 100$		P_f $C = 120$		Velocity	
GPM	L/min	psi/ft	bar/m	psi/ft	bar/m	fps	m/s
130	492.1	0.090	0.0204	0.064	0.0146	8.7	2.7
132	499.7	0.093	0.0210	0.066	0.0150	8.8	2.7
134	507.2	0.095	0.0216	0.068	0.0154	9.0	2.7
136	514.8	0.098	0.0222	0.070	0.0158	9.1	2.8
138	522.4	0.101	0.0228	0.072	0.0163	9.2	2.8
140	530.0	0.103	0.0234	0.074	0.0167	9.4	2.9
142	537.5	0.106	0.0240	0.076	0.0171	9.5	2.9
144	545.1	0.109	0.0246	0.078	0.0176	9.6	2.9
146	552.7	0.112	0.0253	0.080	0.0180	9.8	3.0
148	560.2	0.114	0.0259	0.082	0.0185	9.9	3.0
150	567.8	0.117	0.0266	0.084	0.0190	10.1	3.1
152	575.4	0.120	0.0272	0.086	0.0194	10.2	3.1
154	583.0	0.123	0.0279	0.088	0.0199	10.3	3.1
156	590.5	0.126	0.0286	0.090	0.0204	10.5	3.2
158	598.1	0.129	0.0292	0.092	0.0209	10.6	3.2
160	605.7	0.132	0.0299	0.094	0.0214	10.7	3.3
162	613.2	0.135	0.0306	0.097	0.0219	10.9	3.3
164	620.8	0.138	0.0313	0.099	0.0224	11.0	3.3
166	628.4	0.142	0.0320	0.101	0.0229	11.1	3.4
168	635.9	0.145	0.0328	0.103	0.0234	11.3	3.4
170	643.5	0.148	0.0335	0.106	0.0239	11.4	3.5
172	651.1	0.151	0.0342	0.108	0.0244	11.5	3.5
174	658.7	0.154	0.0350	0.110	0.0250	11.7	3.6
176	666.2	0.158	0.0357	0.113	0.0255	11.8	3.6
178	673.8	0.161	0.0365	0.115	0.0260	11.9	3.6
180	681.4	0.164	0.0372	0.117	0.0266	12.1	3.7
182	688.9	0.168	0.0380	0.120	0.0271	12.2	3.7
184	696.5	0.171	0.0388	0.122	0.0277	12.3	3.8
186	704.1	0.175	0.0396	0.125	0.0282	12.5	3.8
188	711.7	0.178	0.0403	0.127	0.0288	12.6	3.8
190	719.2	0.182	0.0411	0.130	0.0294	12.7	3.9
192	726.8	0.185	0.0419	0.132	0.0299	12.9	3.9
194	734.4	0.189	0.0428	0.135	0.0305	13.0	4.0
196	741.9	0.192	0.0436	0.137	0.0311	13.1	4.0

(continued)

Table B.5 (continued)

Q		P_f $C = 100$		P_f $C = 120$		Velocity	
GPM	L/min	psi/ft	bar/m	psi/ft	bar/m	fps	m/s
198	749.5	0.196	0.0444	0.140	0.0317	13.3	4.0
200	757.1	0.200	0.0452	0.143	0.0323	13.4	4.1
202	764.7	0.203	0.0461	0.145	0.0329	13.5	4.1
204	772.2	0.207	0.0469	0.148	0.0335	13.7	4.2
206	779.8	0.211	0.0478	0.151	0.0341	13.8	4.2
208	787.4	0.215	0.0486	0.153	0.0347	13.9	4.2
210	794.9	0.219	0.0495	0.156	0.0353	14.1	4.3
212	802.5	0.222	0.0504	0.159	0.0360	14.2	4.3
214	810.1	0.226	0.0513	0.162	0.0366	14.3	4.4
216	817.6	0.230	0.0522	0.164	0.0372	14.5	4.4
218	825.2	0.234	0.0531	0.167	0.0379	14.6	4.5
220	832.8	0.238	0.0540	0.170	0.0385	14.7	4.5
222	840.4	0.242	0.0549	0.173	0.0392	14.9	4.5
224	847.9	0.246	0.0558	0.176	0.0398	15.0	4.6
226	855.5	0.250	0.0567	0.179	0.0405	15.1	4.6
228	863.1	0.255	0.0576	0.182	0.0411	15.3	4.7
230	870.6	0.259	0.0586	0.185	0.0418	15.4	4.7
232	878.2	0.263	0.0595	0.188	0.0425	15.5	4.7
234	885.8	0.267	0.0605	0.191	0.0432	15.7	4.8
236	893.4	0.271	0.0614	0.194	0.0439	15.8	4.8
238	900.9	0.276	0.0624	0.197	0.0445	15.9	4.9
240	908.5	0.280	0.0634	0.200	0.0452	16.1	4.9

Table B.6 Water flowing in 3-in. (80 mm) schedule 40 steel pipe I.D. = 3.068 in. (77.9 mm)

Q		P_f $C = 100$		P_f $C = 120$		Velocity	
GPM	L/min	psi/ft	bar/m	psi/ft	bar/m	fps	m/s
50	189.3	0.005	0.0012	0.004	0.0009	2.2	0.7
55	208.2	0.006	0.0014	0.005	0.0010	2.4	0.7
60	227.1	0.007	0.0017	0.005	0.0012	2.6	0.8
65	246.1	0.009	0.0020	0.006	0.0014	2.8	0.9
70	265.0	0.010	0.0023	0.007	0.0016	3.0	0.9
75	283.9	0.011	0.0026	0.008	0.0018	3.2	1.0
80	302.8	0.013	0.0029	0.009	0.0021	3.4	1.0
85	321.8	0.014	0.0032	0.010	0.0023	3.7	1.1
90	340.7	0.016	0.0036	0.011	0.0026	3.9	1.2
95	359.6	0.017	0.0040	0.012	0.0028	4.1	1.2
100	378.5	0.019	0.0044	0.014	0.0031	4.3	1.3
105	397.5	0.021	0.0048	0.015	0.0034	4.5	1.4
110	416.4	0.023	0.0052	0.016	0.0037	4.7	1.4
115	435.3	0.025	0.0056	0.018	0.0040	4.9	1.5
120	454.2	0.027	0.0061	0.019	0.0044	5.2	1.6
125	473.2	0.029	0.0066	0.021	0.0047	5.4	1.6
130	492.1	0.031	0.0071	0.022	0.0051	5.6	1.7
135	511.0	0.034	0.0076	0.024	0.0054	5.8	1.8
140	530.0	0.036	0.0081	0.026	0.0058	6.0	1.8
145	548.9	0.038	0.0087	0.027	0.0062	6.2	1.9
150	567.8	0.041	0.0092	0.029	0.0066	6.5	2.0
155	586.7	0.043	0.0098	0.031	0.0070	6.7	2.0
160	605.7	0.046	0.0104	0.033	0.0074	6.9	2.1
165	624.6	0.049	0.0110	0.035	0.0079	7.1	2.2
170	643.5	0.051	0.0116	0.037	0.0083	7.3	2.2
175	662.4	0.054	0.0123	0.039	0.0088	7.5	2.3
180	681.4	0.057	0.0129	0.041	0.0092	7.7	2.4
185	700.3	0.060	0.0136	0.043	0.0097	8.0	2.4
190	719.2	0.063	0.0143	0.045	0.0102	8.2	2.5
195	738.2	0.066	0.0150	0.047	0.0107	8.4	2.6
200	757.1	0.069	0.0157	0.049	0.0112	8.6	2.6
205	776.0	0.073	0.0165	0.052	0.0117	8.8	2.7
210	794.9	0.076	0.0172	0.054	0.0123	9.0	2.8
215	813.9	0.079	0.0180	0.057	0.0128	9.2	2.8
220	832.8	0.083	0.0187	0.059	0.0134	9.5	2.9

(continued)

Table B.6 (continued)

Q		P_f $C = 100$		P_f $C = 120$		Velocity	
GPM	L/min	psi/ft	bar/m	psi/ft	bar/m	fps	m/s
225	851.7	0.086	0.0195	0.062	0.0139	9.7	2.9
230	870.6	0.090	0.0204	0.064	0.0145	9.9	3.0
235	889.6	0.093	0.0212	0.067	0.0151	10.1	3.1
240	908.5	0.097	0.0220	0.069	0.0157	10.3	3.1
245	927.4	0.101	0.0229	0.072	0.0163	10.5	3.2
250	946.4	0.105	0.0238	0.075	0.0170	10.8	3.3
255	965.3	0.109	0.0246	0.078	0.0176	11.0	3.3
260	984.2	0.113	0.0255	0.080	0.0182	11.2	3.4
265	1003.1	0.117	0.0265	0.083	0.0189	11.4	3.5
270	1022.1	0.121	0.0274	0.086	0.0195	11.6	3.5
275	1041.0	0.125	0.0283	0.089	0.0202	11.8	3.6
280	1059.9	0.129	0.0293	0.092	0.0209	12.0	3.7
285	1078.8	0.134	0.0303	0.095	0.0216	12.3	3.7
290	1097.8	0.138	0.0313	0.098	0.0223	12.5	3.8
295	1116.7	0.142	0.0323	0.102	0.0230	12.7	3.9
300	1135.6	0.147	0.0333	0.105	0.0238	12.9	3.9
305	1154.5	0.151	0.0343	0.108	0.0245	13.1	4.0
310	1173.5	0.156	0.0354	0.111	0.0252	13.3	4.1
315	1192.4	0.161	0.0364	0.115	0.0260	13.5	4.1
320	1211.3	0.165	0.0375	0.118	0.0268	13.8	4.2
325	1230.3	0.170	0.0386	0.122	0.0275	14.0	4.3
330	1249.2	0.175	0.0397	0.125	0.0283	14.2	4.3
335	1268.1	0.180	0.0408	0.129	0.0291	14.4	4.4
340	1287.0	0.185	0.0419	0.132	0.0299	14.6	4.5
345	1306.0	0.190	0.0431	0.136	0.0308	14.8	4.5
350	1324.9	0.195	0.0443	0.139	0.0316	15.1	4.6
355	1343.8	0.200	0.0454	0.143	0.0324	15.3	4.7
360	1362.7	0.206	0.0466	0.147	0.0333	15.5	4.7
365	1381.7	0.211	0.0478	0.151	0.0341	15.7	4.8
370	1400.6	0.216	0.0491	0.154	0.0350	15.9	4.8
375	1419.5	0.222	0.0503	0.158	0.0359	16.1	4.9
380	1438.5	0.227	0.0515	0.162	0.0368	16.3	5.0
385	1457.4	0.233	0.0528	0.166	0.0377	16.6	5.0
390	1476.3	0.239	0.0541	0.170	0.0386	16.8	5.1

(continued)

Table B.6 (continued)

Q		P_f $C = 100$		P_f $C = 120$		Velocity	
GPM	L/min	psi/ft	bar/m	psi/ft	bar/m	fps	m/s
395	1495.2	0.244	0.0554	0.174	0.0395	17.0	5.2
400	1514.2	0.250	0.0567	0.178	0.0404	17.2	5.2
405	1533.1	0.256	0.0580	0.183	0.0414	17.4	5.3
410	1552.0	0.262	0.0593	0.187	0.0423	17.6	5.4
415	1570.9	0.268	0.0607	0.191	0.0433	17.8	5.4
420	1589.9	0.274	0.0620	0.195	0.0443	18.1	5.5
425	1608.8	0.280	0.0634	0.200	0.0452	18.3	5.6
430	1627.7	0.286	0.0648	0.204	0.0462	18.5	5.6
435	1646.6	0.292	0.0662	0.208	0.0472	18.7	5.7
440	1665.6	0.298	0.0676	0.213	0.0482	18.9	5.8
445	1684.5	0.305	0.0690	0.217	0.0493	19.1	5.8
450	1703.4	0.311	0.0705	0.222	0.0503	19.4	5.9
455	1722.4	0.317	0.0719	0.226	0.0513	19.6	6.0
460	1741.3	0.324	0.0734	0.231	0.0524	19.8	6.0
465	1760.2	0.330	0.0749	0.236	0.0534	20.0	6.1
470	1779.1	0.337	0.0764	0.240	0.0545	20.2	6.2
475	1798.1	0.344	0.0779	0.245	0.0556	20.4	6.2
480	1817.0	0.350	0.0794	0.250	0.0567	20.6	6.3
485	1835.9	0.357	0.0809	0.255	0.0578	20.9	6.4
490	1854.8	0.364	0.0825	0.260	0.0589	21.1	6.4
495	1873.8	0.371	0.0840	0.265	0.0600	21.3	6.5
500	1892.7	0.378	0.0856	0.270	0.0611	21.5	6.6

Table B.7 Water flowing in 4-in. (100 mm) schedule 40 steel pipe I.D. = 4.026 in. (102.3 mm)

Q		P_f $C = 100$		P_f $C = 120$		Velocity	
GPM	L/min	psi/ft	bar/m	psi/ft	bar/m	fps	m/s
100	378.5	0.005	0.0012	0.004	0.0008	2.5	0.8
110	416.4	0.006	0.0014	0.004	0.0010	2.8	0.8
120	454.2	0.007	0.0016	0.005	0.0012	3.0	0.9
130	492.1	0.008	0.0019	0.006	0.0013	3.3	1.0
140	530.0	0.010	0.0022	0.007	0.0015	3.5	1.1
150	567.8	0.011	0.0024	0.008	0.0017	3.8	1.1
160	605.7	0.012	0.0028	0.009	0.0020	4.0	1.2
170	643.5	0.014	0.0031	0.010	0.0022	4.3	1.3
180	681.4	0.015	0.0034	0.011	0.0024	4.5	1.4
190	719.2	0.017	0.0038	0.012	0.0027	4.8	1.4
200	757.1	0.018	0.0042	0.013	0.0030	5.0	1.5
210	794.9	0.020	0.0046	0.014	0.0033	5.3	1.6
220	832.8	0.022	0.0050	0.016	0.0035	5.5	1.7
230	870.6	0.024	0.0054	0.017	0.0039	5.8	1.8
240	908.5	0.026	0.0058	0.018	0.0042	6.0	1.8
250	946.4	0.028	0.0063	0.020	0.0045	6.3	1.9
260	984.2	0.030	0.0068	0.021	0.0048	6.5	2.0
270	1022.1	0.032	0.0073	0.023	0.0052	6.8	2.1
280	1059.9	0.034	0.0078	0.025	0.0055	7.0	2.1
290	1097.8	0.037	0.0083	0.026	0.0059	7.3	2.2
300	1135.6	0.039	0.0088	0.028	0.0063	7.5	2.3
310	1173.5	0.042	0.0094	0.030	0.0067	7.8	2.4
320	1211.3	0.044	0.0099	0.031	0.0071	8.0	2.4
330	1249.2	0.047	0.0105	0.033	0.0075	8.3	2.5
340	1287.0	0.049	0.0111	0.035	0.0079	8.5	2.6
350	1324.9	0.052	0.0117	0.037	0.0084	8.8	2.7
360	1362.7	0.055	0.0124	0.039	0.0088	9.0	2.7
370	1400.6	0.058	0.0130	0.041	0.0093	9.3	2.8
380	1438.5	0.061	0.0137	0.043	0.0098	9.5	2.9
390	1476.3	0.064	0.0143	0.045	0.0102	9.8	3.0
400	1514.2	0.067	0.0150	0.048	0.0107	10.0	3.0
410	1552.0	0.070	0.0157	0.050	0.0112	10.3	3.1
420	1589.9	0.073	0.0165	0.052	0.0117	10.5	3.2
430	1627.7	0.076	0.0172	0.054	0.0123	10.8	3.3
440	1665.6	0.079	0.0179	0.057	0.0128	11.0	3.4

(continued)

Table B.7 (continued)

Q		P_f C = 100		P_f C = 120		Velocity	
GPM	L/min	psi/ft	bar/m	psi/ft	bar/m	fps	m/s
450	1703.4	0.083	0.0187	0.059	0.0133	11.3	3.4
460	1741.3	0.086	0.0195	0.062	0.0139	11.5	3.5
470	1779.1	0.090	0.0203	0.064	0.0145	11.8	3.6
480	1817.0	0.093	0.0211	0.067	0.0150	12.0	3.7
490	1854.8	0.097	0.0219	0.069	0.0156	12.3	3.7
500	1892.7	0.101	0.0227	0.072	0.0162	12.5	3.8
510	1930.6	0.104	0.0236	0.074	0.0168	12.8	3.9
520	1968.4	0.108	0.0244	0.077	0.0174	13.0	4.0
530	2006.3	0.112	0.0253	0.080	0.0181	13.3	4.0
540	2044.1	0.116	0.0262	0.083	0.0187	13.5	4.1
550	2082.0	0.120	0.0271	0.086	0.0193	13.8	4.2
560	2119.8	0.124	0.0280	0.089	0.0200	14.0	4.3
570	2157.7	0.128	0.0289	0.091	0.0207	14.3	4.3
580	2195.5	0.132	0.0299	0.094	0.0213	14.5	4.4
590	2233.4	0.137	0.0308	0.097	0.0220	14.8	4.5
600	2271.2	0.141	0.0318	0.101	0.0227	15.0	4.6
610	2309.1	0.145	0.0328	0.104	0.0234	15.3	4.6
620	2346.9	0.150	0.0338	0.107	0.0241	15.5	4.7
630	2384.8	0.154	0.0348	0.110	0.0249	15.8	4.8
640	2422.7	0.159	0.0359	0.113	0.0256	16.0	4.9
650	2460.5	0.163	0.0369	0.117	0.0263	16.3	5.0
660	2498.4	0.168	0.0380	0.120	0.0271	16.5	5.0
670	2536.2	0.173	0.0390	0.123	0.0279	16.8	5.1
680	2574.1	0.178	0.0401	0.127	0.0286	17.0	5.2
690	2611.9	0.183	0.0412	0.130	0.0294	17.3	5.3
700	2649.8	0.187	0.0423	0.134	0.0302	17.5	5.3
710	2687.6	0.192	0.0435	0.137	0.0310	17.8	5.4
720	2725.5	0.197	0.0446	0.141	0.0318	18.0	5.5
730	2763.3	0.203	0.0457	0.145	0.0326	18.3	5.6
740	2801.2	0.208	0.0469	0.148	0.0335	18.5	5.6
750	2839.1	0.213	0.0481	0.152	0.0343	18.8	5.7
760	2876.9	0.218	0.0493	0.156	0.0352	19.0	5.8
770	2914.8	0.224	0.0505	0.160	0.0360	19.3	5.9
780	2952.6	0.229	0.0517	0.163	0.0369	19.5	5.9

(continued)

Table B.7 (continued)

Q		P_f $C = 100$		P_f $C = 120$		Velocity	
GPM	L/min	psi/ft	bar/m	psi/ft	bar/m	fps	m/s
790	2990.5	0.234	0.0529	0.167	0.0378	19.8	6.0
800	3028.3	0.240	0.0542	0.171	0.0387	20.0	6.1
810	3066.2	0.246	0.0554	0.175	0.0396	20.3	6.2
820	3104.0	0.251	0.0567	0.179	0.0405	20.5	6.2
830	3141.9	0.257	0.0580	0.183	0.0414	20.8	6.3
840	3179.7	0.263	0.0593	0.187	0.0423	21.0	6.4
850	3217.6	0.268	0.0606	0.192	0.0433	21.3	6.5
860	3255.4	0.274	0.0619	0.196	0.0442	21.5	6.6
870	3293.3	0.280	0.0633	0.200	0.0452	21.8	6.6
880	3331.2	0.286	0.0646	0.204	0.0461	22.0	6.7
890	3369.0	0.292	0.0660	0.209	0.0471	22.3	6.8
900	3406.9	0.298	0.0674	0.213	0.0481	22.5	6.9
910	3444.7	0.305	0.0688	0.217	0.0491	22.8	6.9
920	3482.6	0.311	0.0702	0.222	0.0501	23.0	7.0
930	3520.4	0.317	0.0716	0.226	0.0511	23.3	7.1
940	3558.3	0.323	0.0730	0.231	0.0521	23.5	7.2
950	3596.1	0.330	0.0745	0.235	0.0531	23.8	7.2
960	3634.0	0.336	0.0759	0.240	0.0542	24.0	7.3
970	3671.8	0.343	0.0774	0.245	0.0552	24.3	7.4
980	3709.7	0.349	0.0789	0.249	0.0563	24.5	7.5
990	3747.5	0.356	0.0804	0.254	0.0574	24.8	7.5
1000	3785.4	0.363	0.0819	0.259	0.0584	25.0	7.6

Table B.8 Water flowing in 6-in. (150 mm) schedule 40 steel pipe I.D. = 6.065 in. (154.1 mm)

Q		P_f $C = 100$		P_f $C = 120$		Velocity	
GPM	L/min	psi/ft	bar/m	psi/ft	bar/m	fps	m/s
300	1135.6	0.005	0.0012	0.004	0.0009	3.3	1.0
325	1230.3	0.006	0.0014	0.004	0.0010	3.6	1.1
350	1324.9	0.007	0.0016	0.005	0.0011	3.9	1.2
375	1419.5	0.008	0.0018	0.006	0.0013	4.1	1.3
400	1514.2	0.009	0.0020	0.006	0.0015	4.4	1.3
425	1608.8	0.010	0.0023	0.007	0.0016	4.7	1.4
450	1703.4	0.011	0.0025	0.008	0.0018	5.0	1.5
475	1798.1	0.012	0.0028	0.009	0.0020	5.2	1.6
500	1892.7	0.014	0.0031	0.010	0.0022	5.5	1.7
525	1987.3	0.015	0.0034	0.011	0.0024	5.8	1.8
550	2082.0	0.016	0.0037	0.012	0.0026	6.1	1.8
575	2176.6	0.018	0.0040	0.013	0.0029	6.3	1.9
600	2271.2	0.019	0.0043	0.014	0.0031	6.6	2.0
625	2365.9	0.021	0.0047	0.015	0.0033	6.9	2.1
650	2460.5	0.022	0.0050	0.016	0.0036	7.2	2.2
675	2555.1	0.024	0.0054	0.017	0.0038	7.4	2.3
700	2649.8	0.025	0.0058	0.018	0.0041	7.7	2.3
725	2744.4	0.027	0.0061	0.019	0.0044	8.0	2.4
750	2839.1	0.029	0.0065	0.021	0.0047	8.3	2.5
775	2933.7	0.031	0.0069	0.022	0.0050	8.5	2.6
800	3028.3	0.033	0.0074	0.023	0.0053	8.8	2.7
825	3123.0	0.035	0.0078	0.025	0.0056	9.1	2.8
850	3217.6	0.036	0.0082	0.026	0.0059	9.4	2.8
875	3312.2	0.039	0.0087	0.027	0.0062	9.6	2.9
900	3406.9	0.041	0.0092	0.029	0.0065	9.9	3.0
925	3501.5	0.043	0.0096	0.030	0.0069	10.2	3.1
950	3596.1	0.045	0.0101	0.032	0.0072	10.5	3.2
975	3690.8	0.047	0.0106	0.034	0.0076	10.7	3.3
1000	3785.4	0.049	0.0111	0.035	0.0079	11.0	3.4
1025	3880.0	0.052	0.0117	0.037	0.0083	11.3	3.4
1050	3974.7	0.054	0.0122	0.039	0.0087	11.6	3.5
1075	4069.3	0.056	0.0127	0.040	0.0091	11.8	3.6
1100	4163.9	0.059	0.0133	0.042	0.0095	12.1	3.7
1125	4258.6	0.061	0.0138	0.044	0.0099	12.4	3.8
1150	4353.2	0.064	0.0144	0.046	0.0103	12.7	3.9

(continued)

Table B.8 (continued)

Q		P_f $C = 100$		P_f $C = 120$		Velocity	
GPM	L/min	psi/ft	bar/m	psi/ft	bar/m	fps	m/s
1175	4447.8	0.066	0.0150	0.047	0.0107	12.9	3.9
1200	4542.5	0.069	0.0156	0.049	0.0111	13.2	4.0
1225	4637.1	0.072	0.0162	0.051	0.0116	13.5	4.1
1250	4731.8	0.074	0.0168	0.053	0.0120	13.8	4.2
1275	4826.4	0.077	0.0175	0.055	0.0125	14.0	4.3
1300	4921.0	0.080	0.0181	0.057	0.0129	14.3	4.4
1325	5015.7	0.083	0.0187	0.059	0.0134	14.6	4.4
1350	5110.3	0.086	0.0194	0.061	0.0138	14.9	4.5
1375	5204.9	0.089	0.0201	0.063	0.0143	15.1	4.6
1400	5299.6	0.092	0.0207	0.066	0.0148	15.4	4.7
1425	5394.2	0.095	0.0214	0.068	0.0153	15.7	4.8
1450	5488.8	0.098	0.0221	0.070	0.0158	16.0	4.9
1475	5583.5	0.101	0.0229	0.072	0.0163	16.2	4.9
1500	5678.1	0.104	0.0236	0.074	0.0168	16.5	5.0
1525	5772.7	0.108	0.0243	0.077	0.0173	16.8	5.1
1550	5867.4	0.111	0.0250	0.079	0.0179	17.1	5.2
1575	5962.0	0.114	0.0258	0.082	0.0184	17.3	5.3
1600	6056.6	0.118	0.0266	0.084	0.0190	17.6	5.4
1625	6151.3	0.121	0.0273	0.086	0.0195	17.9	5.4
1650	6245.9	0.124	0.0281	0.089	0.0201	18.2	5.5
1675	6340.5	0.128	0.0289	0.091	0.0206	18.4	5.6
1700	6435.2	0.132	0.0297	0.094	0.0212	18.7	5.7
1725	6529.8	0.135	0.0305	0.096	0.0218	19.0	5.8
1750	6624.5	0.139	0.0314	0.099	0.0224	19.3	5.9
1775	6719.1	0.142	0.0322	0.102	0.0230	19.5	6.0
1800	6813.7	0.146	0.0330	0.104	0.0236	19.8	6.0
1825	6908.4	0.150	0.0339	0.107	0.0242	20.1	6.1
1850	7003.0	0.154	0.0347	0.110	0.0248	20.4	6.2
1875	7097.6	0.158	0.0356	0.113	0.0254	20.6	6.3
1900	7192.3	0.162	0.0365	0.115	0.0261	20.9	6.4
1925	7286.9	0.166	0.0374	0.118	0.0267	21.2	6.5
1950	7381.5	0.170	0.0383	0.121	0.0273	21.5	6.5
1975	7476.2	0.174	0.0392	0.124	0.0280	21.7	6.6
2000	7570.8	0.178	0.0401	0.127	0.0286	22.0	6.7

(continued)

Table B.8 (continued)

Q		P_f $C = 100$		P_f $C = 120$		Velocity	
GPM	L/min	psi/ft	bar/m	psi/ft	bar/m	fps	m/s
2025	7665.4	0.182	0.0411	0.130	0.0293	22.3	6.8
2050	7760.1	0.186	0.0420	0.133	0.0300	22.6	6.9
2075	7854.7	0.190	0.0430	0.136	0.0307	22.8	7.0
2100	7949.3	0.194	0.0439	0.139	0.0314	23.1	7.0
2125	8044.0	0.199	0.0449	0.142	0.0320	23.4	7.1
2150	8138.6	0.203	0.0459	0.145	0.0327	23.7	7.2
2175	8233.2	0.208	0.0469	0.148	0.0335	23.9	7.3
2200	8327.9	0.212	0.0479	0.151	0.0342	24.2	7.4
2225	8422.5	0.216	0.0489	0.154	0.0349	24.5	7.5
2250	8517.2	0.221	0.0499	0.158	0.0356	24.8	7.5
2275	8611.8	0.226	0.0509	0.161	0.0364	25.0	7.6
2300	8706.4	0.230	0.0520	0.164	0.0371	25.3	7.7
2325	8801.1	0.235	0.0530	0.168	0.0378	25.6	7.8
2350	8895.7	0.239	0.0541	0.171	0.0386	25.9	7.9
2375	8990.3	0.244	0.0552	0.174	0.0394	26.1	8.0
2400	9085.0	0.249	0.0562	0.178	0.0401	26.4	8.0
2425	9179.6	0.254	0.0573	0.181	0.0409	26.7	8.1
2450	9274.2	0.259	0.0584	0.185	0.0417	27.0	8.2
2475	9368.9	0.264	0.0595	0.188	0.0425	27.2	8.3
2500	9463.5	0.268	0.0607	0.192	0.0433	27.5	8.4

Table B.9 Water flowing in 8-in. (200 mm) schedule 40 steel pipe I.D. = 7.981 in. (203 mm)

Q		P_f $C = 100$		P_f $C = 120$		Velocity	
GPM	L/min	psi/ft	bar/m	psi/ft	bar/m	fps	m/s
500	1892.7	0.0036	0.0008	0.0026	0.0006	3.2	1.0
525	1987.3	0.0039	0.0009	0.0028	0.0006	3.4	1.0
550	2082.0	0.0043	0.0010	0.0031	0.0007	3.5	1.1
575	2176.6	0.0047	0.0010	0.0033	0.0007	3.7	1.1
600	2271.2	0.0050	0.0011	0.0036	0.0008	3.8	1.2
625	2365.9	0.0054	0.0012	0.0039	0.0009	4.0	1.2
650	2460.5	0.0058	0.0013	0.0042	0.0009	4.2	1.3
675	2555.1	0.0063	0.0014	0.0045	0.0010	4.3	1.3
700	2649.8	0.0067	0.0015	0.0048	0.0011	4.5	1.4
725	2744.4	0.0071	0.0016	0.0051	0.0011	4.6	1.4
750	2839.1	0.0076	0.0017	0.0054	0.0012	4.8	1.5
775	2933.7	0.0081	0.0018	0.0058	0.0013	5.0	1.5
800	3028.3	0.0086	0.0019	0.0061	0.0014	5.1	1.6
825	3123.0	0.0091	0.0020	0.0065	0.0015	5.3	1.6
850	3217.6	0.0096	0.0022	0.0068	0.0015	5.4	1.7
875	3312.2	0.0101	0.0023	0.0072	0.0016	5.6	1.7
900	3406.9	0.0107	0.0024	0.0076	0.0017	5.8	1.8
925	3501.5	0.0112	0.0025	0.0080	0.0018	5.9	1.8
950	3596.1	0.0118	0.0026	0.0084	0.0019	6.1	1.9
975	3690.8	0.0124	0.0028	0.0088	0.0020	6.2	1.9
1000	3785.4	0.0129	0.0029	0.0092	0.0021	6.4	2.0
1025	3880.0	0.0136	0.0030	0.0097	0.0022	6.6	2.0
1050	3974.7	0.0142	0.0032	0.0101	0.0023	6.7	2.0
1075	4069.3	0.0148	0.0033	0.0106	0.0024	6.9	2.1
1100	4163.9	0.0154	0.0035	0.0110	0.0025	7.0	2.1
1125	4258.6	0.0161	0.0036	0.0115	0.0026	7.2	2.2
1150	4353.2	0.0168	0.0038	0.0120	0.0027	7.4	2.2
1175	4447.8	0.0174	0.0039	0.0125	0.0028	7.5	2.3
1200	4542.5	0.0181	0.0041	0.0129	0.0029	7.7	2.3
1225	4637.1	0.0188	0.0042	0.0134	0.0030	7.8	2.4
1250	4731.8	0.0196	0.0044	0.0140	0.0031	8.0	2.4
1275	4826.4	0.0203	0.0046	0.0145	0.0033	8.2	2.5
1300	4921.0	0.0210	0.0047	0.0150	0.0034	8.3	2.5
1325	5015.7	0.0218	0.0049	0.0156	0.0035	8.5	2.6
1350	5110.3	0.0226	0.0051	0.0161	0.0036	8.6	2.6

(continued)

Table B.9 (continued)

Q		P_f $C = 100$		P_f $C = 120$		Velocity	
GPM	L/min	psi/ft	bar/m	psi/ft	bar/m	fps	m/s
1375	5204.9	0.0233	0.0052	0.0167	0.0037	8.8	2.7
1400	5299.6	0.0241	0.0054	0.0172	0.0039	9.0	2.7
1425	5394.2	0.0249	0.0056	0.0178	0.0040	9.1	2.8
1450	5488.8	0.0257	0.0058	0.0184	0.0041	9.3	2.8
1475	5583.5	0.0266	0.0060	0.0190	0.0043	9.4	2.9
1500	5678.1	0.0274	0.0062	0.0196	0.0044	9.6	2.9
1525	5772.7	0.0283	0.0064	0.0202	0.0045	9.8	3.0
1550	5867.4	0.0291	0.0065	0.0208	0.0047	9.9	3.0
1575	5962.0	0.0300	0.0067	0.0214	0.0048	10.1	3.1
1600	6056.6	0.0309	0.0069	0.0220	0.0050	10.2	3.1
1625	6151.3	0.0318	0.0071	0.0227	0.0051	10.4	3.2
1650	6245.9	0.0327	0.0073	0.0233	0.0052	10.6	3.2
1675	6340.5	0.0336	0.0076	0.0240	0.0054	10.7	3.3
1700	6435.2	0.0346	0.0078	0.0247	0.0055	10.9	3.3
1725	6529.8	0.0355	0.0080	0.0253	0.0057	11.0	3.4
1750	6624.5	0.0365	0.0082	0.0260	0.0058	11.2	3.4
1775	6719.1	0.0374	0.0084	0.0267	0.0060	11.4	3.5
1800	6813.7	0.0384	0.0086	0.0274	0.0062	11.5	3.5
1825	6908.4	0.0394	0.0089	0.0281	0.0063	11.7	3.6
1850	7003.0	0.0404	0.0091	0.0288	0.0065	11.8	3.6
1875	7097.6	0.0414	0.0093	0.0296	0.0066	12.0	3.7
1900	7192.3	0.0424	0.0095	0.0303	0.0068	12.2	3.7
1925	7286.9	0.0435	0.0098	0.0310	0.0070	12.3	3.8
1950	7381.5	0.0445	0.0100	0.0318	0.0071	12.5	3.8
1975	7476.2	0.0456	0.0102	0.0325	0.0073	12.6	3.9
2000	7570.8	0.0467	0.0105	0.0333	0.0075	12.8	3.9
2025	7665.4	0.0478	0.0107	0.0341	0.0077	13.0	4.0
2050	7760.1	0.0488	0.0110	0.0349	0.0078	13.1	4.0
2075	7854.7	0.0500	0.0112	0.0357	0.0080	13.3	4.0
2100	7949.3	0.0511	0.0115	0.0365	0.0082	13.4	4.1
2125	8044.0	0.0522	0.0117	0.0373	0.0084	13.6	4.1
2150	8138.6	0.0533	0.0120	0.0381	0.0086	13.8	4.2
2175	8233.2	0.0545	0.0122	0.0389	0.0087	13.9	4.2
2200	8327.9	0.0557	0.0125	0.0397	0.0089	14.1	4.3

(continued)

Table B.9 (continued)

Q		P_f $C = 100$		P_f $C = 120$		Velocity	
GPM	L/min	psi/ft	bar/m	psi/ft	bar/m	fps	m/s
2225	8422.5	0.0568	0.0128	0.0406	0.0091	14.2	4.3
2250	8517.2	0.0580	0.0130	0.0414	0.0093	14.4	4.4
2275	8611.8	0.0592	0.0133	0.0423	0.0095	14.6	4.4
2300	8706.4	0.0604	0.0136	0.0431	0.0097	14.7	4.5
2325	8801.1	0.0617	0.0139	0.0440	0.0099	14.9	4.5
2350	8895.7	0.0629	0.0141	0.0449	0.0101	15.0	4.6
2375	8990.3	0.0641	0.0144	0.0458	0.0103	15.2	4.6
2400	9085.0	0.0654	0.0147	0.0467	0.0105	15.4	4.7
2425	9179.6	0.0667	0.0150	0.0476	0.0107	15.5	4.7
2450	9274.2	0.0679	0.0153	0.0485	0.0109	15.7	4.8
2475	9368.9	0.0692	0.0156	0.0494	0.0111	15.8	4.8
2500	9463.5	0.0705	0.0158	0.0503	0.0113	16.0	4.9
2525	9558.1	0.0718	0.0161	0.0513	0.0115	16.2	4.9
2550	9652.8	0.0732	0.0164	0.0522	0.0117	16.3	5.0
2575	9747.4	0.0745	0.0167	0.0532	0.0119	16.5	5.0
2600	9842.0	0.0758	0.0170	0.0541	0.0122	16.6	5.1
2625	9936.7	0.0772	0.0173	0.0551	0.0124	16.8	5.1
2650	10,031.3	0.0785	0.0177	0.0561	0.0126	17.0	5.2
2675	10,125.9	0.0799	0.0180	0.0570	0.0128	17.1	5.2
2700	10,220.6	0.0813	0.0183	0.0580	0.0130	17.3	5.3
2725	10,315.2	0.0827	0.0186	0.0590	0.0133	17.4	5.3
2750	10,409.9	0.0841	0.0189	0.0600	0.0135	17.6	5.4

Appendix C
Friction Loss Tables for Schedule 10 Steel Pipe

P. Brock and V. Charter, *Fire Protection Hydraulics and Water Supply Analysis*, The Society of Fire Protection Engineers Series, https://doi.org/10.1007/978-3-032-09877-1

Table C.1 Water flowing in 1-in. (25 mm) schedule 10 steel pipe I.D. = 1.097 in. (27.9 mm)

Q		P_f $C = 100$		P_f $C = 120$		Velocity	
GPM	L/min	psi/ft	bar/m	psi/ft	bar/m	fps	m/s
10	37.9	0.041	0.009	0.029	0.007	3.4	1.0
11	41.6	0.049	0.011	0.035	0.008	3.7	1.1
12	45.4	0.057	0.013	0.041	0.009	4.1	1.2
13	49.2	0.066	0.015	0.047	0.011	4.4	1.3
14	53.0	0.076	0.017	0.054	0.012	4.7	1.4
15	56.8	0.086	0.019	0.061	0.014	5.1	1.5
16	60.6	0.097	0.022	0.069	0.016	5.4	1.7
17	64.4	0.109	0.024	0.077	0.017	5.8	1.8
18	68.1	0.121	0.027	0.086	0.019	6.1	1.9
19	71.9	0.133	0.030	0.095	0.021	6.4	2.0
20	75.7	0.147	0.033	0.105	0.024	6.8	2.1
21	79.5	0.160	0.036	0.115	0.026	7.1	2.2
22	83.3	0.175	0.039	0.125	0.028	7.5	2.3
23	87.1	0.190	0.043	0.136	0.030	7.8	2.4
24	90.8	0.205	0.046	0.147	0.033	8.1	2.5
25	94.6	0.222	0.050	0.158	0.036	8.5	2.6
26	98.4	0.238	0.054	0.170	0.038	8.8	2.7
27	102.2	0.255	0.057	0.182	0.041	9.2	2.8
28	106.0	0.273	0.061	0.195	0.044	9.5	2.9
29	109.8	0.292	0.066	0.208	0.047	9.8	3.0
30	113.6	0.310	0.070	0.222	0.050	10.2	3.1
31	117.3	0.330	0.074	0.235	0.053	10.5	3.2
32	121.1	0.350	0.079	0.250	0.056	10.8	3.3
33	124.9	0.370	0.083	0.264	0.059	11.2	3.4
34	128.7	0.391	0.088	0.279	0.063	11.5	3.5
35	132.5	0.413	0.093	0.295	0.066	11.9	3.6
36	136.3	0.435	0.098	0.310	0.070	12.2	3.7
37	140.1	0.458	0.103	0.327	0.073	12.5	3.8
38	143.8	0.481	0.108	0.343	0.077	12.9	3.9
39	147.6	0.504	0.113	0.360	0.081	13.2	4.0
40	151.4	0.529	0.119	0.377	0.085	13.6	4.1
41	155.2	0.553	0.124	0.395	0.089	13.9	4.2
42	159.0	0.579	0.130	0.413	0.093	14.2	4.3
43	162.8	0.604	0.136	0.431	0.097	14.6	4.4
44	166.6	0.631	0.142	0.450	0.101	14.9	4.5

(continued)

Table C.1 (continued)

Q		P_f $C = 100$		P_f $C = 120$		Velocity	
GPM	L/min	psi/ft	bar/m	psi/ft	bar/m	fps	m/s
45	170.3	0.657	0.148	0.469	0.105	15.3	4.6
46	174.1	0.685	0.154	0.489	0.110	15.6	4.8
47	177.9	0.712	0.160	0.508	0.114	15.9	4.9
48	181.7	0.741	0.167	0.529	0.119	16.3	5.0
49	185.5	0.769	0.173	0.549	0.123	16.6	5.1
50	189.3	0.799	0.180	0.570	0.128	17.0	5.2
51	193.1	0.829	0.186	0.591	0.133	17.3	5.3
52	196.8	0.859	0.193	0.613	0.138	17.6	5.4
53	200.6	0.890	0.200	0.635	0.143	18.0	5.5
54	204.4	0.921	0.207	0.657	0.148	18.3	5.6
55	208.2	0.953	0.214	0.680	0.153	18.6	5.7
56	212.0	0.985	0.221	0.703	0.158	19.0	5.8
57	215.8	1.018	0.229	0.726	0.163	19.3	5.9
58	219.6	1.051	0.236	0.750	0.169	19.7	6.0
59	223.3	1.085	0.244	0.774	0.174	20.0	6.1
60	227.1	1.119	0.252	0.799	0.180	20.3	6.2
61	230.9	1.154	0.259	0.824	0.185	20.7	6.3
62	234.7	1.189	0.267	0.849	0.191	21.0	6.4
63	238.5	1.225	0.275	0.874	0.197	21.4	6.5
64	242.3	1.261	0.284	0.900	0.202	21.7	6.6
65	246.1	1.298	0.292	0.926	0.208	22.0	6.7
66	249.8	1.335	0.300	0.953	0.214	22.4	6.8
67	253.6	1.373	0.309	0.980	0.220	22.7	6.9
68	257.4	1.411	0.317	1.007	0.226	23.1	7.0
69	261.2	1.449	0.326	1.034	0.233	23.4	7.1
70	265.0	1.489	0.335	1.062	0.239	23.7	7.2
71	268.8	1.528	0.344	1.091	0.245	24.1	7.3
72	272.5	1.568	0.353	1.119	0.252	24.4	7.4
73	276.3	1.609	0.362	1.148	0.258	24.7	7.5
74	280.1	1.650	0.371	1.177	0.265	25.1	7.6
75	283.9	1.691	0.380	1.207	0.271	25.4	7.7
76	287.7	1.733	0.390	1.237	0.278	25.8	7.9
77	291.5	1.776	0.399	1.267	0.285	26.1	8.0
78	295.3	1.818	0.409	1.298	0.292	26.4	8.1

(continued)

Table C.1 (continued)

Q		P_f $C = 100$		P_f $C = 120$		Velocity	
GPM	L/min	psi/ft	bar/m	psi/ft	bar/m	fps	m/s
79	299.0	1.862	0.419	1.329	0.299	26.8	8.2
80	302.8	1.906	0.428	1.360	0.306	27.1	8.3
81	306.6	1.950	0.438	1.392	0.313	27.5	8.4
82	310.4	1.995	0.448	1.424	0.320	27.8	8.5
83	314.2	2.040	0.459	1.456	0.327	28.1	8.6
84	318.0	2.086	0.469	1.489	0.335	28.5	8.7
85	321.8	2.132	0.479	1.521	0.342	28.8	8.8
86	325.5	2.178	0.490	1.555	0.350	29.2	8.9
87	329.3	2.226	0.500	1.588	0.357	29.5	9.0
88	333.1	2.273	0.511	1.622	0.365	29.8	9.1
89	336.9	2.321	0.522	1.657	0.372	30.2	9.2
90	340.7	2.370	0.533	1.691	0.380	30.5	9.3
91	344.5	2.419	0.544	1.726	0.388	30.8	9.4
92	348.3	2.468	0.555	1.761	0.396	31.2	9.5
93	352.0	2.518	0.566	1.797	0.404	31.5	9.6
94	355.8	2.568	0.577	1.833	0.412	31.9	9.7
95	359.6	2.619	0.589	1.869	0.420	32.2	9.8
96	363.4	2.670	0.600	1.906	0.428	32.5	9.9
97	367.2	2.722	0.612	1.943	0.437	32.9	10.0
98	371.0	2.774	0.624	1.980	0.445	33.2	10.1
99	374.8	2.827	0.635	2.017	0.454	33.6	10.2
100	378.5	2.880	0.647	2.055	0.462	33.9	10.3

Table C.2 Water flowing in 1¼-in. (32 mm) schedule 10 steel pipe I.D. = 1.442 in. (36.6 mm)

Q		P_f $C = 100$		P_f $C = 120$		Velocity	
GPM	L/min	psi/ft	bar/m	psi/ft	bar/m	fps	m/s
10	37.9	0.011	0.002	0.008	0.002	2.0	0.6
11	41.6	0.013	0.003	0.009	0.002	2.2	0.7
12	45.4	0.015	0.003	0.011	0.002	2.4	0.7
13	49.2	0.017	0.004	0.012	0.003	2.5	0.8
14	53.0	0.020	0.005	0.014	0.003	2.7	0.8
15	56.8	0.023	0.005	0.016	0.004	2.9	0.9
16	60.6	0.026	0.006	0.018	0.004	3.1	1.0
17	64.4	0.029	0.007	0.020	0.005	3.3	1.0
18	68.1	0.032	0.007	0.023	0.005	3.5	1.1
19	71.9	0.035	0.008	0.025	0.006	3.7	1.1
20	75.7	0.039	0.009	0.028	0.006	3.9	1.2
21	79.5	0.042	0.010	0.030	0.007	4.1	1.3
22	83.3	0.046	0.010	0.033	0.007	4.3	1.3
23	87.1	0.050	0.011	0.036	0.008	4.5	1.4
24	90.8	0.054	0.012	0.039	0.009	4.7	1.4
25	94.6	0.059	0.013	0.042	0.009	4.9	1.5
26	98.4	0.063	0.014	0.045	0.010	5.1	1.6
27	102.2	0.067	0.015	0.048	0.011	5.3	1.6
28	106.0	0.072	0.016	0.051	0.012	5.5	1.7
29	109.8	0.077	0.017	0.055	0.012	5.7	1.7
30	113.6	0.082	0.019	0.059	0.013	5.9	1.8
31	117.3	0.087	0.020	0.062	0.014	6.1	1.9
32	121.1	0.092	0.021	0.066	0.015	6.3	1.9
33	124.9	0.098	0.022	0.070	0.016	6.5	2.0
34	128.7	0.103	0.023	0.074	0.017	6.7	2.0
35	132.5	0.109	0.025	0.078	0.018	6.9	2.1
36	136.3	0.115	0.026	0.082	0.019	7.1	2.2
37	140.1	0.121	0.027	0.086	0.020	7.3	2.2
38	143.8	0.127	0.029	0.091	0.021	7.4	2.3
39	147.6	0.133	0.030	0.095	0.022	7.6	2.3
40	151.4	0.140	0.032	0.100	0.023	7.8	2.4
41	155.2	0.146	0.033	0.104	0.024	8.0	2.4
42	159.0	0.153	0.035	0.109	0.025	8.2	2.5
43	162.8	0.160	0.036	0.114	0.026	8.4	2.6
44	166.6	0.166	0.038	0.119	0.027	8.6	2.6

(continued)

Table C.2 (continued)

Q		P_f $C = 100$		P_f $C = 120$		Velocity	
GPM	L/min	psi/ft	bar/m	psi/ft	bar/m	fps	m/s
45	170.3	0.174	0.039	0.124	0.028	8.8	2.7
46	174.1	0.181	0.041	0.129	0.029	9.0	2.7
47	177.9	0.188	0.043	0.134	0.030	9.2	2.8
48	181.7	0.196	0.044	0.140	0.032	9.4	2.9
49	185.5	0.203	0.046	0.145	0.033	9.6	2.9
50	189.3	0.211	0.048	0.151	0.034	9.8	3.0
51	193.1	0.219	0.050	0.156	0.035	10.0	3.0
52	196.8	0.227	0.051	0.162	0.037	10.2	3.1
53	200.6	0.235	0.053	0.168	0.038	10.4	3.2
54	204.4	0.243	0.055	0.174	0.039	10.6	3.2
55	208.2	0.252	0.057	0.180	0.041	10.8	3.3
56	212.0	0.260	0.059	0.186	0.042	11.0	3.3
57	215.8	0.269	0.061	0.192	0.044	11.2	3.4
58	219.6	0.278	0.063	0.198	0.045	11.4	3.5
59	223.3	0.286	0.065	0.204	0.046	11.6	3.5
60	227.1	0.295	0.067	0.211	0.048	11.8	3.6
61	230.9	0.305	0.069	0.217	0.049	12.0	3.6
62	234.7	0.314	0.071	0.224	0.051	12.2	3.7
63	238.5	0.323	0.073	0.231	0.052	12.3	3.8
64	242.3	0.333	0.076	0.238	0.054	12.5	3.8
65	246.1	0.343	0.078	0.245	0.056	12.7	3.9
66	249.8	0.352	0.080	0.252	0.057	12.9	3.9
67	253.6	0.362	0.082	0.259	0.059	13.1	4.0
68	257.4	0.372	0.085	0.266	0.060	13.3	4.1
69	261.2	0.383	0.087	0.273	0.062	13.5	4.1
70	265.0	0.393	0.089	0.280	0.064	13.7	4.2
71	268.8	0.403	0.092	0.288	0.065	13.9	4.2
72	272.5	0.414	0.094	0.295	0.067	14.1	4.3
73	276.3	0.425	0.096	0.303	0.069	14.3	4.4
74	280.1	0.436	0.099	0.311	0.071	14.5	4.4
75	283.9	0.447	0.101	0.319	0.072	14.7	4.5
76	287.7	0.458	0.104	0.327	0.074	14.9	4.5
77	291.5	0.469	0.106	0.335	0.076	15.1	4.6
78	295.3	0.480	0.109	0.343	0.078	15.3	4.7

(continued)

Table C.2 (continued)

Q		P_f $C = 100$		P_f $C = 120$		Velocity	
GPM	L/min	psi/ft	bar/m	psi/ft	bar/m	fps	m/s
79	299.0	0.492	0.112	0.351	0.080	15.5	4.7
80	302.8	0.503	0.114	0.359	0.082	15.7	4.8
81	306.6	0.515	0.117	0.367	0.083	15.9	4.8
82	310.4	0.527	0.120	0.376	0.085	16.1	4.9
83	314.2	0.539	0.122	0.384	0.087	16.3	5.0
84	318.0	0.551	0.125	0.393	0.089	16.5	5.0
85	321.8	0.563	0.128	0.402	0.091	16.7	5.1
86	325.5	0.575	0.131	0.411	0.093	16.9	5.1
87	329.3	0.588	0.133	0.419	0.095	17.1	5.2
88	333.1	0.600	0.136	0.428	0.097	17.2	5.3
89	336.9	0.613	0.139	0.437	0.099	17.4	5.3
90	340.7	0.626	0.142	0.447	0.101	17.6	5.4
91	344.5	0.639	0.145	0.456	0.103	17.8	5.4
92	348.3	0.652	0.148	0.465	0.106	18.0	5.5
93	352.0	0.665	0.151	0.474	0.108	18.2	5.6
94	355.8	0.678	0.154	0.484	0.110	18.4	5.6
95	359.6	0.691	0.157	0.493	0.112	18.6	5.7
96	363.4	0.705	0.160	0.503	0.114	18.8	5.7
97	367.2	0.719	0.163	0.513	0.116	19.0	5.8
98	371.0	0.732	0.166	0.523	0.119	19.2	5.9
99	374.8	0.746	0.169	0.533	0.121	19.4	5.9
100	378.5	0.760	0.173	0.543	0.123	19.6	6.0

Table C.3 Water flowing in 1½-in. (40 mm) schedule 10 steel pipe I.D. = 1.682 in. (42.7 mm)

Q		P_f $C = 100$		P_f $C = 120$		Velocity	
GPM	L/min	psi/ft	bar/m	psi/ft	bar/m	fps	m/s
20	75.7	0.018	0.004	0.013	0.003	2.9	0.9
22	83.3	0.022	0.005	0.016	0.004	3.2	1.0
24	90.8	0.026	0.006	0.018	0.004	3.5	1.1
26	98.4	0.030	0.007	0.021	0.005	3.7	1.1
28	106.0	0.034	0.008	0.024	0.006	4.0	1.2
30	113.6	0.039	0.009	0.028	0.006	4.3	1.3
32	121.1	0.044	0.010	0.031	0.007	4.6	1.4
34	128.7	0.049	0.011	0.035	0.008	4.9	1.5
36	136.3	0.054	0.012	0.039	0.009	5.2	1.6
38	143.8	0.060	0.014	0.043	0.010	5.5	1.7
40	151.4	0.066	0.015	0.047	0.011	5.8	1.8
42	159.0	0.072	0.016	0.052	0.012	6.0	1.8
44	166.6	0.079	0.018	0.056	0.013	6.3	1.9
46	174.1	0.085	0.019	0.061	0.014	6.6	2.0
48	181.7	0.092	0.021	0.066	0.015	6.9	2.1
50	189.3	0.100	0.023	0.071	0.016	7.2	2.2
52	196.8	0.107	0.024	0.076	0.017	7.5	2.3
54	204.4	0.115	0.026	0.082	0.019	7.8	2.4
56	212.0	0.123	0.028	0.088	0.020	8.1	2.5
58	219.6	0.131	0.030	0.094	0.021	8.4	2.5
60	227.1	0.140	0.032	0.100	0.023	8.6	2.6
62	234.7	0.148	0.034	0.106	0.024	8.9	2.7
64	242.3	0.157	0.036	0.112	0.025	9.2	2.8
66	249.8	0.167	0.038	0.119	0.027	9.5	2.9
68	257.4	0.176	0.040	0.126	0.028	9.8	3.0
70	265.0	0.186	0.042	0.133	0.030	10.1	3.1
72	272.5	0.196	0.044	0.140	0.032	10.4	3.2
74	280.1	0.206	0.047	0.147	0.033	10.7	3.2
76	287.7	0.216	0.049	0.154	0.035	10.9	3.3
78	295.3	0.227	0.051	0.162	0.037	11.2	3.4
80	302.8	0.238	0.054	0.170	0.038	11.5	3.5
82	310.4	0.249	0.056	0.178	0.040	11.8	3.6
84	318.0	0.260	0.059	0.186	0.042	12.1	3.7
86	325.5	0.272	0.062	0.194	0.044	12.4	3.8
88	333.1	0.284	0.064	0.202	0.046	12.7	3.9

(continued)

Table C.3 (continued)

Q		P_f $C = 100$		P_f $C = 120$		Velocity	
GPM	L/min	psi/ft	bar/m	psi/ft	bar/m	fps	m/s
90	340.7	0.296	0.067	0.211	0.048	13.0	4.0
92	348.3	0.308	0.070	0.220	0.050	13.2	4.0
94	355.8	0.320	0.073	0.229	0.052	13.5	4.1
96	363.4	0.333	0.076	0.238	0.054	13.8	4.2
98	371.0	0.346	0.078	0.247	0.056	14.1	4.3
100	378.5	0.359	0.081	0.256	0.058	14.4	4.4
102	386.1	0.373	0.085	0.266	0.060	14.7	4.5
104	393.7	0.386	0.088	0.276	0.063	15.0	4.6
106	401.3	0.400	0.091	0.286	0.065	15.3	4.7
108	408.8	0.414	0.094	0.296	0.067	15.6	4.7
110	416.4	0.428	0.097	0.306	0.069	15.8	4.8
112	424.0	0.443	0.100	0.316	0.072	16.1	4.9
114	431.5	0.458	0.104	0.327	0.074	16.4	5.0
116	439.1	0.473	0.107	0.337	0.077	16.7	5.1
118	446.7	0.488	0.111	0.348	0.079	17.0	5.2
120	454.2	0.503	0.114	0.359	0.081	17.3	5.3
122	461.8	0.519	0.118	0.370	0.084	17.6	5.4
124	469.4	0.535	0.121	0.382	0.087	17.9	5.4
126	477.0	0.551	0.125	0.393	0.089	18.1	5.5
128	484.5	0.567	0.129	0.405	0.092	18.4	5.6
130	492.1	0.584	0.132	0.417	0.094	18.7	5.7
132	499.7	0.600	0.136	0.428	0.097	19.0	5.8
134	507.2	0.617	0.140	0.441	0.100	19.3	5.9
136	514.8	0.634	0.144	0.453	0.103	19.6	6.0
138	522.4	0.652	0.148	0.465	0.106	19.9	6.1
140	530.0	0.669	0.152	0.478	0.108	20.2	6.1
142	537.5	0.687	0.156	0.490	0.111	20.4	6.2
144	545.1	0.705	0.160	0.503	0.114	20.7	6.3
146	552.7	0.723	0.164	0.516	0.117	21.0	6.4
148	560.2	0.742	0.168	0.529	0.120	21.3	6.5
150	567.8	0.761	0.173	0.543	0.123	21.6	6.6
152	575.4	0.779	0.177	0.556	0.126	21.9	6.7
154	583.0	0.799	0.181	0.570	0.129	22.2	6.8
156	590.5	0.818	0.186	0.584	0.132	22.5	6.8

(continued)

Table C.3 (continued)

Q		P_f $C = 100$		P_f $C = 120$		Velocity	
GPM	L/min	psi/ft	bar/m	psi/ft	bar/m	fps	m/s
158	598.1	0.837	0.190	0.598	0.136	22.8	6.9
160	605.7	0.857	0.194	0.612	0.139	23.0	7.0
162	613.2	0.877	0.199	0.626	0.142	23.3	7.1
164	620.8	0.897	0.203	0.640	0.145	23.6	7.2
166	628.4	0.917	0.208	0.655	0.149	23.9	7.3
168	635.9	0.938	0.213	0.669	0.152	24.2	7.4
170	643.5	0.959	0.217	0.684	0.155	24.5	7.5
172	651.1	0.980	0.222	0.699	0.159	24.8	7.5
174	658.7	1.001	0.227	0.714	0.162	25.1	7.6
176	666.2	1.022	0.232	0.730	0.165	25.3	7.7
178	673.8	1.044	0.237	0.745	0.169	25.6	7.8
180	681.4	1.066	0.242	0.761	0.173	25.9	7.9
182	688.9	1.088	0.247	0.776	0.176	26.2	8.0
184	696.5	1.110	0.252	0.792	0.180	26.5	8.1
186	704.1	1.132	0.257	0.808	0.183	26.8	8.2
188	711.7	1.155	0.262	0.824	0.187	27.1	8.3
190	719.2	1.178	0.267	0.841	0.191	27.4	8.3
192	726.8	1.201	0.272	0.857	0.194	27.6	8.4
194	734.4	1.224	0.278	0.874	0.198	27.9	8.5
196	741.9	1.247	0.283	0.890	0.202	28.2	8.6
198	749.5	1.271	0.288	0.907	0.206	28.5	8.7
200	757.1	1.295	0.294	0.924	0.210	28.8	8.8

Table C.4 Water flowing in 2-in. (50 mm) schedule 10 steel pipe I.D. = 2.157 in. (54.8 mm)

Q		P_f $C = 100$		P_f $C = 120$		Velocity	
GPM	L/min	psi/ft	bar/m	psi/ft	bar/m	fps	m/s
40	151.4	0.020	0.004	0.014	0.003	3.5	1.1
42	159.0	0.021	0.005	0.015	0.003	3.7	1.1
44	166.6	0.023	0.005	0.017	0.004	3.9	1.2
46	174.1	0.025	0.006	0.018	0.004	4.0	1.2
48	181.7	0.028	0.006	0.020	0.004	4.2	1.3
50	189.3	0.030	0.007	0.021	0.005	4.4	1.3
52	196.8	0.032	0.007	0.023	0.005	4.6	1.4
54	204.4	0.034	0.008	0.024	0.006	4.8	1.4
56	212.0	0.037	0.008	0.026	0.006	4.9	1.5
58	219.6	0.039	0.009	0.028	0.006	5.1	1.6
60	227.1	0.042	0.009	0.030	0.007	5.3	1.6
62	234.7	0.044	0.010	0.032	0.007	5.5	1.7
64	242.3	0.047	0.011	0.033	0.008	5.6	1.7
66	249.8	0.050	0.011	0.035	0.008	5.8	1.8
68	257.4	0.052	0.012	0.037	0.008	6.0	1.8
70	265.0	0.055	0.012	0.039	0.009	6.2	1.9
72	272.5	0.058	0.013	0.042	0.009	6.3	1.9
74	280.1	0.061	0.014	0.044	0.010	6.5	2.0
76	287.7	0.064	0.015	0.046	0.010	6.7	2.0
78	295.3	0.068	0.015	0.048	0.011	6.9	2.1
80	302.8	0.071	0.016	0.051	0.011	7.0	2.1
82	310.4	0.074	0.017	0.053	0.012	7.2	2.2
84	318.0	0.077	0.018	0.055	0.012	7.4	2.3
86	325.5	0.081	0.018	0.058	0.013	7.6	2.3
88	333.1	0.084	0.019	0.060	0.014	7.7	2.4
90	340.7	0.088	0.020	0.063	0.014	7.9	2.4
92	348.3	0.092	0.021	0.065	0.015	8.1	2.5
94	355.8	0.095	0.022	0.068	0.015	8.3	2.5
96	363.4	0.099	0.022	0.071	0.016	8.4	2.6
98	371.0	0.103	0.023	0.074	0.017	8.6	2.6
100	378.5	0.107	0.024	0.076	0.017	8.8	2.7
102	386.1	0.111	0.025	0.079	0.018	9.0	2.7
104	393.7	0.115	0.026	0.082	0.019	9.2	2.8
106	401.3	0.119	0.027	0.085	0.019	9.3	2.8
108	408.8	0.123	0.028	0.088	0.020	9.5	2.9

(continued)

Table C.4 (continued)

Q		P_f $C = 100$		P_f $C = 120$		Velocity	
GPM	L/min	psi/ft	bar/m	psi/ft	bar/m	fps	m/s
110	416.4	0.128	0.029	0.091	0.021	9.7	3.0
112	424.0	0.132	0.030	0.094	0.021	9.9	3.0
114	431.5	0.136	0.031	0.097	0.022	10.0	3.1
116	439.1	0.141	0.032	0.100	0.023	10.2	3.1
118	446.7	0.145	0.033	0.104	0.023	10.4	3.2
120	454.2	0.150	0.034	0.107	0.024	10.6	3.2
122	461.8	0.155	0.035	0.110	0.025	10.7	3.3
124	469.4	0.159	0.036	0.114	0.026	10.9	3.3
126	477.0	0.164	0.037	0.117	0.026	11.1	3.4
128	484.5	0.169	0.038	0.121	0.027	11.3	3.4
130	492.1	0.174	0.039	0.124	0.028	11.4	3.5
132	499.7	0.179	0.040	0.128	0.029	11.6	3.5
134	507.2	0.184	0.042	0.131	0.030	11.8	3.6
136	514.8	0.189	0.043	0.135	0.030	12.0	3.6
138	522.4	0.194	0.044	0.139	0.031	12.1	3.7
140	530.0	0.199	0.045	0.142	0.032	12.3	3.8
142	537.5	0.205	0.046	0.146	0.033	12.5	3.8
144	545.1	0.210	0.047	0.150	0.034	12.7	3.9
146	552.7	0.215	0.049	0.154	0.035	12.8	3.9
148	560.2	0.221	0.050	0.158	0.036	13.0	4.0
150	567.8	0.226	0.051	0.162	0.037	13.2	4.0
152	575.4	0.232	0.052	0.166	0.037	13.4	4.1
154	583.0	0.238	0.054	0.170	0.038	13.6	4.1
156	590.5	0.244	0.055	0.174	0.039	13.7	4.2
158	598.1	0.249	0.056	0.178	0.040	13.9	4.2
160	605.7	0.255	0.058	0.182	0.041	14.1	4.3
162	613.2	0.261	0.059	0.186	0.042	14.3	4.3
164	620.8	0.267	0.060	0.191	0.043	14.4	4.4
166	628.4	0.273	0.062	0.195	0.044	14.6	4.5
168	635.9	0.279	0.063	0.199	0.045	14.8	4.5
170	643.5	0.286	0.065	0.204	0.046	15.0	4.6
172	651.1	0.292	0.066	0.208	0.047	15.1	4.6
174	658.7	0.298	0.067	0.213	0.048	15.3	4.7
176	666.2	0.304	0.069	0.217	0.049	15.5	4.7

(continued)

Table C.4 (continued)

Q		P_f $C = 100$		P_f $C = 120$		Velocity	
GPM	L/min	psi/ft	bar/m	psi/ft	bar/m	fps	m/s
178	673.8	0.311	0.070	0.222	0.050	15.7	4.8
180	681.4	0.317	0.072	0.226	0.051	15.8	4.8
182	688.9	0.324	0.073	0.231	0.052	16.0	4.9
184	696.5	0.331	0.075	0.236	0.053	16.2	4.9
186	704.1	0.337	0.076	0.241	0.054	16.4	5.0
188	711.7	0.344	0.078	0.245	0.055	16.5	5.0
190	719.2	0.351	0.079	0.250	0.057	16.7	5.1
192	726.8	0.358	0.081	0.255	0.058	16.9	5.1
194	734.4	0.365	0.082	0.260	0.059	17.1	5.2
196	741.9	0.372	0.084	0.265	0.060	17.2	5.3
198	749.5	0.379	0.086	0.270	0.061	17.4	5.3
200	757.1	0.386	0.087	0.275	0.062	17.6	5.4
202	764.7	0.393	0.089	0.280	0.063	17.8	5.4
204	772.2	0.400	0.090	0.286	0.065	18.0	5.5
206	779.8	0.407	0.092	0.291	0.066	18.1	5.5
208	787.4	0.415	0.094	0.296	0.067	18.3	5.6
210	794.9	0.422	0.095	0.301	0.068	18.5	5.6
212	802.5	0.430	0.097	0.307	0.069	18.7	5.7
214	810.1	0.437	0.099	0.312	0.070	18.8	5.7
216	817.6	0.445	0.100	0.317	0.072	19.0	5.8
218	825.2	0.452	0.102	0.323	0.073	19.2	5.8
220	832.8	0.460	0.104	0.328	0.074	19.4	5.9

Table C.5 Water flowing in 2½-in. (65 mm) schedule 10 steel pipe I.D. = 2.635 in. (66.9 mm)

Q		P_f $C = 100$		P_f $C = 120$		Velocity	
GPM	L/min	psi/ft	bar/m	psi/ft	bar/m	fps	m/s
60	227.1	0.016	0.0036	0.011	0.0025	3.5	1.1
62	234.7	0.017	0.0038	0.012	0.0027	3.7	1.1
64	242.3	0.018	0.0040	0.013	0.0029	3.8	1.2
66	249.8	0.019	0.0042	0.013	0.0030	3.9	1.2
68	257.4	0.020	0.0045	0.014	0.0032	4.0	1.2
70	265.0	0.021	0.0047	0.015	0.0034	4.1	1.3
72	272.5	0.022	0.0050	0.016	0.0036	4.2	1.3
74	280.1	0.023	0.0052	0.017	0.0037	4.4	1.3
76	287.7	0.024	0.0055	0.017	0.0039	4.5	1.4
78	295.3	0.025	0.0058	0.018	0.0041	4.6	1.4
80	302.8	0.027	0.0061	0.019	0.0043	4.7	1.4
82	310.4	0.028	0.0063	0.020	0.0045	4.8	1.5
84	318.0	0.029	0.0066	0.021	0.0047	5.0	1.5
86	325.5	0.031	0.0069	0.022	0.0049	5.1	1.5
88	333.1	0.032	0.0072	0.023	0.0052	5.2	1.6
90	340.7	0.033	0.0075	0.024	0.0054	5.3	1.6
92	348.3	0.035	0.0078	0.025	0.0056	5.4	1.7
94	355.8	0.036	0.0082	0.026	0.0058	5.5	1.7
96	363.4	0.037	0.0085	0.027	0.0061	5.7	1.7
98	371.0	0.039	0.0088	0.028	0.0063	5.8	1.8
100	378.5	0.040	0.0092	0.029	0.0065	5.9	1.8
102	386.1	0.042	0.0095	0.030	0.0068	6.0	1.8
104	393.7	0.043	0.0098	0.031	0.0070	6.1	1.9
106	401.3	0.045	0.0102	0.032	0.0073	6.3	1.9
108	408.8	0.047	0.0106	0.033	0.0075	6.4	1.9
110	416.4	0.048	0.0109	0.034	0.0078	6.5	2.0
112	424.0	0.050	0.0113	0.036	0.0081	6.6	2.0
114	431.5	0.051	0.0117	0.037	0.0083	6.7	2.1
116	439.1	0.053	0.0120	0.038	0.0086	6.8	2.1
118	446.7	0.055	0.0124	0.039	0.0089	7.0	2.1
120	454.2	0.057	0.0128	0.040	0.0092	7.1	2.2
122	461.8	0.058	0.0132	0.042	0.0094	7.2	2.2
124	469.4	0.060	0.0136	0.043	0.0097	7.3	2.2
126	477.0	0.062	0.0140	0.044	0.0100	7.4	2.3
128	484.5	0.064	0.0144	0.045	0.0103	7.6	2.3

(continued)

Table C.5 (continued)

Q		P_f $C = 100$		P_f $C = 120$		Velocity	
GPM	L/min	psi/ft	bar/m	psi/ft	bar/m	fps	m/s
130	492.1	0.066	0.0149	0.047	0.0106	7.7	2.3
132	499.7	0.067	0.0153	0.048	0.0109	7.8	2.4
134	507.2	0.069	0.0157	0.049	0.0112	7.9	2.4
136	514.8	0.071	0.0162	0.051	0.0115	8.0	2.4
138	522.4	0.073	0.0166	0.052	0.0119	8.1	2.5
140	530.0	0.075	0.0171	0.054	0.0122	8.3	2.5
142	537.5	0.077	0.0175	0.055	0.0125	8.4	2.6
144	545.1	0.079	0.0180	0.057	0.0128	8.5	2.6
146	552.7	0.081	0.0184	0.058	0.0132	8.6	2.6
148	560.2	0.083	0.0189	0.059	0.0135	8.7	2.7
150	567.8	0.085	0.0194	0.061	0.0138	8.9	2.7
152	575.4	0.088	0.0199	0.062	0.0142	9.0	2.7
154	583.0	0.090	0.0203	0.064	0.0145	9.1	2.8
156	590.5	0.092	0.0208	0.066	0.0149	9.2	2.8
158	598.1	0.094	0.0213	0.067	0.0152	9.3	2.8
160	605.7	0.096	0.0218	0.069	0.0156	9.4	2.9
162	613.2	0.099	0.0223	0.070	0.0159	9.6	2.9
164	620.8	0.101	0.0229	0.072	0.0163	9.7	2.9
166	628.4	0.103	0.0234	0.074	0.0167	9.8	3.0
168	635.9	0.105	0.0239	0.075	0.0171	9.9	3.0
170	643.5	0.108	0.0244	0.077	0.0174	10.0	3.1
172	651.1	0.110	0.0250	0.079	0.0178	10.1	3.1
174	658.7	0.112	0.0255	0.080	0.0182	10.3	3.1
176	666.2	0.115	0.0260	0.082	0.0186	10.4	3.2
178	673.8	0.117	0.0266	0.084	0.0190	10.5	3.2
180	681.4	0.120	0.0271	0.085	0.0194	10.6	3.2
182	688.9	0.122	0.0277	0.087	0.0198	10.7	3.3
184	696.5	0.125	0.0283	0.089	0.0202	10.9	3.3
186	704.1	0.127	0.0288	0.091	0.0206	11.0	3.3
188	711.7	0.130	0.0294	0.093	0.0210	11.1	3.4
190	719.2	0.132	0.0300	0.094	0.0214	11.2	3.4
192	726.8	0.135	0.0306	0.096	0.0218	11.3	3.5
194	734.4	0.138	0.0312	0.098	0.0223	11.4	3.5
196	741.9	0.140	0.0318	0.100	0.0227	11.6	3.5

(continued)

Table C.5 (continued)

Q		P_f $C = 100$		P_f $C = 120$		Velocity	
GPM	L/min	psi/ft	bar/m	psi/ft	bar/m	fps	m/s
198	749.5	0.143	0.0324	0.102	0.0231	11.7	3.6
200	757.1	0.145	0.0330	0.104	0.0235	11.8	3.6
202	764.7	0.148	0.0336	0.106	0.0240	11.9	3.6
204	772.2	0.151	0.0342	0.108	0.0244	12.0	3.7
206	779.8	0.154	0.0348	0.110	0.0249	12.2	3.7
208	787.4	0.156	0.0355	0.112	0.0253	12.3	3.7
210	794.9	0.159	0.0361	0.114	0.0258	12.4	3.8
212	802.5	0.162	0.0367	0.116	0.0262	12.5	3.8
214	810.1	0.165	0.0374	0.118	0.0267	12.6	3.8
216	817.6	0.168	0.0380	0.120	0.0271	12.7	3.9
218	825.2	0.171	0.0387	0.122	0.0276	12.9	3.9
220	832.8	0.174	0.0393	0.124	0.0281	13.0	4.0
222	840.4	0.176	0.0400	0.126	0.0286	13.1	4.0
224	847.9	0.179	0.0407	0.128	0.0290	13.2	4.0
226	855.5	0.182	0.0414	0.130	0.0295	13.3	4.1
228	863.1	0.185	0.0420	0.132	0.0300	13.5	4.1
230	870.6	0.188	0.0427	0.134	0.0305	13.6	4.1
232	878.2	0.191	0.0434	0.137	0.0310	13.7	4.2
234	885.8	0.195	0.0441	0.139	0.0315	13.8	4.2
236	893.4	0.198	0.0448	0.141	0.0320	13.9	4.2
238	900.9	0.201	0.0455	0.143	0.0325	14.0	4.3
240	908.5	0.204	0.0462	0.145	0.0330	14.2	4.3

Table C.6 Water flowing in 3-in. (80 mm) schedule 10 steel pipe I.D. = 3.26 in. (82.8 mm)

Q		P_f $C = 100$		P_f $C = 120$		Velocity	
GPM	L/min	psi/ft	bar/m	psi/ft	bar/m	fps	m/s
50	189.3	0.004	0.0009	0.003	0.0006	1.9	0.6
55	208.2	0.005	0.0011	0.003	0.0008	2.1	0.6
60	227.1	0.006	0.0013	0.004	0.0009	2.3	0.7
65	246.1	0.006	0.0015	0.005	0.0010	2.5	0.8
70	265.0	0.007	0.0017	0.005	0.0012	2.7	0.8
75	283.9	0.008	0.0019	0.006	0.0014	2.9	0.9
80	302.8	0.009	0.0021	0.007	0.0015	3.1	0.9
85	321.8	0.011	0.0024	0.008	0.0017	3.3	1.0
90	340.7	0.012	0.0027	0.008	0.0019	3.5	1.1
95	359.6	0.013	0.0029	0.009	0.0021	3.6	1.1
100	378.5	0.014	0.0032	0.010	0.0023	3.8	1.2
105	397.5	0.016	0.0035	0.011	0.0025	4.0	1.2
110	416.4	0.017	0.0039	0.012	0.0028	4.2	1.3
115	435.3	0.019	0.0042	0.013	0.0030	4.4	1.3
120	454.2	0.020	0.0045	0.014	0.0032	4.6	1.4
125	473.2	0.022	0.0049	0.015	0.0035	4.8	1.5
130	492.1	0.023	0.0053	0.017	0.0038	5.0	1.5
135	511.0	0.025	0.0056	0.018	0.0040	5.2	1.6
140	530.0	0.027	0.0060	0.019	0.0043	5.4	1.6
145	548.9	0.028	0.0064	0.020	0.0046	5.6	1.7
150	567.8	0.030	0.0069	0.022	0.0049	5.8	1.8
155	586.7	0.032	0.0073	0.023	0.0052	6.0	1.8
160	605.7	0.034	0.0077	0.024	0.0055	6.1	1.9
165	624.6	0.036	0.0082	0.026	0.0058	6.3	1.9
170	643.5	0.038	0.0086	0.027	0.0062	6.5	2.0
175	662.4	0.040	0.0091	0.029	0.0065	6.7	2.0
180	681.4	0.042	0.0096	0.030	0.0069	6.9	2.1
185	700.3	0.045	0.0101	0.032	0.0072	7.1	2.2
190	719.2	0.047	0.0106	0.033	0.0076	7.3	2.2
195	738.2	0.049	0.0111	0.035	0.0080	7.5	2.3
200	757.1	0.052	0.0117	0.037	0.0083	7.7	2.3
205	776.0	0.054	0.0122	0.039	0.0087	7.9	2.4
210	794.9	0.056	0.0128	0.040	0.0091	8.1	2.5
215	813.9	0.059	0.0133	0.042	0.0095	8.3	2.5
220	832.8	0.062	0.0139	0.044	0.0099	8.4	2.6

(continued)

Table C.6 (continued)

Q		P_f $C = 100$		P_f $C = 120$		Velocity	
GPM	L/min	psi/ft	bar/m	psi/ft	bar/m	fps	m/s
225	851.7	0.064	0.0145	0.046	0.0104	8.6	2.6
230	870.6	0.067	0.0151	0.048	0.0108	8.8	2.7
235	889.6	0.070	0.0157	0.050	0.0112	9.0	2.8
240	908.5	0.072	0.0164	0.052	0.0117	9.2	2.8
245	927.4	0.075	0.0170	0.054	0.0121	9.4	2.9
250	946.4	0.078	0.0176	0.056	0.0126	9.6	2.9
255	965.3	0.081	0.0183	0.058	0.0131	9.8	3.0
260	984.2	0.084	0.0190	0.060	0.0135	10.0	3.0
265	1003.1	0.087	0.0197	0.062	0.0140	10.2	3.1
270	1022.1	0.090	0.0203	0.064	0.0145	10.4	3.2
275	1041.0	0.093	0.0210	0.066	0.0150	10.6	3.2
280	1059.9	0.096	0.0218	0.069	0.0155	10.8	3.3
285	1078.8	0.099	0.0225	0.071	0.0160	10.9	3.3
290	1097.8	0.103	0.0232	0.073	0.0166	11.1	3.4
295	1116.7	0.106	0.0240	0.076	0.0171	11.3	3.5
300	1135.6	0.109	0.0247	0.078	0.0176	11.5	3.5
305	1154.5	0.113	0.0255	0.080	0.0182	11.7	3.6
310	1173.5	0.116	0.0263	0.083	0.0187	11.9	3.6
315	1192.4	0.120	0.0271	0.085	0.0193	12.1	3.7
320	1211.3	0.123	0.0279	0.088	0.0199	12.3	3.7
325	1230.3	0.127	0.0287	0.090	0.0205	12.5	3.8
330	1249.2	0.130	0.0295	0.093	0.0210	12.7	3.9
335	1268.1	0.134	0.0303	0.096	0.0216	12.9	3.9
340	1287.0	0.138	0.0312	0.098	0.0222	13.1	4.0
345	1306.0	0.141	0.0320	0.101	0.0229	13.2	4.0
350	1324.9	0.145	0.0329	0.104	0.0235	13.4	4.1
355	1343.8	0.149	0.0338	0.106	0.0241	13.6	4.2
360	1362.7	0.153	0.0346	0.109	0.0247	13.8	4.2
365	1381.7	0.157	0.0355	0.112	0.0254	14.0	4.3
370	1400.6	0.161	0.0364	0.115	0.0260	14.2	4.3
375	1419.5	0.165	0.0374	0.118	0.0267	14.4	4.4
380	1438.5	0.169	0.0383	0.121	0.0273	14.6	4.4
385	1457.4	0.173	0.0392	0.124	0.0280	14.8	4.5
390	1476.3	0.178	0.0402	0.127	0.0287	15.0	4.6

(continued)

Table C.6 (continued)

Q		P_f $C = 100$		P_f $C = 120$		Velocity	
GPM	L/min	psi/ft	bar/m	psi/ft	bar/m	fps	m/s
395	1495.2	0.182	0.0411	0.130	0.0294	15.2	4.6
400	1514.2	0.186	0.0421	0.133	0.0300	15.4	4.7
405	1533.1	0.190	0.0431	0.136	0.0307	15.6	4.7
410	1552.0	0.195	0.0441	0.139	0.0314	15.7	4.8
415	1570.9	0.199	0.0451	0.142	0.0322	15.9	4.9
420	1589.9	0.204	0.0461	0.145	0.0329	16.1	4.9
425	1608.8	0.208	0.0471	0.149	0.0336	16.3	5.0
430	1627.7	0.213	0.0481	0.152	0.0343	16.5	5.0
435	1646.6	0.217	0.0492	0.155	0.0351	16.7	5.1
440	1665.6	0.222	0.0502	0.158	0.0358	16.9	5.1
445	1684.5	0.227	0.0513	0.162	0.0366	17.1	5.2
450	1703.4	0.231	0.0523	0.165	0.0374	17.3	5.3
455	1722.4	0.236	0.0534	0.169	0.0381	17.5	5.3
460	1741.3	0.241	0.0545	0.172	0.0389	17.7	5.4
465	1760.2	0.246	0.0556	0.175	0.0397	17.9	5.4
470	1779.1	0.251	0.0567	0.179	0.0405	18.0	5.5
475	1798.1	0.256	0.0579	0.182	0.0413	18.2	5.6
480	1817.0	0.261	0.0590	0.186	0.0421	18.4	5.6
485	1835.9	0.266	0.0601	0.190	0.0429	18.6	5.7
490	1854.8	0.271	0.0613	0.193	0.0437	18.8	5.7
495	1873.8	0.276	0.0624	0.197	0.0446	19.0	5.8
500	1892.7	0.281	0.0636	0.201	0.0454	19.2	5.9

Table C.7 Water flowing in 4-in. (100 mm) schedule 10 steel pipe I.D. = 4.26 in. (108.2 mm)

Q		P_f $C = 100$		P_f $C = 120$		Velocity	
GPM	L/min	psi/ft	bar/m	psi/ft	bar/m	fps	m/s
100	378.5	0.004	0.0009	0.003	0.0006	2.3	0.7
110	416.4	0.005	0.0010	0.003	0.0007	2.5	0.8
120	454.2	0.005	0.0012	0.004	0.0009	2.7	0.8
130	492.1	0.006	0.0014	0.005	0.0010	2.9	0.9
140	530.0	0.007	0.0016	0.005	0.0012	3.2	1.0
150	567.8	0.008	0.0019	0.006	0.0013	3.4	1.0
160	605.7	0.009	0.0021	0.007	0.0015	3.6	1.1
170	643.5	0.010	0.0023	0.007	0.0017	3.8	1.2
180	681.4	0.012	0.0026	0.008	0.0019	4.1	1.2
190	719.2	0.013	0.0029	0.009	0.0021	4.3	1.3
200	757.1	0.014	0.0032	0.010	0.0023	4.5	1.4
210	794.9	0.015	0.0035	0.011	0.0025	4.7	1.4
220	832.8	0.017	0.0038	0.012	0.0027	5.0	1.5
230	870.6	0.018	0.0041	0.013	0.0029	5.2	1.6
240	908.5	0.020	0.0044	0.014	0.0032	5.4	1.6
250	946.4	0.021	0.0048	0.015	0.0034	5.6	1.7
260	984.2	0.023	0.0052	0.016	0.0037	5.9	1.8
270	1022.1	0.024	0.0055	0.017	0.0039	6.1	1.9
280	1059.9	0.026	0.0059	0.019	0.0042	6.3	1.9
290	1097.8	0.028	0.0063	0.020	0.0045	6.5	2.0
300	1135.6	0.030	0.0067	0.021	0.0048	6.8	2.1
310	1173.5	0.032	0.0071	0.023	0.0051	7.0	2.1
320	1211.3	0.033	0.0076	0.024	0.0054	7.2	2.2
330	1249.2	0.035	0.0080	0.025	0.0057	7.4	2.3
340	1287.0	0.037	0.0085	0.027	0.0060	7.7	2.3
350	1324.9	0.039	0.0089	0.028	0.0064	7.9	2.4
360	1362.7	0.042	0.0094	0.030	0.0067	8.1	2.5
370	1400.6	0.044	0.0099	0.031	0.0071	8.3	2.5
380	1438.5	0.046	0.0104	0.033	0.0074	8.6	2.6
390	1476.3	0.048	0.0109	0.034	0.0078	8.8	2.7
400	1514.2	0.051	0.0114	0.036	0.0082	9.0	2.7
410	1552.0	0.053	0.0120	0.038	0.0085	9.2	2.8
420	1589.9	0.055	0.0125	0.039	0.0089	9.5	2.9
430	1627.7	0.058	0.0131	0.041	0.0093	9.7	2.9
440	1665.6	0.060	0.0136	0.043	0.0097	9.9	3.0

(continued)

Table C.7 (continued)

Q		P_f $C = 100$		P_f $C = 120$		Velocity	
GPM	L/min	psi/ft	bar/m	psi/ft	bar/m	fps	m/s
450	1703.4	0.063	0.0142	0.045	0.0102	10.1	3.1
460	1741.3	0.065	0.0148	0.047	0.0106	10.4	3.2
470	1779.1	0.068	0.0154	0.049	0.0110	10.6	3.2
480	1817.0	0.071	0.0160	0.051	0.0114	10.8	3.3
490	1854.8	0.074	0.0167	0.053	0.0119	11.0	3.4
500	1892.7	0.076	0.0173	0.055	0.0123	11.3	3.4
510	1930.6	0.079	0.0179	0.057	0.0128	11.5	3.5
520	1968.4	0.082	0.0186	0.059	0.0133	11.7	3.6
530	2006.3	0.085	0.0193	0.061	0.0137	11.9	3.6
540	2044.1	0.088	0.0199	0.063	0.0142	12.2	3.7
550	2082.0	0.091	0.0206	0.065	0.0147	12.4	3.8
560	2119.8	0.094	0.0213	0.067	0.0152	12.6	3.8
570	2157.7	0.097	0.0220	0.069	0.0157	12.8	3.9
580	2195.5	0.101	0.0227	0.072	0.0162	13.1	4.0
590	2233.4	0.104	0.0235	0.074	0.0168	13.3	4.0
600	2271.2	0.107	0.0242	0.076	0.0173	13.5	4.1
610	2309.1	0.110	0.0250	0.079	0.0178	13.7	4.2
620	2346.9	0.114	0.0257	0.081	0.0184	14.0	4.3
630	2384.8	0.117	0.0265	0.084	0.0189	14.2	4.3
640	2422.7	0.121	0.0273	0.086	0.0195	14.4	4.4
650	2460.5	0.124	0.0281	0.089	0.0200	14.6	4.5
660	2498.4	0.128	0.0289	0.091	0.0206	14.9	4.5
670	2536.2	0.131	0.0297	0.094	0.0212	15.1	4.6
680	2574.1	0.135	0.0305	0.096	0.0218	15.3	4.7
690	2611.9	0.139	0.0314	0.099	0.0224	15.5	4.7
700	2649.8	0.142	0.0322	0.102	0.0230	15.8	4.8
710	2687.6	0.146	0.0331	0.104	0.0236	16.0	4.9
720	2725.5	0.150	0.0339	0.107	0.0242	16.2	4.9
730	2763.3	0.154	0.0348	0.110	0.0248	16.4	5.0
740	2801.2	0.158	0.0357	0.113	0.0255	16.7	5.1
750	2839.1	0.162	0.0366	0.115	0.0261	16.9	5.1
760	2876.9	0.166	0.0375	0.118	0.0268	17.1	5.2
770	2914.8	0.170	0.0384	0.121	0.0274	17.3	5.3
780	2952.6	0.174	0.0394	0.124	0.0281	17.6	5.3

(continued)

Table C.7 (continued)

Q		P_f $C = 100$		P_f $C = 120$		Velocity	
GPM	L/min	psi/ft	bar/m	psi/ft	bar/m	fps	m/s
790	2990.5	0.178	0.0403	0.127	0.0288	17.8	5.4
800	3028.3	0.182	0.0412	0.130	0.0294	18.0	5.5
810	3066.2	0.186	0.0422	0.133	0.0301	18.2	5.6
820	3104.0	0.191	0.0432	0.136	0.0308	18.5	5.6
830	3141.9	0.195	0.0441	0.139	0.0315	18.7	5.7
840	3179.7	0.199	0.0451	0.142	0.0322	18.9	5.8
850	3217.6	0.204	0.0461	0.145	0.0329	19.1	5.8
860	3255.4	0.208	0.0471	0.149	0.0336	19.4	5.9
870	3293.3	0.213	0.0482	0.152	0.0344	19.6	6.0
880	3331.2	0.217	0.0492	0.155	0.0351	19.8	6.0
890	3369.0	0.222	0.0502	0.158	0.0358	20.0	6.1
900	3406.9	0.227	0.0513	0.162	0.0366	20.3	6.2
910	3444.7	0.231	0.0523	0.165	0.0374	20.5	6.2
920	3482.6	0.236	0.0534	0.168	0.0381	20.7	6.3
930	3520.4	0.241	0.0545	0.172	0.0389	20.9	6.4
940	3558.3	0.246	0.0556	0.175	0.0397	21.2	6.4
950	3596.1	0.250	0.0567	0.179	0.0404	21.4	6.5
960	3634.0	0.255	0.0578	0.182	0.0412	21.6	6.6
970	3671.8	0.260	0.0589	0.186	0.0420	21.8	6.7
980	3709.7	0.265	0.0600	0.189	0.0428	22.1	6.7
990	3747.5	0.270	0.0612	0.193	0.0437	22.3	6.8
1000	3785.4	0.275	0.0623	0.197	0.0445	22.5	6.9

Table C.8 Water flowing in 6-in. (150 mm) schedule 10 steel pipe I.D. = 6.357 in. (161.5 mm)

Q		P_f $C = 100$		P_f $C = 120$		Velocity	
GPM	L/min	psi/ft	bar/m	psi/ft	bar/m	fps	m/s
300	1135.6	0.004	0.0010	0.003	0.0007	3.0	0.9
325	1230.3	0.005	0.0011	0.003	0.0008	3.3	1.0
350	1324.9	0.006	0.0013	0.004	0.0009	3.5	1.1
375	1419.5	0.006	0.0014	0.005	0.0010	3.8	1.2
400	1514.2	0.007	0.0016	0.005	0.0012	4.0	1.2
425	1608.8	0.008	0.0018	0.006	0.0013	4.3	1.3
450	1703.4	0.009	0.0020	0.006	0.0014	4.5	1.4
475	1798.1	0.010	0.0022	0.007	0.0016	4.8	1.5
500	1892.7	0.011	0.0025	0.008	0.0018	5.1	1.5
525	1987.3	0.012	0.0027	0.008	0.0019	5.3	1.6
550	2082.0	0.013	0.0029	0.009	0.0021	5.6	1.7
575	2176.6	0.014	0.0032	0.010	0.0023	5.8	1.8
600	2271.2	0.015	0.0034	0.011	0.0025	6.1	1.8
625	2365.9	0.016	0.0037	0.012	0.0027	6.3	1.9
650	2460.5	0.018	0.0040	0.013	0.0029	6.6	2.0
675	2555.1	0.019	0.0043	0.014	0.0031	6.8	2.1
700	2649.8	0.020	0.0046	0.014	0.0033	7.1	2.2
725	2744.4	0.022	0.0049	0.015	0.0035	7.3	2.2
750	2839.1	0.023	0.0052	0.016	0.0037	7.6	2.3
775	2933.7	0.024	0.0055	0.017	0.0039	7.8	2.4
800	3028.3	0.026	0.0059	0.019	0.0042	8.1	2.5
825	3123.0	0.027	0.0062	0.020	0.0044	8.3	2.5
850	3217.6	0.029	0.0066	0.021	0.0047	8.6	2.6
875	3312.2	0.031	0.0069	0.022	0.0049	8.8	2.7
900	3406.9	0.032	0.0073	0.023	0.0052	9.1	2.8
925	3501.5	0.034	0.0077	0.024	0.0055	9.3	2.8
950	3596.1	0.036	0.0081	0.025	0.0058	9.6	2.9
975	3690.8	0.037	0.0085	0.027	0.0060	9.8	3.0
1000	3785.4	0.039	0.0089	0.028	0.0063	10.1	3.1
1025	3880.0	0.041	0.0093	0.029	0.0066	10.4	3.2
1050	3974.7	0.043	0.0097	0.031	0.0069	10.6	3.2
1075	4069.3	0.045	0.0101	0.032	0.0072	10.9	3.3
1100	4163.9	0.047	0.0106	0.033	0.0075	11.1	3.4
1125	4258.6	0.049	0.0110	0.035	0.0079	11.4	3.5
1150	4353.2	0.051	0.0115	0.036	0.0082	11.6	3.5

(continued)

Table C.8 (continued)

Q		P_f $C = 100$		P_f $C = 120$		Velocity	
GPM	L/min	psi/ft	bar/m	psi/ft	bar/m	fps	m/s
1175	4447.8	0.053	0.0119	0.038	0.0085	11.9	3.6
1200	4542.5	0.055	0.0124	0.039	0.0089	12.1	3.7
1225	4637.1	0.057	0.0129	0.041	0.0092	12.4	3.8
1250	4731.8	0.059	0.0134	0.042	0.0096	12.6	3.8
1275	4826.4	0.061	0.0139	0.044	0.0099	12.9	3.9
1300	4921.0	0.064	0.0144	0.045	0.0103	13.1	4.0
1325	5015.7	0.066	0.0149	0.047	0.0106	13.4	4.1
1350	5110.3	0.068	0.0154	0.049	0.0110	13.6	4.2
1375	5204.9	0.071	0.0160	0.050	0.0114	13.9	4.2
1400	5299.6	0.073	0.0165	0.052	0.0118	14.1	4.3
1425	5394.2	0.075	0.0171	0.054	0.0122	14.4	4.4
1450	5488.8	0.078	0.0176	0.056	0.0126	14.6	4.5
1475	5583.5	0.080	0.0182	0.057	0.0130	14.9	4.5
1500	5678.1	0.083	0.0188	0.059	0.0134	15.2	4.6
1525	5772.7	0.086	0.0193	0.061	0.0138	15.4	4.7
1550	5867.4	0.088	0.0199	0.063	0.0142	15.7	4.8
1575	5962.0	0.091	0.0205	0.065	0.0147	15.9	4.8
1600	6056.6	0.094	0.0211	0.067	0.0151	16.2	4.9
1625	6151.3	0.096	0.0218	0.069	0.0155	16.4	5.0
1650	6245.9	0.099	0.0224	0.071	0.0160	16.7	5.1
1675	6340.5	0.102	0.0230	0.073	0.0164	16.9	5.2
1700	6435.2	0.105	0.0236	0.075	0.0169	17.2	5.2
1725	6529.8	0.107	0.0243	0.077	0.0173	17.4	5.3
1750	6624.5	0.110	0.0250	0.079	0.0178	17.7	5.4
1775	6719.1	0.113	0.0256	0.081	0.0183	17.9	5.5
1800	6813.7	0.116	0.0263	0.083	0.0188	18.2	5.5
1825	6908.4	0.119	0.0270	0.085	0.0192	18.4	5.6
1850	7003.0	0.122	0.0277	0.087	0.0197	18.7	5.7
1875	7097.6	0.125	0.0283	0.090	0.0202	18.9	5.8
1900	7192.3	0.129	0.0291	0.092	0.0207	19.2	5.8
1925	7286.9	0.132	0.0298	0.094	0.0212	19.4	5.9
1950	7381.5	0.135	0.0305	0.096	0.0218	19.7	6.0
1975	7476.2	0.138	0.0312	0.099	0.0223	19.9	6.1
2000	7570.8	0.141	0.0319	0.101	0.0228	20.2	6.2

(continued)

Table C.8 (continued)

Q		P_f $C = 100$		P_f $C = 120$		Velocity	
GPM	L/min	psi/ft	bar/m	psi/ft	bar/m	fps	m/s
2025	7665.4	0.145	0.0327	0.103	0.0233	20.5	6.2
2050	7760.1	0.148	0.0334	0.106	0.0239	20.7	6.3
2075	7854.7	0.151	0.0342	0.108	0.0244	21.0	6.4
2100	7949.3	0.155	0.0350	0.110	0.0250	21.2	6.5
2125	8044.0	0.158	0.0357	0.113	0.0255	21.5	6.5
2150	8138.6	0.162	0.0365	0.115	0.0261	21.7	6.6
2175	8233.2	0.165	0.0373	0.118	0.0266	22.0	6.7
2200	8327.9	0.169	0.0381	0.120	0.0272	22.2	6.8
2225	8422.5	0.172	0.0389	0.123	0.0278	22.5	6.8
2250	8517.2	0.176	0.0397	0.125	0.0283	22.7	6.9
2275	8611.8	0.179	0.0405	0.128	0.0289	23.0	7.0
2300	8706.4	0.183	0.0414	0.131	0.0295	23.2	7.1
2325	8801.1	0.187	0.0422	0.133	0.0301	23.5	7.2
2350	8895.7	0.190	0.0430	0.136	0.0307	23.7	7.2
2375	8990.3	0.194	0.0439	0.139	0.0313	24.0	7.3
2400	9085.0	0.198	0.0448	0.141	0.0319	24.2	7.4
2425	9179.6	0.202	0.0456	0.144	0.0326	24.5	7.5
2450	9274.2	0.206	0.0465	0.147	0.0332	24.7	7.5
2475	9368.9	0.210	0.0474	0.150	0.0338	25.0	7.6
2500	9463.5	0.214	0.0483	0.152	0.0344	25.3	7.7

Table C.9 Water flowing in 8-in. (200 mm) schedule 10 steel pipe I.D. = 8.249 in. (209.5 mm)

Q		P_f $C = 100$		P_f $C = 120$		Velocity	
GPM	L/min	psi/ft	bar/m	GPM	L/min	psi/ft	bar/m
500	1892.7	0.0031	0.0007	0.0022	0.0005	3.0	0.9
525	1987.3	0.0033	0.0008	0.0024	0.0005	3.2	1.0
550	2082.0	0.0036	0.0008	0.0026	0.0006	3.3	1.0
575	2176.6	0.0040	0.0009	0.0028	0.0006	3.5	1.1
600	2271.2	0.0043	0.0010	0.0031	0.0007	3.6	1.1
625	2365.9	0.0046	0.0010	0.0033	0.0007	3.8	1.1
650	2460.5	0.0050	0.0011	0.0035	0.0008	3.9	1.2
675	2555.1	0.0053	0.0012	0.0038	0.0009	4.1	1.2
700	2649.8	0.0057	0.0013	0.0041	0.0009	4.2	1.3
725	2744.4	0.0061	0.0014	0.0043	0.0010	4.4	1.3
750	2839.1	0.0065	0.0015	0.0046	0.0010	4.5	1.4
775	2933.7	0.0069	0.0016	0.0049	0.0011	4.7	1.4
800	3028.3	0.0073	0.0017	0.0052	0.0012	4.8	1.5
825	3123.0	0.0077	0.0017	0.0055	0.0012	5.0	1.5
850	3217.6	0.0082	0.0018	0.0058	0.0013	5.1	1.6
875	3312.2	0.0086	0.0019	0.0061	0.0014	5.3	1.6
900	3406.9	0.0091	0.0021	0.0065	0.0015	5.4	1.6
925	3501.5	0.0095	0.0022	0.0068	0.0015	5.6	1.7
950	3596.1	0.0100	0.0023	0.0072	0.0016	5.7	1.7
975	3690.8	0.0105	0.0024	0.0075	0.0017	5.9	1.8
1000	3785.4	0.0110	0.0025	0.0079	0.0018	6.0	1.8
1025	3880.0	0.0115	0.0026	0.0082	0.0019	6.2	1.9
1050	3974.7	0.0121	0.0027	0.0086	0.0019	6.3	1.9
1075	4069.3	0.0126	0.0029	0.0090	0.0020	6.5	2.0
1100	4163.9	0.0131	0.0030	0.0094	0.0021	6.6	2.0
1125	4258.6	0.0137	0.0031	0.0098	0.0022	6.8	2.1
1150	4353.2	0.0143	0.0032	0.0102	0.0023	6.9	2.1
1175	4447.8	0.0149	0.0034	0.0106	0.0024	7.1	2.1
1200	4542.5	0.0154	0.0035	0.0110	0.0025	7.2	2.2
1225	4637.1	0.0160	0.0036	0.0115	0.0026	7.4	2.2
1250	4731.8	0.0167	0.0038	0.0119	0.0027	7.5	2.3
1275	4826.4	0.0173	0.0039	0.0123	0.0028	7.7	2.3
1300	4921.0	0.0179	0.0041	0.0128	0.0029	7.8	2.4
1325	5015.7	0.0186	0.0042	0.0132	0.0030	8.0	2.4
1350	5110.3	0.0192	0.0043	0.0137	0.0031	8.1	2.5

(continued)

Table C.9 (continued)

Q		P_f $C = 100$		P_f $C = 120$		Velocity	
GPM	L/min	psi/ft	bar/m	GPM	L/min	psi/ft	bar/m
1375	5204.9	0.0199	0.0045	0.0142	0.0032	8.3	2.5
1400	5299.6	0.0205	0.0046	0.0147	0.0033	8.4	2.6
1425	5394.2	0.0212	0.0048	0.0151	0.0034	8.6	2.6
1450	5488.8	0.0219	0.0050	0.0156	0.0035	8.7	2.7
1475	5583.5	0.0226	0.0051	0.0161	0.0037	8.9	2.7
1500	5678.1	0.0233	0.0053	0.0167	0.0038	9.0	2.7
1525	5772.7	0.0241	0.0054	0.0172	0.0039	9.2	2.8
1550	5867.4	0.0248	0.0056	0.0177	0.0040	9.3	2.8
1575	5962.0	0.0255	0.0058	0.0182	0.0041	9.5	2.9
1600	6056.6	0.0263	0.0060	0.0188	0.0042	9.6	2.9
1625	6151.3	0.0271	0.0061	0.0193	0.0044	9.8	3.0
1650	6245.9	0.0278	0.0063	0.0199	0.0045	9.9	3.0
1675	6340.5	0.0286	0.0065	0.0204	0.0046	10.1	3.1
1700	6435.2	0.0294	0.0067	0.0210	0.0048	10.2	3.1
1725	6529.8	0.0302	0.0068	0.0216	0.0049	10.4	3.2
1750	6624.5	0.0310	0.0070	0.0222	0.0050	10.5	3.2
1775	6719.1	0.0319	0.0072	0.0227	0.0051	10.7	3.2
1800	6813.7	0.0327	0.0074	0.0233	0.0053	10.8	3.3
1825	6908.4	0.0335	0.0076	0.0239	0.0054	11.0	3.3
1850	7003.0	0.0344	0.0078	0.0245	0.0056	11.1	3.4
1875	7097.6	0.0353	0.0080	0.0252	0.0057	11.3	3.4
1900	7192.3	0.0361	0.0082	0.0258	0.0058	11.4	3.5
1925	7286.9	0.0370	0.0084	0.0264	0.0060	11.6	3.5
1950	7381.5	0.0379	0.0086	0.0271	0.0061	11.7	3.6
1975	7476.2	0.0388	0.0088	0.0277	0.0063	11.9	3.6
2000	7570.8	0.0397	0.0090	0.0284	0.0064	12.0	3.7
2025	7665.4	0.0407	0.0092	0.0290	0.0066	12.2	3.7
2050	7760.1	0.0416	0.0094	0.0297	0.0067	12.3	3.7
2075	7854.7	0.0425	0.0096	0.0304	0.0069	12.5	3.8
2100	7949.3	0.0435	0.0098	0.0310	0.0070	12.6	3.8
2125	8044.0	0.0445	0.0101	0.0317	0.0072	12.8	3.9
2150	8138.6	0.0454	0.0103	0.0324	0.0073	12.9	3.9
2175	8233.2	0.0464	0.0105	0.0331	0.0075	13.1	4.0
2200	8327.9	0.0474	0.0107	0.0338	0.0077	13.2	4.0

(continued)

Table C.9 (continued)

Q		P_f $C = 100$		P_f $C = 120$		Velocity	
GPM	L/min	psi/ft	bar/m	GPM	L/min	psi/ft	bar/m
2225	8422.5	0.0484	0.0110	0.0345	0.0078	13.4	4.1
2250	8517.2	0.0494	0.0112	0.0353	0.0080	13.5	4.1
2275	8611.8	0.0504	0.0114	0.0360	0.0081	13.7	4.2
2300	8706.4	0.0515	0.0116	0.0367	0.0083	13.8	4.2
2325	8801.1	0.0525	0.0119	0.0375	0.0085	14.0	4.3
2350	8895.7	0.0535	0.0121	0.0382	0.0087	14.1	4.3
2375	8990.3	0.0546	0.0124	0.0390	0.0088	14.3	4.3
2400	9085.0	0.0557	0.0126	0.0397	0.0090	14.4	4.4
2425	9179.6	0.0568	0.0128	0.0405	0.0092	14.6	4.4
2450	9274.2	0.0578	0.0131	0.0413	0.0093	14.7	4.5
2475	9368.9	0.0589	0.0133	0.0421	0.0095	14.9	4.5
2500	9463.5	0.0600	0.0136	0.0429	0.0097	15.0	4.6
2525	9558.1	0.0612	0.0138	0.0436	0.0099	15.2	4.6
2550	9652.8	0.0623	0.0141	0.0445	0.0101	15.3	4.7
2575	9747.4	0.0634	0.0144	0.0453	0.0102	15.5	4.7
2600	9842.0	0.0646	0.0146	0.0461	0.0104	15.6	4.8
2625	9936.7	0.0657	0.0149	0.0469	0.0106	15.8	4.8
2650	10,031.3	0.0669	0.0151	0.0477	0.0108	15.9	4.8
2675	10,125.9	0.0680	0.0154	0.0486	0.0110	16.1	4.9
2700	10,220.6	0.0692	0.0157	0.0494	0.0112	16.2	4.9
2725	10,315.2	0.0704	0.0159	0.0503	0.0114	16.4	5.0
2750	10,409.9	0.0716	0.0162	0.0511	0.0116	16.5	5.0

Appendix D
Friction Loss Tables for CPVC Plastic Pipe

P. Brock and V. Charter, *Fire Protection Hydraulics and Water Supply Analysis*, The Society of Fire Protection Engineers Series, https://doi.org/10.1007/978-3-032-09877-1

Table D.1 Water flowing in ¾-in. (20 mm) CPVC plastic pipe $C = 150$ I.D. = 0.884 in. (22.4 mm)

Q		P_f		Velocity		Q		P_f		Velocity	
GPM	L/min	psi/ft	bar/m	fps	m/s	GPM	L/min	psi/ft	bar/m	fps	m/s
10	37.9	0.055	0.013	5.2	1.6	56	212.0	1.331	0.305	29.2	8.9
11	41.6	0.066	0.015	5.7	1.8	57	215.8	1.376	0.315	29.8	9.1
12	45.4	0.077	0.018	6.3	1.9	58	219.6	1.421	0.325	30.3	9.2
13	49.2	0.089	0.020	6.8	2.1	59	223.3	1.466	0.336	30.8	9.4
14	53.0	0.102	0.023	7.3	2.2	60	227.1	1.513	0.346	31.3	9.5
15	56.8	0.116	0.027	7.8	2.4	61	230.9	1.560	0.357	31.8	9.7
16	60.6	0.131	0.030	8.4	2.5	62	234.7	1.607	0.368	32.4	9.9
17	64.4	0.147	0.034	8.9	2.7	63	238.5	1.655	0.379	32.9	10.0
18	68.1	0.163	0.037	9.4	2.9	64	242.3	1.704	0.390	33.4	10.2
19	71.9	0.180	0.041	9.9	3.0	65	246.1	1.754	0.401	33.9	10.3
20	75.7	0.198	0.045	10.4	3.2	66	249.8	1.804	0.413	34.5	10.5
21	79.5	0.217	0.050	11.0	3.3	67	253.6	1.855	0.425	35.0	10.7
22	83.3	0.236	0.054	11.5	3.5	68	257.4	1.907	0.436	35.5	10.8
23	87.1	0.257	0.059	12.0	3.7	69	261.2	1.959	0.448	36.0	11.0
24	90.8	0.278	0.064	12.5	3.8	70	265.0	2.012	0.460	36.5	11.1
25	94.6	0.299	0.069	13.1	4.0	71	268.8	2.065	0.473	37.1	11.3
26	98.4	0.322	0.074	13.6	4.1	72	272.5	2.119	0.485	37.6	11.5
27	102.2	0.345	0.079	14.1	4.3	73	276.3	2.174	0.498	38.1	11.6
28	106.0	0.369	0.085	14.6	4.5	74	280.1	2.230	0.510	38.6	11.8
29	109.8	0.394	0.090	15.1	4.6	75	283.9	2.286	0.523	39.2	11.9
30	113.6	0.420	0.096	15.7	4.8	76	287.7	2.342	0.536	39.7	12.1
31	117.3	0.446	0.102	16.2	4.9	77	291.5	2.400	0.549	40.2	12.3
32	121.1	0.473	0.108	16.7	5.1	78	295.3	2.458	0.563	40.7	12.4
33	124.9	0.500	0.115	17.2	5.3	79	299.0	2.516	0.576	41.2	12.6
34	128.7	0.529	0.121	17.7	5.4	80	302.8	2.576	0.590	41.8	12.7
35	132.5	0.558	0.128	18.3	5.6	81	306.6	2.635	0.603	42.3	12.9
36	136.3	0.588	0.135	18.8	5.7	82	310.4	2.696	0.617	42.8	13.0
37	140.1	0.618	0.142	19.3	5.9	83	314.2	2.757	0.631	43.3	13.2
38	143.8	0.650	0.149	19.8	6.0	84	318.0	2.819	0.645	43.8	13.4
39	147.6	0.682	0.156	20.4	6.2	85	321.8	2.881	0.659	44.4	13.5
40	151.4	0.714	0.164	20.9	6.4	86	325.5	2.944	0.674	44.9	13.7
41	155.2	0.748	0.171	21.4	6.5	87	329.3	3.008	0.688	45.4	13.8
42	159.0	0.782	0.179	21.9	6.7	88	333.1	3.072	0.703	45.9	14.0
43	162.8	0.817	0.187	22.4	6.8	89	336.9	3.137	0.718	46.5	14.2
44	166.6	0.852	0.195	23.0	7.0	90	340.7	3.203	0.733	47.0	14.3
45	170.3	0.888	0.203	23.5	7.2	91	344.5	3.269	0.748	47.5	14.5

(continued)

Table D.1 (continued)

Q		P_f		Velocity		Q		P_f		Velocity	
GPM	L/min	psi/ft	bar/m	fps	m/s	GPM	L/min	psi/ft	bar/m	fps	m/s
46	174.1	0.925	0.212	24.0	7.3	92	348.3	3.335	0.763	48.0	14.6
47	177.9	0.963	0.220	24.5	7.5	93	352.0	3.403	0.779	48.5	14.8
48	181.7	1.001	0.229	25.1	7.6	94	355.8	3.471	0.794	49.1	15.0
49	185.5	1.040	0.238	25.6	7.8	95	359.6	3.539	0.810	49.6	15.1
50	189.3	1.080	0.247	26.1	8.0	96	363.4	3.609	0.826	50.1	15.3
51	193.1	1.120	0.256	26.6	8.1	97	367.2	3.679	0.842	50.6	15.4
52	196.8	1.161	0.266	27.1	8.3	98	371.0	3.749	0.858	51.2	15.6
53	200.6	1.202	0.275	27.7	8.4	99	374.8	3.820	0.874	51.7	15.8
54	204.4	1.245	0.285	28.2	8.6	100	378.5	3.892	0.891	52.2	15.9
55	208.2	1.288	0.295	28.7	8.8	101	382.3	3.964	0.907	52.7	16.1

Table D.2 Water flowing in 1-in. (25 mm) CPVC plastic pipe $C = 150$ I.D. = 1.109 in. (28.2 mm)

Q		P_f		Velocity		Q		P_f		Velocity	
GPM	L/min	psi/ft	bar/m	fps	m/s	GPM	L/min	psi/ft	bar/m	fps	m/s
10	37.9	0.018	0.004	3.3	1.0	56	212.0	0.441	0.099	18.6	5.7
11	41.6	0.022	0.005	3.7	1.1	57	215.8	0.456	0.103	18.9	5.8
12	45.4	0.026	0.006	4.0	1.2	58	219.6	0.471	0.106	19.3	5.9
13	49.2	0.030	0.007	4.3	1.3	59	223.3	0.486	0.109	19.6	6.0
14	53.0	0.034	0.008	4.6	1.4	60	227.1	0.501	0.113	19.9	6.1
15	56.8	0.039	0.009	5.0	1.5	61	230.9	0.517	0.116	20.3	6.2
16	60.6	0.043	0.010	5.3	1.6	62	234.7	0.533	0.120	20.6	6.3
17	64.4	0.049	0.011	5.6	1.7	63	238.5	0.549	0.123	20.9	6.4
18	68.1	0.054	0.012	6.0	1.8	64	242.3	0.565	0.127	21.2	6.5
19	71.9	0.060	0.013	6.3	1.9	65	246.1	0.581	0.131	21.6	6.6
20	75.7	0.066	0.015	6.6	2.0	66	249.8	0.598	0.135	21.9	6.7
21	79.5	0.072	0.016	7.0	2.1	67	253.6	0.615	0.138	22.2	6.8
22	83.3	0.078	0.018	7.3	2.2	68	257.4	0.632	0.142	22.6	6.9
23	87.1	0.085	0.019	7.6	2.3	69	261.2	0.649	0.146	22.9	7.0
24	90.8	0.092	0.021	8.0	2.4	70	265.0	0.667	0.150	23.2	7.1
25	94.6	0.099	0.022	8.3	2.5	71	268.8	0.685	0.154	23.6	7.2
26	98.4	0.107	0.024	8.6	2.6	72	272.5	0.702	0.158	23.9	7.3
27	102.2	0.114	0.026	9.0	2.7	73	276.3	0.721	0.162	24.2	7.4
28	106.0	0.122	0.028	9.3	2.8	74	280.1	0.739	0.166	24.6	7.5
29	109.8	0.131	0.029	9.6	2.9	75	283.9	0.758	0.170	24.9	7.6
30	113.6	0.139	0.031	10.0	3.0	76	287.7	0.776	0.175	25.2	7.7
31	117.3	0.148	0.033	10.3	3.1	77	291.5	0.795	0.179	25.6	7.8
32	121.1	0.157	0.035	10.6	3.2	78	295.3	0.815	0.183	25.9	7.9
33	124.9	0.166	0.037	11.0	3.3	79	299.0	0.834	0.188	26.2	8.0
34	128.7	0.175	0.039	11.3	3.4	80	302.8	0.854	0.192	26.6	8.1
35	132.5	0.185	0.042	11.6	3.5	81	306.6	0.873	0.197	26.9	8.2
36	136.3	0.195	0.044	12.0	3.6	82	310.4	0.894	0.201	27.2	8.3
37	140.1	0.205	0.046	12.3	3.7	83	314.2	0.914	0.206	27.6	8.4
38	143.8	0.215	0.048	12.6	3.8	84	318.0	0.934	0.210	27.9	8.5
39	147.6	0.226	0.051	12.9	3.9	85	321.8	0.955	0.215	28.2	8.6
40	151.4	0.237	0.053	13.3	4.0	86	325.5	0.976	0.220	28.6	8.7
41	155.2	0.248	0.056	13.6	4.1	87	329.3	0.997	0.224	28.9	8.8
42	159.0	0.259	0.058	13.9	4.3	88	333.1	1.018	0.229	29.2	8.9
43	162.8	0.271	0.061	14.3	4.4	89	336.9	1.040	0.234	29.5	9.0
44	166.6	0.282	0.064	14.6	4.5	90	340.7	1.061	0.239	29.9	9.1
45	170.3	0.294	0.066	14.9	4.6	91	344.5	1.083	0.244	30.2	9.2

(continued)

Table D.2 (continued)

Q		P_f		Velocity		*Q*		P_f		Velocity	
GPM	L/min	psi/ft	bar/m	fps	m/s	GPM	L/min	psi/ft	bar/m	fps	m/s
46	174.1	0.307	0.069	15.3	4.7	92	348.3	1.106	0.249	30.5	9.3
47	177.9	0.319	0.072	15.6	4.8	93	352.0	1.128	0.254	30.9	9.4
48	181.7	0.332	0.075	15.9	4.9	94	355.8	1.150	0.259	31.2	9.5
49	185.5	0.345	0.078	16.3	5.0	95	359.6	1.173	0.264	31.5	9.6
50	189.3	0.358	0.081	16.6	5.1	96	363.4	1.196	0.269	31.9	9.7
51	193.1	0.371	0.084	16.9	5.2	97	367.2	1.219	0.274	32.2	9.8
52	196.8	0.385	0.087	17.3	5.3	98	371.0	1.243	0.280	32.5	9.9
53	200.6	0.399	0.090	17.6	5.4	99	374.8	1.266	0.285	32.9	10.0
54	204.4	0.413	0.093	17.9	5.5	100	378.5	1.290	0.290	33.2	10.1
55	208.2	0.427	0.096	18.3	5.6	101	382.3	1.314	0.296	33.5	10.2

Table D.3 Water flowing in 1¼-in. (32 mm) CPVC plastic pipe $C = 150$ I.D. = 1.40 in. (35.6 mm)

Q		P_f		Velocity		Q		P_f		Velocity	
GPM	L/min	psi/ft	bar/m	fps	m/s	GPM	L/min	psi/ft	bar/m	fps	m/s
20	75.7	0.021	0.005	4.2	1.3	112	424.0	0.511	0.115	23.3	7.1
22	83.3	0.025	0.006	4.6	1.4	114	431.5	0.528	0.119	23.7	7.2
24	90.8	0.030	0.007	5.0	1.5	116	439.1	0.546	0.123	24.2	7.4
26	98.4	0.034	0.008	5.4	1.7	118	446.7	0.563	0.127	24.6	7.5
28	106.0	0.039	0.009	5.8	1.8	120	454.2	0.581	0.131	25.0	7.6
30	113.6	0.045	0.010	6.2	1.9	122	461.8	0.599	0.135	25.4	7.7
32	121.1	0.050	0.011	6.7	2.0	124	469.4	0.617	0.139	25.8	7.9
34	128.7	0.056	0.013	7.1	2.2	126	477.0	0.636	0.143	26.2	8.0
36	136.3	0.063	0.014	7.5	2.3	128	484.5	0.655	0.147	26.7	8.1
38	143.8	0.069	0.016	7.9	2.4	130	492.1	0.674	0.152	27.1	8.3
40	151.4	0.076	0.017	8.3	2.5	132	499.7	0.693	0.156	27.5	8.4
42	159.0	0.083	0.019	8.7	2.7	134	507.2	0.713	0.160	27.9	8.5
44	166.6	0.091	0.020	9.2	2.8	136	514.8	0.732	0.165	28.3	8.6
46	174.1	0.099	0.022	9.6	2.9	138	522.4	0.752	0.169	28.7	8.8
48	181.7	0.107	0.024	10.0	3.0	140	530.0	0.773	0.174	29.2	8.9
50	189.3	0.115	0.026	10.4	3.2	142	537.5	0.793	0.179	29.6	9.0
52	196.8	0.124	0.028	10.8	3.3	144	545.1	0.814	0.183	30.0	9.1
54	204.4	0.133	0.030	11.2	3.4	146	552.7	0.835	0.188	30.4	9.3
56	212.0	0.142	0.032	11.7	3.6	148	560.2	0.856	0.193	30.8	9.4
58	219.6	0.151	0.034	12.1	3.7	150	567.8	0.878	0.198	31.2	9.5
60	227.1	0.161	0.036	12.5	3.8	152	575.4	0.900	0.202	31.7	9.7
62	234.7	0.171	0.039	12.9	3.9	154	583.0	0.922	0.207	32.1	9.8
64	242.3	0.182	0.041	13.3	4.1	156	590.5	0.944	0.212	32.5	9.9
66	249.8	0.192	0.043	13.7	4.2	158	598.1	0.967	0.217	32.9	10.0
68	257.4	0.203	0.046	14.2	4.3	160	605.7	0.989	0.223	33.3	10.2
70	265.0	0.214	0.048	14.6	4.4	162	613.2	1.012	0.228	33.7	10.3
72	272.5	0.226	0.051	15.0	4.6	164	620.8	1.036	0.233	34.2	10.4
74	280.1	0.238	0.053	15.4	4.7	166	628.4	1.059	0.238	34.6	10.5
76	287.7	0.250	0.056	15.8	4.8	168	635.9	1.083	0.244	35.0	10.7
78	295.3	0.262	0.059	16.2	5.0	170	643.5	1.107	0.249	35.4	10.8
80	302.8	0.274	0.062	16.7	5.1	172	651.1	1.131	0.254	35.8	10.9
82	310.4	0.287	0.065	17.1	5.2	174	658.7	1.155	0.260	36.2	11.0
84	318.0	0.300	0.068	17.5	5.3	176	666.2	1.180	0.266	36.7	11.2
86	325.5	0.314	0.071	17.9	5.5	178	673.8	1.205	0.271	37.1	11.3
88	333.1	0.327	0.074	18.3	5.6	180	681.4	1.230	0.277	37.5	11.4
90	340.7	0.341	0.077	18.7	5.7	182	688.9	1.256	0.283	37.9	11.6

(continued)

Table D.3 (continued)

Q		P_f		Velocity		Q		P_f		Velocity	
GPM	L/min	psi/ft	bar/m	fps	m/s	GPM	L/min	psi/ft	bar/m	fps	m/s
92	348.3	0.355	0.080	19.2	5.8	184	696.5	1.281	0.288	38.3	11.7
94	355.8	0.370	0.083	19.6	6.0	186	704.1	1.307	0.294	38.7	11.8
96	363.4	0.385	0.087	20.0	6.1	188	711.7	1.333	0.300	39.2	11.9
98	371.0	0.399	0.090	20.4	6.2	190	719.2	1.360	0.306	39.6	12.1
100	378.5	0.415	0.093	20.8	6.3	192	726.8	1.386	0.312	40.0	12.2
102	386.1	0.430	0.097	21.2	6.5	194	734.4	1.413	0.318	40.4	12.3
104	393.7	0.446	0.100	21.7	6.6	196	741.9	1.440	0.324	40.8	12.4
106	401.3	0.462	0.104	22.1	6.7	198	749.5	1.467	0.330	41.2	12.6
108	408.8	0.478	0.108	22.5	6.9	200	757.1	1.495	0.336	41.7	12.7
110	416.4	0.495	0.111	22.9	7.0	202	764.7	1.523	0.343	42.1	12.8

Table D.4 Water flowing in 1½-in. (40 mm) CPVC plastic pipe $C = 150$ I.D. = 1.602 in. (40.7 mm)

Q		P_f		Velocity		Q		P_f		Velocity	
GPM	L/min	psi/ft	bar/m	fps	m/s	GPM	L/min	psi/ft	bar/m	fps	m/s
20	75.7	0.011	0.0025	3.2	1.0	112	424.0	0.265	0.0600	17.8	5.4
22	83.3	0.013	0.0030	3.5	1.1	114	431.5	0.274	0.0620	18.1	5.5
24	90.8	0.015	0.0035	3.8	1.2	116	439.1	0.283	0.0640	18.4	5.6
26	98.4	0.018	0.0040	4.1	1.3	118	446.7	0.292	0.0660	18.8	5.7
28	106.0	0.020	0.0046	4.5	1.4	120	454.2	0.301	0.0681	19.1	5.8
30	113.6	0.023	0.0052	4.8	1.5	122	461.8	0.311	0.0702	19.4	5.9
32	121.1	0.026	0.0059	5.1	1.6	124	469.4	0.320	0.0724	19.7	6.0
34	128.7	0.029	0.0066	5.4	1.6	126	477.0	0.330	0.0746	20.0	6.1
36	136.3	0.032	0.0073	5.7	1.7	128	484.5	0.340	0.0768	20.4	6.2
38	143.8	0.036	0.0081	6.0	1.8	130	492.1	0.350	0.0790	20.7	6.3
40	151.4	0.039	0.0089	6.4	1.9	132	499.7	0.360	0.0813	21.0	6.4
42	159.0	0.043	0.0098	6.7	2.0	134	507.2	0.370	0.0835	21.3	6.5
44	166.6	0.047	0.0106	7.0	2.1	136	514.8	0.380	0.0859	21.6	6.6
46	174.1	0.051	0.0116	7.3	2.2	138	522.4	0.390	0.0882	21.9	6.7
48	181.7	0.055	0.0125	7.6	2.3	140	530.0	0.401	0.0906	22.3	6.8
50	189.3	0.060	0.0135	8.0	2.4	142	537.5	0.412	0.0930	22.6	6.9
52	196.8	0.064	0.0145	8.3	2.5	144	545.1	0.422	0.0954	22.9	7.0
54	204.4	0.069	0.0155	8.6	2.6	146	552.7	0.433	0.0979	23.2	7.1
56	212.0	0.074	0.0166	8.9	2.7	148	560.2	0.444	0.1004	23.5	7.2
58	219.6	0.079	0.0177	9.2	2.8	150	567.8	0.455	0.1029	23.9	7.3
60	227.1	0.084	0.0189	9.5	2.9	152	575.4	0.467	0.1055	24.2	7.4
62	234.7	0.089	0.0201	9.9	3.0	154	583.0	0.478	0.1081	24.5	7.5
64	242.3	0.094	0.0213	10.2	3.1	156	590.5	0.490	0.1107	24.8	7.6
66	249.8	0.100	0.0225	10.5	3.2	158	598.1	0.501	0.1133	25.1	7.7
68	257.4	0.105	0.0238	10.8	3.3	160	605.7	0.513	0.1160	25.4	7.8
70	265.0	0.111	0.0251	11.1	3.4	162	613.2	0.525	0.1187	25.8	7.9
72	272.5	0.117	0.0265	11.4	3.5	164	620.8	0.537	0.1214	26.1	7.9
74	280.1	0.123	0.0279	11.8	3.6	166	628.4	0.549	0.1242	26.4	8.0
76	287.7	0.129	0.0293	12.1	3.7	168	635.9	0.562	0.1269	26.7	8.1
78	295.3	0.136	0.0307	12.4	3.8	170	643.5	0.574	0.1297	27.0	8.2
80	302.8	0.142	0.0322	12.7	3.9	172	651.1	0.587	0.1326	27.3	8.3
82	310.4	0.149	0.0337	13.0	4.0	174	658.7	0.599	0.1355	27.7	8.4
84	318.0	0.156	0.0352	13.4	4.1	176	666.2	0.612	0.1383	28.0	8.5
86	325.5	0.163	0.0368	13.7	4.2	178	673.8	0.625	0.1413	28.3	8.6
88	333.1	0.170	0.0384	14.0	4.3	180	681.4	0.638	0.1442	28.6	8.7
90	340.7	0.177	0.0400	14.3	4.4	182	688.9	0.651	0.1472	28.9	8.8
92	348.3	0.184	0.0417	14.6	4.5	184	696.5	0.665	0.1502	29.3	8.9
94	355.8	0.192	0.0434	14.9	4.6	186	704.1	0.678	0.1532	29.6	9.0

(continued)

Table D.4 (continued)

Q		P_f		Velocity		Q		P_f		Velocity	
GPM	L/min	psi/ft	bar/m	fps	m/s	GPM	L/min	psi/ft	bar/m	fps	m/s
96	363.4	0.199	0.0451	15.3	4.7	188	711.7	0.692	0.1563	29.9	9.1
98	371.0	0.207	0.0468	15.6	4.7	190	719.2	0.705	0.1594	30.2	9.2
100	378.5	0.215	0.0486	15.9	4.8	192	726.8	0.719	0.1625	30.5	9.3
102	386.1	0.223	0.0504	16.2	4.9	194	734.4	0.733	0.1657	30.8	9.4
104	393.7	0.231	0.0523	16.5	5.0	196	741.9	0.747	0.1688	31.2	9.5
106	401.3	0.240	0.0541	16.9	5.1	198	749.5	0.761	0.1720	31.5	9.6
108	408.8	0.248	0.0561	17.2	5.2	200	757.1	0.775	0.1753	31.8	9.7
110	416.4	0.257	0.0580	17.5	5.3	202	764.7	0.790	0.1785	32.1	9.8

Table D.5 Water flowing in 2-in. (50 mm) CPVC plastic pipe $C = 150$ I.D. = 2.003 in. (50.9 mm)

Q		P_f		Velocity		Q		P_f		Velocity	
GPM	L/min	psi/ft	bar/m	fps	m/s	GPM	L/min	psi/ft	bar/m	fps	m/s
40	151.4	0.013	0.0030	4.1	1.2	132	499.7	0.121	0.0273	13.5	4.1
42	159.0	0.015	0.0033	4.3	1.3	134	507.2	0.125	0.0281	13.7	4.2
44	166.6	0.016	0.0036	4.5	1.4	136	514.8	0.128	0.0289	13.9	4.2
46	174.1	0.017	0.0039	4.7	1.4	138	522.4	0.132	0.0297	14.1	4.3
48	181.7	0.019	0.0042	4.9	1.5	140	530.0	0.135	0.0305	14.3	4.4
50	189.3	0.020	0.0045	5.1	1.6	142	537.5	0.139	0.0313	14.5	4.4
52	196.8	0.022	0.0049	5.3	1.6	144	545.1	0.142	0.0321	14.7	4.5
54	204.4	0.023	0.0052	5.5	1.7	146	552.7	0.146	0.0329	14.9	4.5
56	212.0	0.025	0.0056	5.7	1.7	148	560.2	0.150	0.0338	15.1	4.6
58	219.6	0.026	0.0060	5.9	1.8	150	567.8	0.153	0.0346	15.3	4.7
60	227.1	0.028	0.0064	6.1	1.9	152	575.4	0.157	0.0355	15.5	4.7
62	234.7	0.030	0.0068	6.3	1.9	154	583.0	0.161	0.0364	15.7	4.8
64	242.3	0.032	0.0072	6.5	2.0	156	590.5	0.165	0.0372	15.9	4.8
66	249.8	0.034	0.0076	6.7	2.1	158	598.1	0.169	0.0381	16.1	4.9
68	257.4	0.036	0.0080	6.9	2.1	160	605.7	0.173	0.0390	16.3	5.0
70	265.0	0.037	0.0085	7.1	2.2	162	613.2	0.177	0.0399	16.5	5.0
72	272.5	0.039	0.0089	7.3	2.2	164	620.8	0.181	0.0409	16.7	5.1
74	280.1	0.042	0.0094	7.5	2.3	166	628.4	0.185	0.0418	16.9	5.2
76	287.7	0.044	0.0098	7.8	2.4	168	635.9	0.189	0.0427	17.1	5.2
78	295.3	0.046	0.0103	8.0	2.4	170	643.5	0.193	0.0437	17.3	5.3
80	302.8	0.048	0.0108	8.2	2.5	172	651.1	0.198	0.0446	17.5	5.3
82	310.4	0.050	0.0113	8.4	2.5	174	658.7	0.202	0.0456	17.7	5.4
84	318.0	0.052	0.0118	8.6	2.6	176	666.2	0.206	0.0466	18.0	5.5
86	325.5	0.055	0.0124	8.8	2.7	178	673.8	0.211	0.0475	18.2	5.5
88	333.1	0.057	0.0129	9.0	2.7	180	681.4	0.215	0.0485	18.4	5.6
90	340.7	0.060	0.0135	9.2	2.8	182	688.9	0.219	0.0495	18.6	5.7
92	348.3	0.062	0.0140	9.4	2.9	184	696.5	0.224	0.0505	18.8	5.7
94	355.8	0.065	0.0146	9.6	2.9	186	704.1	0.228	0.0516	19.0	5.8
96	363.4	0.067	0.0152	9.8	3.0	188	711.7	0.233	0.0526	19.2	5.8
98	371.0	0.070	0.0158	10.0	3.0	190	719.2	0.238	0.0536	19.4	5.9
100	378.5	0.072	0.0164	10.2	3.1	192	726.8	0.242	0.0547	19.6	6.0
102	386.1	0.075	0.0170	10.4	3.2	194	734.4	0.247	0.0557	19.8	6.0
104	393.7	0.078	0.0176	10.6	3.2	196	741.9	0.252	0.0568	20.0	6.1
106	401.3	0.081	0.0182	10.8	3.3	198	749.5	0.256	0.0579	20.2	6.2
108	408.8	0.084	0.0189	11.0	3.4	200	757.1	0.261	0.0590	20.4	6.2
110	416.4	0.086	0.0195	11.2	3.4	202	764.7	0.266	0.0601	20.6	6.3

(continued)

Table D.5 (continued)

Q		P_f		Velocity		Q		P_f		Velocity	
GPM	L/min	psi/ft	bar/m	fps	m/s	GPM	L/min	psi/ft	bar/m	fps	m/s
112	424.0	0.089	0.0202	11.4	3.5	204	772.2	0.271	0.0612	20.8	6.3
114	431.5	0.092	0.0208	11.6	3.5	206	779.8	0.276	0.0623	21.0	6.4
116	439.1	0.095	0.0215	11.8	3.6	208	787.4	0.281	0.0634	21.2	6.5
118	446.7	0.098	0.0222	12.0	3.7	210	794.9	0.286	0.0645	21.4	6.5
120	454.2	0.102	0.0229	12.2	3.7	212	802.5	0.291	0.0657	21.6	6.6
122	461.8	0.105	0.0236	12.4	3.8	214	810.1	0.296	0.0668	21.8	6.7
124	469.4	0.108	0.0244	12.6	3.9	216	817.6	0.301	0.0680	22.0	6.7
126	477.0	0.111	0.0251	12.9	3.9	218	825.2	0.306	0.0692	22.2	6.8
128	484.5	0.114	0.0258	13.1	4.0	220	832.8	0.312	0.0703	22.4	6.8
130	492.1	0.118	0.0266	13.3	4.0	222	840.4	0.317	0.0715	22.6	6.9

Table D.6 Water flowing in 2½-in. (65 mm) CPVC plastic pipe $C = 150$ I.D. = 2.423 in. (61.5 mm)

Q		P_f		Velocity		Q		P_f		Velocity	
GPM	L/min	psi/ft	bar/m	fps	m/s	GPM	L/min	psi/ft	bar/m	fps	m/s
60	227.1	0.011	0.0025	4.2	1.3	152	575.4	0.062	0.0141	10.6	3.2
62	234.7	0.012	0.0027	4.3	1.3	154	583.0	0.064	0.0145	10.7	3.3
64	242.3	0.013	0.0029	4.4	1.4	156	590.5	0.065	0.0148	10.8	3.3
66	249.8	0.013	0.0030	4.6	1.4	158	598.1	0.067	0.0152	11.0	3.3
68	257.4	0.014	0.0032	4.7	1.4	160	605.7	0.068	0.0155	11.1	3.4
70	265.0	0.015	0.0034	4.9	1.5	162	613.2	0.070	0.0159	11.3	3.4
72	272.5	0.016	0.0035	5.0	1.5	164	620.8	0.072	0.0163	11.4	3.5
74	280.1	0.016	0.0037	5.1	1.6	166	628.4	0.073	0.0166	11.5	3.5
76	287.7	0.017	0.0039	5.3	1.6	168	635.9	0.075	0.0170	11.7	3.6
78	295.3	0.018	0.0041	5.4	1.7	170	643.5	0.077	0.0174	11.8	3.6
80	302.8	0.019	0.0043	5.6	1.7	172	651.1	0.078	0.0178	12.0	3.6
82	310.4	0.020	0.0045	5.7	1.7	174	658.7	0.080	0.0181	12.1	3.7
84	318.0	0.021	0.0047	5.8	1.8	176	666.2	0.082	0.0185	12.2	3.7
86	325.5	0.022	0.0049	6.0	1.8	178	673.8	0.083	0.0189	12.4	3.8
88	333.1	0.023	0.0051	6.1	1.9	180	681.4	0.085	0.0193	12.5	3.8
90	340.7	0.024	0.0054	6.3	1.9	182	688.9	0.087	0.0197	12.6	3.9
92	348.3	0.025	0.0056	6.4	1.9	184	696.5	0.089	0.0201	12.8	3.9
94	355.8	0.026	0.0058	6.5	2.0	186	704.1	0.090	0.0205	12.9	3.9
96	363.4	0.027	0.0060	6.7	2.0	188	711.7	0.092	0.0209	13.1	4.0
98	371.0	0.028	0.0063	6.8	2.1	190	719.2	0.094	0.0213	13.2	4.0
100	378.5	0.029	0.0065	7.0	2.1	192	726.8	0.096	0.0218	13.3	4.1
102	386.1	0.030	0.0068	7.1	2.2	194	734.4	0.098	0.0222	13.5	4.1
104	393.7	0.031	0.0070	7.2	2.2	196	741.9	0.100	0.0226	13.6	4.2
106	401.3	0.032	0.0073	7.4	2.2	198	749.5	0.101	0.0230	13.8	4.2
108	408.8	0.033	0.0075	7.5	2.3	200	757.1	0.103	0.0235	13.9	4.2
110	416.4	0.034	0.0078	7.6	2.3	202	764.7	0.105	0.0239	14.0	4.3
112	424.0	0.035	0.0080	7.8	2.4	204	772.2	0.107	0.0243	14.2	4.3
114	431.5	0.037	0.0083	7.9	2.4	206	779.8	0.109	0.0248	14.3	4.4
116	439.1	0.038	0.0086	8.1	2.5	208	787.4	0.111	0.0252	14.5	4.4
118	446.7	0.039	0.0088	8.2	2.5	210	794.9	0.113	0.0257	14.6	4.4
120	454.2	0.040	0.0091	8.3	2.5	212	802.5	0.115	0.0261	14.7	4.5
122	461.8	0.041	0.0094	8.5	2.6	214	810.1	0.117	0.0266	14.9	4.5
124	469.4	0.043	0.0097	8.6	2.6	216	817.6	0.119	0.0271	15.0	4.6
126	477.0	0.044	0.0100	8.8	2.7	218	825.2	0.121	0.0275	15.2	4.6
128	484.5	0.045	0.0103	8.9	2.7	220	832.8	0.123	0.0280	15.3	4.7
130	492.1	0.047	0.0106	9.0	2.8	222	840.4	0.125	0.0285	15.4	4.7

(continued)

Table D.6 (continued)

Q		P_f		Velocity		Q		P_f		Velocity	
GPM	L/min	psi/ft	bar/m	fps	m/s	GPM	L/min	psi/ft	bar/m	fps	m/s
132	499.7	0.048	0.0109	9.2	2.8	224	847.9	0.128	0.0289	15.6	4.7
134	507.2	0.049	0.0112	9.3	2.8	226	855.5	0.130	0.0294	15.7	4.8
136	514.8	0.051	0.0115	9.5	2.9	228	863.1	0.132	0.0299	15.8	4.8
138	522.4	0.052	0.0118	9.6	2.9	230	870.6	0.134	0.0304	16.0	4.9
140	530.0	0.053	0.0121	9.7	3.0	232	878.2	0.136	0.0309	16.1	4.9
142	537.5	0.055	0.0125	9.9	3.0	234	885.8	0.138	0.0314	16.3	5.0
144	545.1	0.056	0.0128	10.0	3.1	236	893.4	0.140	0.0319	16.4	5.0
146	552.7	0.058	0.0131	10.1	3.1	238	900.9	0.143	0.0324	16.5	5.0
148	560.2	0.059	0.0134	10.3	3.1	240	908.5	0.145	0.0329	16.7	5.1
150	567.8	0.061	0.0138	10.4	3.2	242	916.1	0.147	0.0334	16.8	5.1

Table D.7 Water flowing in 3-in. (80 mm) CPVC plastic pipe $C = 150$ I.D. = 2.951 in. (75 mm)

Q		P_f		Velocity		Q		P_f		Velocity	
GPM	L/min	psi/ft	bar/m	fps	m/s	GPM	L/min	psi/ft	bar/m	fps	m/s
65	246.1	0.005	0.0011	3.1	0.9	295	1116.7	0.081	0.0183	13.9	4.2
70	265.0	0.006	0.0013	3.3	1.0	300	1135.6	0.084	0.0189	14.1	4.3
75	283.9	0.006	0.0015	3.5	1.1	305	1154.5	0.086	0.0195	14.3	4.4
80	302.8	0.007	0.0016	3.8	1.1	310	1173.5	0.089	0.0201	14.6	4.4
85	321.8	0.008	0.0018	4.0	1.2	315	1192.4	0.092	0.0207	14.8	4.5
90	340.7	0.009	0.0020	4.2	1.3	320	1211.3	0.094	0.0213	15.0	4.6
95	359.6	0.010	0.0023	4.5	1.4	325	1230.3	0.097	0.0219	15.3	4.7
100	378.5	0.011	0.0025	4.7	1.4	330	1249.2	0.100	0.0226	15.5	4.7
105	397.5	0.012	0.0027	4.9	1.5	335	1268.1	0.103	0.0232	15.7	4.8
110	416.4	0.013	0.0030	5.2	1.6	340	1287.0	0.106	0.0238	16.0	4.9
115	435.3	0.014	0.0032	5.4	1.6	345	1306.0	0.109	0.0245	16.2	4.9
120	454.2	0.015	0.0035	5.6	1.7	350	1324.9	0.111	0.0251	16.5	5.0
125	473.2	0.017	0.0037	5.9	1.8	355	1343.8	0.114	0.0258	16.7	5.1
130	492.1	0.018	0.0040	6.1	1.9	360	1362.7	0.117	0.0265	16.9	5.2
135	511.0	0.019	0.0043	6.3	1.9	365	1381.7	0.120	0.0272	17.2	5.2
140	530.0	0.020	0.0046	6.6	2.0	370	1400.6	0.124	0.0279	17.4	5.3
145	548.9	0.022	0.0049	6.8	2.1	375	1419.5	0.127	0.0286	17.6	5.4
150	567.8	0.023	0.0052	7.1	2.1	380	1438.5	0.130	0.0293	17.9	5.4
155	586.7	0.025	0.0056	7.3	2.2	385	1457.4	0.133	0.0300	18.1	5.5
160	605.7	0.026	0.0059	7.5	2.3	390	1476.3	0.136	0.0307	18.3	5.6
165	624.6	0.028	0.0063	7.8	2.4	395	1495.2	0.139	0.0315	18.6	5.7
170	643.5	0.029	0.0066	8.0	2.4	400	1514.2	0.143	0.0322	18.8	5.7
175	662.4	0.031	0.0070	8.2	2.5	405	1533.1	0.146	0.0329	19.0	5.8
180	681.4	0.033	0.0073	8.5	2.6	410	1552.0	0.149	0.0337	19.3	5.9
185	700.3	0.034	0.0077	8.7	2.7	415	1570.9	0.153	0.0345	19.5	5.9
190	719.2	0.036	0.0081	8.9	2.7	420	1589.9	0.156	0.0352	19.7	6.0
195	738.2	0.038	0.0085	9.2	2.8	425	1608.8	0.160	0.0360	20.0	6.1
200	757.1	0.040	0.0089	9.4	2.9	430	1627.7	0.163	0.0368	20.2	6.2
205	776.0	0.041	0.0093	9.6	2.9	435	1646.6	0.167	0.0376	20.4	6.2
210	794.9	0.043	0.0098	9.9	3.0	440	1665.6	0.170	0.0384	20.7	6.3
215	813.9	0.045	0.0102	10.1	3.1	445	1684.5	0.174	0.0392	20.9	6.4
220	832.8	0.047	0.0107	10.3	3.2	450	1703.4	0.177	0.0400	21.2	6.4
225	851.7	0.049	0.0111	10.6	3.2	455	1722.4	0.181	0.0409	21.4	6.5
230	870.6	0.051	0.0116	10.8	3.3	460	1741.3	0.185	0.0417	21.6	6.6
235	889.6	0.053	0.0120	11.0	3.4	465	1760.2	0.189	0.0425	21.9	6.7
240	908.5	0.055	0.0125	11.3	3.4	470	1779.1	0.192	0.0434	22.1	6.7
245	927.4	0.058	0.0130	11.5	3.5	475	1798.1	0.196	0.0442	22.3	6.8
250	946.4	0.060	0.0135	11.8	3.6	480	1817.0	0.200	0.0451	22.6	6.9

(continued)

Table D.7 (continued)

Q		P_f		Velocity		Q		P_f		Velocity	
GPM	L/min	psi/ft	bar/m	fps	m/s	GPM	L/min	psi/ft	bar/m	fps	m/s
255	965.3	0.062	0.0140	12.0	3.7	485	1835.9	0.204	0.0460	22.8	6.9
260	984.2	0.064	0.0145	12.2	3.7	490	1854.8	0.208	0.0469	23.0	7.0
265	1003.1	0.067	0.0150	12.5	3.8	495	1873.8	0.212	0.0477	23.3	7.1
270	1022.1	0.069	0.0156	12.7	3.9	500	1892.7	0.216	0.0486	23.5	7.2
275	1041.0	0.071	0.0161	12.9	3.9	505	1911.6	0.220	0.0495	23.7	7.2
280	1059.9	0.074	0.0166	13.2	4.0	510	1930.6	0.224	0.0505	24.0	7.3
285	1078.8	0.076	0.0172	13.4	4.1	515	1949.5	0.228	0.0514	24.2	7.4
290	1097.8	0.079	0.0178	13.6	4.2	520	1968.4	0.232	0.0523	24.4	7.4

Appendix E
Friction Loss Tables for Ductile-Iron Underground Pipe

P. Brock and V. Charter, *Fire Protection Hydraulics and Water Supply Analysis*, The Society of Fire Protection Engineers Series, https://doi.org/10.1007/978-3-032-09877-1

Table E.1 Water flowing in 4-in. (100 mm) unlined class 51 ductile iron pipe $C = 100$ I.D. = 4.28 in. (108.7 mm)

Q		P_f		Velocity		Q		P_f		Velocity	
GPM	L/min	psi/ft	bar/m	fps	m/s	GPM	L/min	psi/ft	bar/m	fps	m/s
100	378.5	0.004	0.0009	2.2	0.7	560	2119.8	0.092	0.0208	12.5	3.8
110	416.4	0.005	0.0010	2.5	0.7	570	2157.7	0.095	0.0215	12.7	3.9
120	454.2	0.005	0.0012	2.7	0.8	580	2195.5	0.098	0.0222	12.9	3.9
130	492.1	0.006	0.0014	2.9	0.9	590	2233.4	0.101	0.0230	13.2	4.0
140	530.0	0.007	0.0016	3.1	1.0	600	2271.2	0.105	0.0237	13.4	4.1
150	567.8	0.008	0.0018	3.3	1.0	610	2309.1	0.108	0.0244	13.6	4.1
160	605.7	0.009	0.0021	3.6	1.1	620	2346.9	0.111	0.0252	13.8	4.2
170	643.5	0.010	0.0023	3.8	1.2	630	2384.8	0.114	0.0259	14.0	4.3
180	681.4	0.011	0.0026	4.0	1.2	640	2422.7	0.118	0.0267	14.3	4.3
190	719.2	0.012	0.0028	4.2	1.3	650	2460.5	0.121	0.0275	14.5	4.4
200	757.1	0.014	0.0031	4.5	1.4	660	2498.4	0.125	0.0282	14.7	4.5
210	794.9	0.015	0.0034	4.7	1.4	670	2536.2	0.128	0.0290	14.9	4.6
220	832.8	0.016	0.0037	4.9	1.5	680	2574.1	0.132	0.0299	15.2	4.6
230	870.6	0.018	0.0040	5.1	1.6	690	2611.9	0.135	0.0307	15.4	4.7
240	908.5	0.019	0.0043	5.3	1.6	700	2649.8	0.139	0.0315	15.6	4.8
250	946.4	0.021	0.0047	5.6	1.7	710	2687.6	0.143	0.0323	15.8	4.8
260	984.2	0.022	0.0050	5.8	1.8	720	2725.5	0.147	0.0332	16.0	4.9
270	1022.1	0.024	0.0054	6.0	1.8	730	2763.3	0.150	0.0340	16.3	5.0
280	1059.9	0.026	0.0058	6.2	1.9	740	2801.2	0.154	0.0349	16.5	5.0
290	1097.8	0.027	0.0062	6.5	2.0	750	2839.1	0.158	0.0358	16.7	5.1
300	1135.6	0.029	0.0066	6.7	2.0	760	2876.9	0.162	0.0367	16.9	5.2
310	1173.5	0.031	0.0070	6.9	2.1	770	2914.8	0.166	0.0376	17.2	5.2
320	1211.3	0.033	0.0074	7.1	2.2	780	2952.6	0.170	0.0385	17.4	5.3
330	1249.2	0.035	0.0078	7.4	2.2	790	2990.5	0.174	0.0394	17.6	5.4
340	1287.0	0.037	0.0083	7.6	2.3	800	3028.3	0.178	0.0403	17.8	5.4
350	1324.9	0.039	0.0087	7.8	2.4	810	3066.2	0.182	0.0413	18.1	5.5
360	1362.7	0.041	0.0092	8.0	2.4	820	3104.0	0.186	0.0422	18.3	5.6
370	1400.6	0.043	0.0097	8.2	2.5	830	3141.9	0.191	0.0432	18.5	5.6
380	1438.5	0.045	0.0102	8.5	2.6	840	3179.7	0.195	0.0441	18.7	5.7
390	1476.3	0.047	0.0107	8.7	2.6	850	3217.6	0.199	0.0451	18.9	5.8
400	1514.2	0.049	0.0112	8.9	2.7	860	3255.4	0.204	0.0461	19.2	5.8
410	1552.0	0.052	0.0117	9.1	2.8	870	3293.3	0.208	0.0471	19.4	5.9
420	1589.9	0.054	0.0122	9.4	2.9	880	3331.2	0.212	0.0481	19.6	6.0
430	1627.7	0.056	0.0128	9.6	2.9	890	3369.0	0.217	0.0491	19.8	6.0
440	1665.6	0.059	0.0133	9.8	3.0	900	3406.9	0.221	0.0501	20.1	6.1

(continued)

Table E.1 (continued)

Q		P_f		Velocity		Q		P_f		Velocity	
GPM	L/min	psi/ft	bar/m	fps	m/s	GPM	L/min	psi/ft	bar/m	fps	m/s
450	1703.4	0.061	0.0139	10.0	3.1	910	3444.7	0.226	0.0512	20.3	6.2
460	1741.3	0.064	0.0145	10.3	3.1	920	3482.6	0.231	0.0522	20.5	6.3
470	1779.1	0.067	0.0151	10.5	3.2	930	3520.4	0.235	0.0533	20.7	6.3
480	1817.0	0.069	0.0157	10.7	3.3	940	3558.3	0.240	0.0543	21.0	6.4
490	1854.8	0.072	0.0163	10.9	3.3	950	3596.1	0.245	0.0554	21.2	6.5
500	1892.7	0.075	0.0169	11.1	3.4	960	3634.0	0.250	0.0565	21.4	6.5
510	1930.6	0.077	0.0175	11.4	3.5	970	3671.8	0.254	0.0576	21.6	6.6
520	1968.4	0.080	0.0182	11.6	3.5	980	3709.7	0.259	0.0587	21.8	6.7
530	2006.3	0.083	0.0188	11.8	3.6	990	3747.5	0.264	0.0598	22.1	6.7
540	2044.1	0.086	0.0195	12.0	3.7	1000	3785.4	0.269	0.0609	22.3	6.8
550	2082.0	0.089	0.0202	12.3	3.7	1010	3823.3	0.274	0.0621	22.5	6.9

Table E.2 Water flowing in 4-in. (100 mm) unlined class 52 ductile iron pipe $C = 100$ I.D. = 4.155 in. (105.5 mm)

Q		P_f		Velocity		Q		P_f		Velocity	
GPM	L/min	psi/ft	bar/m	fps	m/s	GPM	L/min	psi/ft	bar/m	fps	m/s
100	378.5	0.004	0.0010	2.4	0.7	560	2119.8	0.106	0.0241	13.2	4.0
110	416.4	0.005	0.0012	2.6	0.8	570	2157.7	0.110	0.0249	13.5	4.1
120	454.2	0.006	0.0014	2.8	0.9	580	2195.5	0.114	0.0257	13.7	4.2
130	492.1	0.007	0.0016	3.1	0.9	590	2233.4	0.117	0.0266	14.0	4.3
140	530.0	0.008	0.0019	3.3	1.0	600	2271.2	0.121	0.0274	14.2	4.3
150	567.8	0.009	0.0021	3.5	1.1	610	2309.1	0.125	0.0282	14.4	4.4
160	605.7	0.010	0.0024	3.8	1.2	620	2346.9	0.128	0.0291	14.7	4.5
170	643.5	0.012	0.0027	4.0	1.2	630	2384.8	0.132	0.0300	14.9	4.5
180	681.4	0.013	0.0030	4.3	1.3	640	2422.7	0.136	0.0309	15.1	4.6
190	719.2	0.014	0.0033	4.5	1.4	650	2460.5	0.140	0.0318	15.4	4.7
200	757.1	0.016	0.0036	4.7	1.4	660	2498.4	0.144	0.0327	15.6	4.8
210	794.9	0.017	0.0039	5.0	1.5	670	2536.2	0.148	0.0336	15.8	4.8
220	832.8	0.019	0.0043	5.2	1.6	680	2574.1	0.152	0.0345	16.1	4.9
230	870.6	0.021	0.0046	5.4	1.7	690	2611.9	0.157	0.0355	16.3	5.0
240	908.5	0.022	0.0050	5.7	1.7	700	2649.8	0.161	0.0364	16.6	5.0
250	946.4	0.024	0.0054	5.9	1.8	710	2687.6	0.165	0.0374	16.8	5.1
260	984.2	0.026	0.0058	6.1	1.9	720	2725.5	0.169	0.0384	17.0	5.2
270	1022.1	0.028	0.0063	6.4	1.9	730	2763.3	0.174	0.0394	17.3	5.3
280	1059.9	0.030	0.0067	6.6	2.0	740	2801.2	0.178	0.0404	17.5	5.3
290	1097.8	0.031	0.0071	6.9	2.1	750	2839.1	0.183	0.0414	17.7	5.4
300	1135.6	0.034	0.0076	7.1	2.2	760	2876.9	0.187	0.0424	18.0	5.5
310	1173.5	0.036	0.0081	7.3	2.2	770	2914.8	0.192	0.0435	18.2	5.6
320	1211.3	0.038	0.0086	7.6	2.3	780	2952.6	0.196	0.0445	18.4	5.6
330	1249.2	0.040	0.0091	7.8	2.4	790	2990.5	0.201	0.0456	18.7	5.7
340	1287.0	0.042	0.0096	8.0	2.5	800	3028.3	0.206	0.0466	18.9	5.8
350	1324.9	0.045	0.0101	8.3	2.5	810	3066.2	0.211	0.0477	19.2	5.8
360	1362.7	0.047	0.0106	8.5	2.6	820	3104.0	0.215	0.0488	19.4	5.9
370	1400.6	0.049	0.0112	8.8	2.7	830	3141.9	0.220	0.0499	19.6	6.0
380	1438.5	0.052	0.0118	9.0	2.7	840	3179.7	0.225	0.0510	19.9	6.1
390	1476.3	0.054	0.0123	9.2	2.8	850	3217.6	0.230	0.0522	20.1	6.1
400	1514.2	0.057	0.0129	9.5	2.9	860	3255.4	0.235	0.0533	20.3	6.2
410	1552.0	0.060	0.0135	9.7	3.0	870	3293.3	0.240	0.0545	20.6	6.3
420	1589.9	0.062	0.0142	9.9	3.0	880	3331.2	0.245	0.0556	20.8	6.3
430	1627.7	0.065	0.0148	10.2	3.1	890	3369.0	0.251	0.0568	21.0	6.4
440	1665.6	0.068	0.0154	10.4	3.2	900	3406.9	0.256	0.0580	21.3	6.5

(continued)

Table E.2 (continued)

Q		P_f		Velocity		Q		P_f		Velocity	
GPM	L/min	psi/ft	bar/m	fps	m/s	GPM	L/min	psi/ft	bar/m	fps	m/s
450	1703.4	0.071	0.0161	10.6	3.2	910	3444.7	0.261	0.0592	21.5	6.6
460	1741.3	0.074	0.0168	10.9	3.3	920	3482.6	0.267	0.0604	21.8	6.6
470	1779.1	0.077	0.0174	11.1	3.4	930	3520.4	0.272	0.0616	22.0	6.7
480	1817.0	0.080	0.0181	11.4	3.5	940	3558.3	0.277	0.0629	22.2	6.8
490	1854.8	0.083	0.0188	11.6	3.5	950	3596.1	0.283	0.0641	22.5	6.8
500	1892.7	0.086	0.0195	11.8	3.6	960	3634.0	0.288	0.0653	22.7	6.9
510	1930.6	0.089	0.0203	12.1	3.7	970	3671.8	0.294	0.0666	22.9	7.0
520	1968.4	0.093	0.0210	12.3	3.7	980	3709.7	0.300	0.0679	23.2	7.1
530	2006.3	0.096	0.0218	12.5	3.8	990	3747.5	0.305	0.0692	23.4	7.1
540	2044.1	0.099	0.0225	12.8	3.9	1000	3785.4	0.311	0.0705	23.7	7.2
550	2082.0	0.103	0.0233	13.0	4.0	1010	3823.3	0.317	0.0718	23.9	7.3

Table E.3 Water flowing in 4-in. (100 mm) cement lined class 51 ductile iron pipe $C = 140$ I.D. = 4.16 in. (105.7 mm)

Q		P_f		Velocity		Q		P_f		Velocity	
GPM	L/min	psi/ft	bar/m	fps	m/s	GPM	L/min	psi/ft	bar/m	fps	m/s
100	378.5	0.002	0.0005	2.4	0.7	560	2119.8	0.057	0.0128	13.2	4.0
110	416.4	0.003	0.0006	2.6	0.8	570	2157.7	0.059	0.0132	13.4	4.1
120	454.2	0.003	0.0007	2.8	0.9	580	2195.5	0.061	0.0137	13.7	4.2
130	492.1	0.004	0.0009	3.1	0.9	590	2233.4	0.063	0.0141	13.9	4.2
140	530.0	0.004	0.0010	3.3	1.0	600	2271.2	0.064	0.0146	14.2	4.3
150	567.8	0.005	0.0011	3.5	1.1	610	2309.1	0.066	0.0150	14.4	4.4
160	605.7	0.006	0.0013	3.8	1.2	620	2346.9	0.069	0.0155	14.6	4.5
170	643.5	0.006	0.0014	4.0	1.2	630	2384.8	0.071	0.0159	14.9	4.5
180	681.4	0.007	0.0016	4.2	1.3	640	2422.7	0.073	0.0164	15.1	4.6
190	719.2	0.008	0.0017	4.5	1.4	650	2460.5	0.075	0.0169	15.3	4.7
200	757.1	0.008	0.0019	4.7	1.4	660	2498.4	0.077	0.0174	15.6	4.7
210	794.9	0.009	0.0021	5.0	1.5	670	2536.2	0.079	0.0179	15.8	4.8
220	832.8	0.010	0.0023	5.2	1.6	680	2574.1	0.081	0.0184	16.0	4.9
230	870.6	0.011	0.0025	5.4	1.7	690	2611.9	0.084	0.0189	16.3	5.0
240	908.5	0.012	0.0027	5.7	1.7	700	2649.8	0.086	0.0194	16.5	5.0
250	946.4	0.013	0.0029	5.9	1.8	710	2687.6	0.088	0.0199	16.8	5.1
260	984.2	0.014	0.0031	6.1	1.9	720	2725.5	0.090	0.0204	17.0	5.2
270	1022.1	0.015	0.0033	6.4	1.9	730	2763.3	0.093	0.0209	17.2	5.2
280	1059.9	0.016	0.0036	6.6	2.0	740	2801.2	0.095	0.0215	17.5	5.3
290	1097.8	0.017	0.0038	6.8	2.1	750	2839.1	0.097	0.0220	17.7	5.4
300	1135.6	0.018	0.0040	7.1	2.2	760	2876.9	0.100	0.0226	17.9	5.5
310	1173.5	0.019	0.0043	7.3	2.2	770	2914.8	0.102	0.0231	18.2	5.5
320	1211.3	0.020	0.0046	7.5	2.3	780	2952.6	0.105	0.0237	18.4	5.6
330	1249.2	0.021	0.0048	7.8	2.4	790	2990.5	0.107	0.0242	18.6	5.7
340	1287.0	0.023	0.0051	8.0	2.4	800	3028.3	0.110	0.0248	18.9	5.8
350	1324.9	0.024	0.0054	8.3	2.5	810	3066.2	0.112	0.0254	19.1	5.8
360	1362.7	0.025	0.0057	8.5	2.6	820	3104.0	0.115	0.0260	19.3	5.9
370	1400.6	0.026	0.0060	8.7	2.7	830	3141.9	0.118	0.0265	19.6	6.0
380	1438.5	0.028	0.0063	9.0	2.7	840	3179.7	0.120	0.0271	19.8	6.0
390	1476.3	0.029	0.0066	9.2	2.8	850	3217.6	0.123	0.0277	20.1	6.1
400	1514.2	0.030	0.0069	9.4	2.9	860	3255.4	0.125	0.0283	20.3	6.2
410	1552.0	0.032	0.0072	9.7	2.9	870	3293.3	0.128	0.0290	20.5	6.3
420	1589.9	0.033	0.0075	9.9	3.0	880	3331.2	0.131	0.0296	20.8	6.3
430	1627.7	0.035	0.0079	10.1	3.1	890	3369.0	0.134	0.0302	21.0	6.4
440	1665.6	0.036	0.0082	10.4	3.2	900	3406.9	0.137	0.0308	21.2	6.5

(continued)

Table E.3 (continued)

Q		P_f		Velocity		Q		P_f		Velocity	
GPM	L/min	psi/ft	bar/m	fps	m/s	GPM	L/min	psi/ft	bar/m	fps	m/s
450	1703.4	0.038	0.0086	10.6	3.2	910	3444.7	0.139	0.0315	21.5	6.5
460	1741.3	0.039	0.0089	10.9	3.3	920	3482.6	0.142	0.0321	21.7	6.6
470	1779.1	0.041	0.0093	11.1	3.4	930	3520.4	0.145	0.0328	21.9	6.7
480	1817.0	0.043	0.0096	11.3	3.5	940	3558.3	0.148	0.0334	22.2	6.8
490	1854.8	0.044	0.0100	11.6	3.5	950	3596.1	0.151	0.0341	22.4	6.8
500	1892.7	0.046	0.0104	11.8	3.6	960	3634.0	0.154	0.0347	22.6	6.9
510	1930.6	0.048	0.0108	12.0	3.7	970	3671.8	0.157	0.0354	22.9	7.0
520	1968.4	0.049	0.0112	12.3	3.7	980	3709.7	0.160	0.0361	23.1	7.0
530	2006.3	0.051	0.0116	12.5	3.8	990	3747.5	0.163	0.0368	23.4	7.1
540	2044.1	0.053	0.0120	12.7	3.9	1000	3785.4	0.166	0.0375	23.6	7.2
550	2082.0	0.055	0.0124	13.0	4.0	1010	3823.3	0.169	0.0382	23.8	7.3

Table E.4 Water flowing in 4-in. (100 mm) cement lined class 52 ductile iron pipe $C = 140$ I.D. = 4.10 in. (104.1 mm)

Q		P_f		Velocity		Q		P_f		Velocity	
GPM	L/min	psi/ft	bar/m	fps	m/s	GPM	L/min	psi/ft	bar/m	fps	m/s
100	378.5	0.0025	0.0006	2.4	0.7	560	2119.8	0.0609	0.0138	13.6	4.1
110	416.4	0.0030	0.0007	2.7	0.8	570	2157.7	0.0629	0.0143	13.8	4.2
120	454.2	0.0035	0.0008	2.9	0.9	580	2195.5	0.0650	0.0147	14.1	4.3
130	492.1	0.0041	0.0009	3.2	1.0	590	2233.4	0.0671	0.0152	14.3	4.4
140	530.0	0.0047	0.0011	3.4	1.0	600	2271.2	0.0692	0.0157	14.6	4.4
150	567.8	0.0053	0.0012	3.6	1.1	610	2309.1	0.0714	0.0162	14.8	4.5
160	605.7	0.0060	0.0014	3.9	1.2	620	2346.9	0.0735	0.0167	15.1	4.6
170	643.5	0.0067	0.0015	4.1	1.3	630	2384.8	0.0757	0.0172	15.3	4.7
180	681.4	0.0075	0.0017	4.4	1.3	640	2422.7	0.0780	0.0177	15.5	4.7
190	719.2	0.0082	0.0019	4.6	1.4	650	2460.5	0.0802	0.0182	15.8	4.8
200	757.1	0.0091	0.0021	4.9	1.5	660	2498.4	0.0825	0.0187	16.0	4.9
210	794.9	0.0099	0.0022	5.1	1.6	670	2536.2	0.0849	0.0192	16.3	5.0
220	832.8	0.0108	0.0025	5.3	1.6	680	2574.1	0.0872	0.0198	16.5	5.0
230	870.6	0.0117	0.0027	5.6	1.7	690	2611.9	0.0896	0.0203	16.8	5.1
240	908.5	0.0127	0.0029	5.8	1.8	700	2649.8	0.0920	0.0209	17.0	5.2
250	946.4	0.0137	0.0031	6.1	1.9	710	2687.6	0.0945	0.0214	17.2	5.3
260	984.2	0.0147	0.0033	6.3	1.9	720	2725.5	0.0970	0.0220	17.5	5.3
270	1022.1	0.0158	0.0036	6.6	2.0	730	2763.3	0.0995	0.0225	17.7	5.4
280	1059.9	0.0169	0.0038	6.8	2.1	740	2801.2	0.1020	0.0231	18.0	5.5
290	1097.8	0.0180	0.0041	7.0	2.1	750	2839.1	0.1046	0.0237	18.2	5.6
300	1135.6	0.0192	0.0044	7.3	2.2	760	2876.9	0.1072	0.0243	18.5	5.6
310	1173.5	0.0204	0.0046	7.5	2.3	770	2914.8	0.1098	0.0249	18.7	5.7
320	1211.3	0.0216	0.0049	7.8	2.4	780	2952.6	0.1124	0.0255	18.9	5.8
330	1249.2	0.0229	0.0052	8.0	2.4	790	2990.5	0.1151	0.0261	19.2	5.8
340	1287.0	0.0242	0.0055	8.3	2.5	800	3028.3	0.1178	0.0267	19.4	5.9
350	1324.9	0.0255	0.0058	8.5	2.6	810	3066.2	0.1206	0.0273	19.7	6.0
360	1362.7	0.0269	0.0061	8.7	2.7	820	3104.0	0.1233	0.0280	19.9	6.1
370	1400.6	0.0283	0.0064	9.0	2.7	830	3141.9	0.1261	0.0286	20.2	6.1
380	1438.5	0.0297	0.0067	9.2	2.8	840	3179.7	0.1290	0.0292	20.4	6.2
390	1476.3	0.0312	0.0071	9.5	2.9	850	3217.6	0.1318	0.0299	20.6	6.3
400	1514.2	0.0327	0.0074	9.7	3.0	860	3255.4	0.1347	0.0305	20.9	6.4
410	1552.0	0.0342	0.0078	10.0	3.0	870	3293.3	0.1376	0.0312	21.1	6.4
420	1589.9	0.0358	0.0081	10.2	3.1	880	3331.2	0.1406	0.0319	21.4	6.5
430	1627.7	0.0374	0.0085	10.4	3.2	890	3369.0	0.1435	0.0325	21.6	6.6
440	1665.6	0.0390	0.0088	10.7	3.3	900	3406.9	0.1465	0.0332	21.9	6.7

(continued)

Table E.4 (continued)

Q		P_f		Velocity		Q		P_f		Velocity	
GPM	L/min	psi/ft	bar/m	fps	m/s	GPM	L/min	psi/ft	bar/m	fps	m/s
450	1703.4	0.0406	0.0092	10.9	3.3	910	3444.7	0.1495	0.0339	22.1	6.7
460	1741.3	0.0423	0.0096	11.2	3.4	920	3482.6	0.1526	0.0346	22.3	6.8
470	1779.1	0.0440	0.0100	11.4	3.5	930	3520.4	0.1557	0.0353	22.6	6.9
480	1817.0	0.0458	0.0104	11.7	3.6	940	3558.3	0.1588	0.0360	22.8	7.0
490	1854.8	0.0476	0.0108	11.9	3.6	950	3596.1	0.1619	0.0367	23.1	7.0
500	1892.7	0.0494	0.0112	12.1	3.7	960	3634.0	0.1651	0.0374	23.3	7.1
510	1930.6	0.0512	0.0116	12.4	3.8	970	3671.8	0.1683	0.0381	23.6	7.2
520	1968.4	0.0531	0.0120	12.6	3.8	980	3709.7	0.1715	0.0389	23.8	7.3
530	2006.3	0.0550	0.0125	12.9	3.9	990	3747.5	0.1748	0.0396	24.0	7.3
540	2044.1	0.0569	0.0129	13.1	4.0	1000	3785.4	0.1781	0.0404	24.3	7.4
550	2082.0	0.0589	0.0134	13.4	4.1	1010	3823.3	0.1814	0.0411	24.5	7.5

Table E.5 Water flowing in 6-in. (150 mm) unlined class 51 ductile iron pipe $C = 100$ I.D. = 6.34 in. (161.0 mm)

Q		P_f		Velocity		*Q*		P_f		Velocity	
GPM	L/min	psi/ft	bar/m	fps	m/s	GPM	L/min	psi/ft	bar/m	fps	m/s
200	757.1	0.0020	0.0005	2.0	0.6	660	2498.4	0.0184	0.0042	6.7	2.0
210	794.9	0.0022	0.0005	2.1	0.7	670	2536.2	0.0189	0.0043	6.8	2.1
220	832.8	0.0024	0.0005	2.2	0.7	680	2574.1	0.0195	0.0044	6.9	2.1
230	870.6	0.0026	0.0006	2.3	0.7	690	2611.9	0.0200	0.0045	7.0	2.1
240	908.5	0.0028	0.0006	2.4	0.7	700	2649.8	0.0205	0.0046	7.1	2.2
250	946.4	0.0031	0.0007	2.5	0.8	710	2687.6	0.0211	0.0048	7.2	2.2
260	984.2	0.0033	0.0007	2.6	0.8	720	2725.5	0.0216	0.0049	7.3	2.2
270	1022.1	0.0035	0.0008	2.7	0.8	730	2763.3	0.0222	0.0050	7.4	2.3
280	1059.9	0.0038	0.0009	2.8	0.9	740	2801.2	0.0228	0.0052	7.5	2.3
290	1097.8	0.0040	0.0009	2.9	0.9	750	2839.1	0.0233	0.0053	7.6	2.3
300	1135.6	0.0043	0.0010	3.0	0.9	760	2876.9	0.0239	0.0054	7.7	2.4
310	1173.5	0.0045	0.0010	3.1	1.0	770	2914.8	0.0245	0.0055	7.8	2.4
320	1211.3	0.0048	0.0011	3.3	1.0	780	2952.6	0.0251	0.0057	7.9	2.4
330	1249.2	0.0051	0.0012	3.4	1.0	790	2990.5	0.0257	0.0058	8.0	2.4
340	1287.0	0.0054	0.0012	3.5	1.1	800	3028.3	0.0263	0.0060	8.1	2.5
350	1324.9	0.0057	0.0013	3.6	1.1	810	3066.2	0.0269	0.0061	8.2	2.5
360	1362.7	0.0060	0.0014	3.7	1.1	820	3104.0	0.0275	0.0062	8.3	2.5
370	1400.6	0.0063	0.0014	3.8	1.1	830	3141.9	0.0281	0.0064	8.4	2.6
380	1438.5	0.0066	0.0015	3.9	1.2	840	3179.7	0.0288	0.0065	8.5	2.6
390	1476.3	0.0070	0.0016	4.0	1.2	850	3217.6	0.0294	0.0067	8.6	2.6
400	1514.2	0.0073	0.0017	4.1	1.2	860	3255.4	0.0300	0.0068	8.7	2.7
410	1552.0	0.0076	0.0017	4.2	1.3	870	3293.3	0.0307	0.0070	8.8	2.7
420	1589.9	0.0080	0.0018	4.3	1.3	880	3331.2	0.0314	0.0071	8.9	2.7
430	1627.7	0.0083	0.0019	4.4	1.3	890	3369.0	0.0320	0.0073	9.0	2.8
440	1665.6	0.0087	0.0020	4.5	1.4	900	3406.9	0.0327	0.0074	9.1	2.8
450	1703.4	0.0091	0.0021	4.6	1.4	910	3444.7	0.0334	0.0076	9.2	2.8
460	1741.3	0.0094	0.0021	4.7	1.4	920	3482.6	0.0340	0.0077	9.3	2.8
470	1779.1	0.0098	0.0022	4.8	1.5	930	3520.4	0.0347	0.0079	9.4	2.9
480	1817.0	0.0102	0.0023	4.9	1.5	940	3558.3	0.0354	0.0080	9.5	2.9
490	1854.8	0.0106	0.0024	5.0	1.5	950	3596.1	0.0361	0.0082	9.6	2.9
500	1892.7	0.0110	0.0025	5.1	1.5	960	3634.0	0.0368	0.0083	9.8	3.0
510	1930.6	0.0114	0.0026	5.2	1.6	970	3671.8	0.0375	0.0085	9.9	3.0
520	1968.4	0.0118	0.0027	5.3	1.6	980	3709.7	0.0383	0.0087	10.0	3.0
530	2006.3	0.0123	0.0028	5.4	1.6	990	3747.5	0.0390	0.0088	10.1	3.1
540	2044.1	0.0127	0.0029	5.5	1.7	1000	3785.4	0.0397	0.0090	10.2	3.1

(continued)

Table E.5 (continued)

Q		P_f		Velocity		Q		P_f		Velocity	
GPM	L/min	psi/ft	bar/m	fps	m/s	GPM	L/min	psi/ft	bar/m	fps	m/s
550	2082.0	0.0131	0.0030	5.6	1.7	1010	3823.3	0.0405	0.0092	10.3	3.1
560	2119.8	0.0136	0.0031	5.7	1.7	1020	3861.1	0.0412	0.0093	10.4	3.2
570	2157.7	0.0140	0.0032	5.8	1.8	1030	3899.0	0.0419	0.0095	10.5	3.2
580	2195.5	0.0145	0.0033	5.9	1.8	1040	3936.8	0.0427	0.0097	10.6	3.2
590	2233.4	0.0150	0.0034	6.0	1.8	1050	3974.7	0.0435	0.0098	10.7	3.3
600	2271.2	0.0154	0.0035	6.1	1.9	1060	4012.5	0.0442	0.0100	10.8	3.3
610	2309.1	0.0159	0.0036	6.2	1.9	1070	4050.4	0.0450	0.0102	10.9	3.3
620	2346.9	0.0164	0.0037	6.3	1.9	1080	4088.2	0.0458	0.0104	11.0	3.3
630	2384.8	0.0169	0.0038	6.4	2.0	1090	4126.1	0.0466	0.0106	11.1	3.4
640	2422.7	0.0174	0.0039	6.5	2.0	1100	4163.9	0.0474	0.0107	11.2	3.4
650	2460.5	0.0179	0.0041	6.6	2.0	1110	4201.8	0.0482	0.0109	11.3	3.4

Table E.6 Water flowing in 6-in. (150 mm) unlined class 52 ductile iron pipe $C = 100$ I.D. = 6.275 in. (159.4 mm)

Q		P_f		Velocity		Q		P_f		Velocity	
GPM	L/min	psi/ft	bar/m	fps	m/s	GPM	L/min	psi/ft	bar/m	fps	m/s
200	757.1	0.0021	0.0005	2.1	0.6	660	2498.4	0.0194	0.0044	6.8	2.1
210	794.9	0.0023	0.0005	2.2	0.7	670	2536.2	0.0199	0.0045	6.9	2.1
220	832.8	0.0025	0.0006	2.3	0.7	680	2574.1	0.0205	0.0046	7.1	2.1
230	870.6	0.0028	0.0006	2.4	0.7	690	2611.9	0.0210	0.0048	7.2	2.2
240	908.5	0.0030	0.0007	2.5	0.8	700	2649.8	0.0216	0.0049	7.3	2.2
250	946.4	0.0032	0.0007	2.6	0.8	710	2687.6	0.0222	0.0050	7.4	2.2
260	984.2	0.0035	0.0008	2.7	0.8	720	2725.5	0.0227	0.0051	7.5	2.3
270	1022.1	0.0037	0.0008	2.8	0.9	730	2763.3	0.0233	0.0053	7.6	2.3
280	1059.9	0.0040	0.0009	2.9	0.9	740	2801.2	0.0239	0.0054	7.7	2.3
290	1097.8	0.0042	0.0010	3.0	0.9	750	2839.1	0.0245	0.0055	7.8	2.4
300	1135.6	0.0045	0.0010	3.1	0.9	760	2876.9	0.0251	0.0057	7.9	2.4
310	1173.5	0.0048	0.0011	3.2	1.0	770	2914.8	0.0257	0.0058	8.0	2.4
320	1211.3	0.0051	0.0011	3.3	1.0	780	2952.6	0.0264	0.0060	8.1	2.5
330	1249.2	0.0054	0.0012	3.4	1.0	790	2990.5	0.0270	0.0061	8.2	2.5
340	1287.0	0.0057	0.0013	3.5	1.1	800	3028.3	0.0276	0.0062	8.3	2.5
350	1324.9	0.0060	0.0014	3.6	1.1	810	3066.2	0.0283	0.0064	8.4	2.6
360	1362.7	0.0063	0.0014	3.7	1.1	820	3104.0	0.0289	0.0065	8.5	2.6
370	1400.6	0.0066	0.0015	3.8	1.2	830	3141.9	0.0296	0.0067	8.6	2.6
380	1438.5	0.0070	0.0016	3.9	1.2	840	3179.7	0.0302	0.0068	8.7	2.7
390	1476.3	0.0073	0.0017	4.0	1.2	850	3217.6	0.0309	0.0070	8.8	2.7
400	1514.2	0.0077	0.0017	4.1	1.3	860	3255.4	0.0316	0.0071	8.9	2.7
410	1552.0	0.0080	0.0018	4.3	1.3	870	3293.3	0.0323	0.0073	9.0	2.7
420	1589.9	0.0084	0.0019	4.4	1.3	880	3331.2	0.0330	0.0075	9.1	2.8
430	1627.7	0.0088	0.0020	4.5	1.4	890	3369.0	0.0337	0.0076	9.2	2.8
440	1665.6	0.0091	0.0021	4.6	1.4	900	3406.9	0.0344	0.0078	9.3	2.8
450	1703.4	0.0095	0.0022	4.7	1.4	910	3444.7	0.0351	0.0079	9.4	2.9
460	1741.3	0.0099	0.0022	4.8	1.5	920	3482.6	0.0358	0.0081	9.5	2.9
470	1779.1	0.0103	0.0023	4.9	1.5	930	3520.4	0.0365	0.0083	9.6	2.9
480	1817.0	0.0107	0.0024	5.0	1.5	940	3558.3	0.0372	0.0084	9.7	3.0
490	1854.8	0.0112	0.0025	5.1	1.5	950	3596.1	0.0380	0.0086	9.9	3.0
500	1892.7	0.0116	0.0026	5.2	1.6	960	3634.0	0.0387	0.0088	10.0	3.0
510	1930.6	0.0120	0.0027	5.3	1.6	970	3671.8	0.0395	0.0089	10.1	3.1
520	1968.4	0.0125	0.0028	5.4	1.6	980	3709.7	0.0402	0.0091	10.2	3.1
530	2006.3	0.0129	0.0029	5.5	1.7	990	3747.5	0.0410	0.0093	10.3	3.1
540	2044.1	0.0134	0.0030	5.6	1.7	1000	3785.4	0.0418	0.0094	10.4	3.2

(continued)

Table E.6 (continued)

Q		P_f		Velocity		Q		P_f		Velocity	
GPM	L/min	psi/ft	bar/m	fps	m/s	GPM	L/min	psi/ft	bar/m	fps	m/s
550	2082.0	0.0138	0.0031	5.7	1.7	1010	3823.3	0.0425	0.0096	10.5	3.2
560	2119.8	0.0143	0.0032	5.8	1.8	1020	3861.1	0.0433	0.0098	10.6	3.2
570	2157.7	0.0148	0.0033	5.9	1.8	1030	3899.0	0.0441	0.0100	10.7	3.3
580	2195.5	0.0152	0.0034	6.0	1.8	1040	3936.8	0.0449	0.0102	10.8	3.3
590	2233.4	0.0157	0.0036	6.1	1.9	1050	3974.7	0.0457	0.0103	10.9	3.3
600	2271.2	0.0162	0.0037	6.2	1.9	1060	4012.5	0.0465	0.0105	11.0	3.4
610	2309.1	0.0167	0.0038	6.3	1.9	1070	4050.4	0.0473	0.0107	11.1	3.4
620	2346.9	0.0172	0.0039	6.4	2.0	1080	4088.2	0.0481	0.0109	11.2	3.4
630	2384.8	0.0178	0.0040	6.5	2.0	1090	4126.1	0.0490	0.0111	11.3	3.4
640	2422.7	0.0183	0.0041	6.6	2.0	1100	4163.9	0.0498	0.0113	11.4	3.5
650	2460.5	0.0188	0.0043	6.7	2.1	1110	4201.8	0.0507	0.0115	11.5	3.5

Table E.7 Water flowing in 6-in. (150 mm) cement lined class 51 ductile iron pipe $C = 140$ I.D. = 6.22 in. (158 mm)

Q		P_f		Velocity		Q		P_f		Velocity	
GPM	L/min	psi/ft	bar/m	fps	m/s	GPM	L/min	psi/ft	bar/m	fps	m/s
200	757.1	0.0012	0.0003	2.1	0.6	660	2498.4	0.0108	0.0025	7.0	2.1
210	794.9	0.0013	0.0003	2.2	0.7	670	2536.2	0.0112	0.0025	7.1	2.2
220	832.8	0.0014	0.0003	2.3	0.7	680	2574.1	0.0115	0.0026	7.2	2.2
230	870.6	0.0015	0.0003	2.4	0.7	690	2611.9	0.0118	0.0027	7.3	2.2
240	908.5	0.0017	0.0004	2.5	0.8	700	2649.8	0.0121	0.0027	7.4	2.3
250	946.4	0.0018	0.0004	2.6	0.8	710	2687.6	0.0124	0.0028	7.5	2.3
260	984.2	0.0019	0.0004	2.7	0.8	720	2725.5	0.0127	0.0029	7.6	2.3
270	1022.1	0.0021	0.0005	2.8	0.9	730	2763.3	0.0131	0.0030	7.7	2.3
280	1059.9	0.0022	0.0005	3.0	0.9	740	2801.2	0.0134	0.0030	7.8	2.4
290	1097.8	0.0024	0.0005	3.1	0.9	750	2839.1	0.0137	0.0031	7.9	2.4
300	1135.6	0.0025	0.0006	3.2	1.0	760	2876.9	0.0141	0.0032	8.0	2.4
310	1173.5	0.0027	0.0006	3.3	1.0	770	2914.8	0.0144	0.0033	8.1	2.5
320	1211.3	0.0028	0.0006	3.4	1.0	780	2952.6	0.0148	0.0033	8.2	2.5
330	1249.2	0.0030	0.0007	3.5	1.1	790	2990.5	0.0151	0.0034	8.3	2.5
340	1287.0	0.0032	0.0007	3.6	1.1	800	3028.3	0.0155	0.0035	8.4	2.6
350	1324.9	0.0034	0.0008	3.7	1.1	810	3066.2	0.0158	0.0036	8.5	2.6
360	1362.7	0.0035	0.0008	3.8	1.2	820	3104.0	0.0162	0.0037	8.7	2.6
370	1400.6	0.0037	0.0008	3.9	1.2	830	3141.9	0.0166	0.0037	8.8	2.7
380	1438.5	0.0039	0.0009	4.0	1.2	840	3179.7	0.0169	0.0038	8.9	2.7
390	1476.3	0.0041	0.0009	4.1	1.3	850	3217.6	0.0173	0.0039	9.0	2.7
400	1514.2	0.0043	0.0010	4.2	1.3	860	3255.4	0.0177	0.0040	9.1	2.8
410	1552.0	0.0045	0.0010	4.3	1.3	870	3293.3	0.0181	0.0041	9.2	2.8
420	1589.9	0.0047	0.0011	4.4	1.4	880	3331.2	0.0185	0.0042	9.3	2.8
430	1627.7	0.0049	0.0011	4.5	1.4	890	3369.0	0.0189	0.0043	9.4	2.9
440	1665.6	0.0051	0.0012	4.6	1.4	900	3406.9	0.0192	0.0044	9.5	2.9
450	1703.4	0.0053	0.0012	4.7	1.4	910	3444.7	0.0196	0.0044	9.6	2.9
460	1741.3	0.0056	0.0013	4.9	1.5	920	3482.6	0.0200	0.0045	9.7	3.0
470	1779.1	0.0058	0.0013	5.0	1.5	930	3520.4	0.0205	0.0046	9.8	3.0
480	1817.0	0.0060	0.0014	5.1	1.5	940	3558.3	0.0209	0.0047	9.9	3.0
490	1854.8	0.0063	0.0014	5.2	1.6	950	3596.1	0.0213	0.0048	10.0	3.1
500	1892.7	0.0065	0.0015	5.3	1.6	960	3634.0	0.0217	0.0049	10.1	3.1
510	1930.6	0.0067	0.0015	5.4	1.6	970	3671.8	0.0221	0.0050	10.2	3.1
520	1968.4	0.0070	0.0016	5.5	1.7	980	3709.7	0.0225	0.0051	10.3	3.2
530	2006.3	0.0072	0.0016	5.6	1.7	990	3747.5	0.0230	0.0052	10.4	3.2
540	2044.1	0.0075	0.0017	5.7	1.7	1000	3785.4	0.0234	0.0053	10.6	3.2

(continued)

Table E.7 (continued)

Q		P_f		Velocity		Q		P_f		Velocity	
GPM	L/min	psi/ft	bar/m	fps	m/s	GPM	L/min	psi/ft	bar/m	fps	m/s
550	2082.0	0.0077	0.0018	5.8	1.8	1010	3823.3	0.0238	0.0054	10.7	3.2
560	2119.8	0.0080	0.0018	5.9	1.8	1020	3861.1	0.0243	0.0055	10.8	3.3
570	2157.7	0.0083	0.0019	6.0	1.8	1030	3899.0	0.0247	0.0056	10.9	3.3
580	2195.5	0.0085	0.0019	6.1	1.9	1040	3936.8	0.0252	0.0057	11.0	3.3
590	2233.4	0.0088	0.0020	6.2	1.9	1050	3974.7	0.0256	0.0058	11.1	3.4
600	2271.2	0.0091	0.0021	6.3	1.9	1060	4012.5	0.0261	0.0059	11.2	3.4
610	2309.1	0.0094	0.0021	6.4	2.0	1070	4050.4	0.0265	0.0060	11.3	3.4
620	2346.9	0.0097	0.0022	6.5	2.0	1080	4088.2	0.0270	0.0061	11.4	3.5
630	2384.8	0.0099	0.0023	6.6	2.0	1090	4126.1	0.0274	0.0062	11.5	3.5
640	2422.7	0.0102	0.0023	6.8	2.1	1100	4163.9	0.0279	0.0063	11.6	3.5
650	2460.5	0.0105	0.0024	6.9	2.1	1110	4201.8	0.0284	0.0064	11.7	3.6

Table E.8 Water flowing in 6-in. (150 mm) cement lined class 52 ductile iron pipe $C = 140$ I.D. = 6.16 in. (156.5 mm)

Q		P_f		Velocity		Q		P_f		Velocity	
GPM	L/min	psi/ft	bar/m	fps	m/s	GPM	L/min	psi/ft	bar/m	fps	m/s
200	757.1	0.0012	0.0003	2.2	0.7	660	2498.4	0.0114	0.0026	7.1	2.2
210	794.9	0.0014	0.0003	2.3	0.7	670	2536.2	0.0117	0.0026	7.2	2.2
220	832.8	0.0015	0.0003	2.4	0.7	680	2574.1	0.0120	0.0027	7.3	2.2
230	870.6	0.0016	0.0004	2.5	0.8	690	2611.9	0.0123	0.0028	7.4	2.3
240	908.5	0.0017	0.0004	2.6	0.8	700	2649.8	0.0127	0.0029	7.5	2.3
250	946.4	0.0019	0.0004	2.7	0.8	710	2687.6	0.0130	0.0029	7.6	2.3
260	984.2	0.0020	0.0005	2.8	0.9	720	2725.5	0.0134	0.0030	7.7	2.4
270	1022.1	0.0022	0.0005	2.9	0.9	730	2763.3	0.0137	0.0031	7.9	2.4
280	1059.9	0.0023	0.0005	3.0	0.9	740	2801.2	0.0140	0.0032	8.0	2.4
290	1097.8	0.0025	0.0006	3.1	1.0	750	2839.1	0.0144	0.0033	8.1	2.5
300	1135.6	0.0026	0.0006	3.2	1.0	760	2876.9	0.0148	0.0033	8.2	2.5
310	1173.5	0.0028	0.0006	3.3	1.0	770	2914.8	0.0151	0.0034	8.3	2.5
320	1211.3	0.0030	0.0007	3.4	1.0	780	2952.6	0.0155	0.0035	8.4	2.6
330	1249.2	0.0032	0.0007	3.6	1.1	790	2990.5	0.0159	0.0036	8.5	2.6
340	1287.0	0.0033	0.0008	3.7	1.1	800	3028.3	0.0162	0.0037	8.6	2.6
350	1324.9	0.0035	0.0008	3.8	1.1	810	3066.2	0.0166	0.0038	8.7	2.7
360	1362.7	0.0037	0.0008	3.9	1.2	820	3104.0	0.0170	0.0038	8.8	2.7
370	1400.6	0.0039	0.0009	4.0	1.2	830	3141.9	0.0174	0.0039	8.9	2.7
380	1438.5	0.0041	0.0009	4.1	1.2	840	3179.7	0.0178	0.0040	9.0	2.8
390	1476.3	0.0043	0.0010	4.2	1.3	850	3217.6	0.0182	0.0041	9.1	2.8
400	1514.2	0.0045	0.0010	4.3	1.3	860	3255.4	0.0186	0.0042	9.3	2.8
410	1552.0	0.0047	0.0011	4.4	1.3	870	3293.3	0.0190	0.0043	9.4	2.9
420	1589.9	0.0049	0.0011	4.5	1.4	880	3331.2	0.0194	0.0044	9.5	2.9
430	1627.7	0.0051	0.0012	4.6	1.4	890	3369.0	0.0198	0.0045	9.6	2.9
440	1665.6	0.0054	0.0012	4.7	1.4	900	3406.9	0.0202	0.0046	9.7	3.0
450	1703.4	0.0056	0.0013	4.8	1.5	910	3444.7	0.0206	0.0047	9.8	3.0
460	1741.3	0.0058	0.0013	4.9	1.5	920	3482.6	0.0210	0.0047	9.9	3.0
470	1779.1	0.0061	0.0014	5.1	1.5	930	3520.4	0.0214	0.0048	10.0	3.1
480	1817.0	0.0063	0.0014	5.2	1.6	940	3558.3	0.0219	0.0049	10.1	3.1
490	1854.8	0.0066	0.0015	5.3	1.6	950	3596.1	0.0223	0.0050	10.2	3.1
500	1892.7	0.0068	0.0015	5.4	1.6	960	3634.0	0.0227	0.0051	10.3	3.1
510	1930.6	0.0071	0.0016	5.5	1.7	970	3671.8	0.0232	0.0052	10.4	3.2
520	1968.4	0.0073	0.0017	5.6	1.7	980	3709.7	0.0236	0.0053	10.5	3.2
530	2006.3	0.0076	0.0017	5.7	1.7	990	3747.5	0.0241	0.0054	10.7	3.2
540	2044.1	0.0078	0.0018	5.8	1.8	1000	3785.4	0.0245	0.0055	10.8	3.3

(continued)

Table E.8 (continued)

Q		P_f		Velocity		Q		P_f		Velocity	
GPM	L/min	psi/ft	bar/m	fps	m/s	GPM	L/min	psi/ft	bar/m	fps	m/s
550	2082.0	0.0081	0.0018	5.9	1.8	1010	3823.3	0.0250	0.0056	10.9	3.3
560	2119.8	0.0084	0.0019	6.0	1.8	1020	3861.1	0.0254	0.0057	11.0	3.3
570	2157.7	0.0087	0.0020	6.1	1.9	1030	3899.0	0.0259	0.0059	11.1	3.4
580	2195.5	0.0090	0.0020	6.2	1.9	1040	3936.8	0.0264	0.0060	11.2	3.4
590	2233.4	0.0092	0.0021	6.3	1.9	1050	3974.7	0.0268	0.0061	11.3	3.4
600	2271.2	0.0095	0.0022	6.5	2.0	1060	4012.5	0.0273	0.0062	11.4	3.5
610	2309.1	0.0098	0.0022	6.6	2.0	1070	4050.4	0.0278	0.0063	11.5	3.5
620	2346.9	0.0101	0.0023	6.7	2.0	1080	4088.2	0.0283	0.0064	11.6	3.5
630	2384.8	0.0104	0.0024	6.8	2.1	1090	4126.1	0.0288	0.0065	11.7	3.6
640	2422.7	0.0107	0.0024	6.9	2.1	1100	4163.9	0.0292	0.0066	11.8	3.6
650	2460.5	0.0111	0.0025	7.0	2.1	1110	4201.8	0.0297	0.0067	11.9	3.6

Table E.9 Water flowing in 8-in. (200 mm) unlined class 51 ductile iron pipe $C = 100$ I.D. = 8.45 in. (214.6 mm)

Q		P_f		Velocity		Q		P_f		Velocity	
GPM	L/min	psi/ft	bar/m	fps	m/s	GPM	L/min	psi/ft	bar/m	fps	m/s
500	1892.7	0.0027	0.0006	2.9	0.9	1650	6245.9	0.0248	0.0056	9.4	2.9
525	1987.3	0.0030	0.0007	3.0	0.9	1675	6340.5	0.0255	0.0058	9.6	2.9
550	2082.0	0.0032	0.0007	3.1	1.0	1700	6435.2	0.0262	0.0059	9.7	3.0
575	2176.6	0.0035	0.0008	3.3	1.0	1725	6529.8	0.0269	0.0061	9.9	3.0
600	2271.2	0.0038	0.0009	3.4	1.0	1750	6624.5	0.0276	0.0062	10.0	3.1
625	2365.9	0.0041	0.0009	3.6	1.1	1775	6719.1	0.0283	0.0064	10.1	3.1
650	2460.5	0.0044	0.0010	3.7	1.1	1800	6813.7	0.0291	0.0066	10.3	3.1
675	2555.1	0.0047	0.0011	3.9	1.2	1825	6908.4	0.0298	0.0068	10.4	3.2
700	2649.8	0.0051	0.0011	4.0	1.2	1850	7003.0	0.0306	0.0069	10.6	3.2
725	2744.4	0.0054	0.0012	4.1	1.3	1875	7097.6	0.0314	0.0071	10.7	3.3
750	2839.1	0.0058	0.0013	4.3	1.3	1900	7192.3	0.0321	0.0073	10.9	3.3
775	2933.7	0.0061	0.0014	4.4	1.4	1925	7286.9	0.0329	0.0075	11.0	3.4
800	3028.3	0.0065	0.0015	4.6	1.4	1950	7381.5	0.0337	0.0076	11.2	3.4
825	3123.0	0.0069	0.0016	4.7	1.4	1975	7476.2	0.0345	0.0078	11.3	3.4
850	3217.6	0.0073	0.0016	4.9	1.5	2000	7570.8	0.0353	0.0080	11.4	3.5
875	3312.2	0.0077	0.0017	5.0	1.5	2025	7665.4	0.0362	0.0082	11.6	3.5
900	3406.9	0.0081	0.0018	5.1	1.6	2050	7760.1	0.0370	0.0084	11.7	3.6
925	3501.5	0.0085	0.0019	5.3	1.6	2075	7854.7	0.0378	0.0086	11.9	3.6
950	3596.1	0.0089	0.0020	5.4	1.7	2100	7949.3	0.0387	0.0088	12.0	3.7
975	3690.8	0.0094	0.0021	5.6	1.7	2125	8044.0	0.0395	0.0090	12.2	3.7
1000	3785.4	0.0098	0.0022	5.7	1.7	2150	8138.6	0.0404	0.0091	12.3	3.7
1025	3880.0	0.0103	0.0023	5.9	1.8	2175	8233.2	0.0413	0.0093	12.4	3.8
1050	3974.7	0.0107	0.0024	6.0	1.8	2200	8327.9	0.0422	0.0095	12.6	3.8
1075	4069.3	0.0112	0.0025	6.1	1.9	2225	8422.5	0.0430	0.0097	12.7	3.9
1100	4163.9	0.0117	0.0026	6.3	1.9	2250	8517.2	0.0439	0.0099	12.9	3.9
1125	4258.6	0.0122	0.0028	6.4	2.0	2275	8611.8	0.0448	0.0102	13.0	4.0
1150	4353.2	0.0127	0.0029	6.6	2.0	2300	8706.4	0.0458	0.0104	13.2	4.0
1175	4447.8	0.0132	0.0030	6.7	2.0	2325	8801.1	0.0467	0.0106	13.3	4.1
1200	4542.5	0.0137	0.0031	6.9	2.1	2350	8895.7	0.0476	0.0108	13.4	4.1
1225	4637.1	0.0143	0.0032	7.0	2.1	2375	8990.3	0.0486	0.0110	13.6	4.1
1250	4731.8	0.0148	0.0034	7.1	2.2	2400	9085.0	0.0495	0.0112	13.7	4.2
1275	4826.4	0.0154	0.0035	7.3	2.2	2425	9179.6	0.0505	0.0114	13.9	4.2
1300	4921.0	0.0159	0.0036	7.4	2.3	2450	9274.2	0.0514	0.0116	14.0	4.3
1325	5015.7	0.0165	0.0037	7.6	2.3	2475	9368.9	0.0524	0.0119	14.2	4.3
1350	5110.3	0.0171	0.0039	7.7	2.4	2500	9463.5	0.0534	0.0121	14.3	4.4

(continued)

Table E.9 (continued)

Q		P_f		Velocity		Q		P_f		Velocity	
GPM	L/min	psi/ft	bar/m	fps	m/s	GPM	L/min	psi/ft	bar/m	fps	m/s
1375	5204.9	0.0177	0.0040	7.9	2.4	2525	9558.1	0.0544	0.0123	14.4	4.4
1400	5299.6	0.0183	0.0041	8.0	2.4	2550	9652.8	0.0554	0.0125	14.6	4.4
1425	5394.2	0.0189	0.0043	8.1	2.5	2575	9747.4	0.0564	0.0128	14.7	4.5
1450	5488.8	0.0195	0.0044	8.3	2.5	2600	9842.0	0.0574	0.0130	14.9	4.5
1475	5583.5	0.0201	0.0046	8.4	2.6	2625	9936.7	0.0584	0.0132	15.0	4.6
1500	5678.1	0.0208	0.0047	8.6	2.6	2650	10,031.3	0.0595	0.0135	15.2	4.6
1525	5772.7	0.0214	0.0048	8.7	2.7	2675	10,125.9	0.0605	0.0137	15.3	4.7
1550	5867.4	0.0221	0.0050	8.9	2.7	2700	10,220.6	0.0616	0.0139	15.4	4.7
1575	5962.0	0.0227	0.0051	9.0	2.7	2725	10,315.2	0.0626	0.0142	15.6	4.7
1600	6056.6	0.0234	0.0053	9.1	2.8	2750	10,409.9	0.0637	0.0144	15.7	4.8
1625	6151.3	0.0241	0.0054	9.3	2.8	2775	10,504.5	0.0648	0.0147	15.9	4.8

Table E.10 Water flowing in 8-in. (200 mm) unlined class 52 ductile iron pipe $C = 100$ I.D. = 8.385 in. (213.0 mm)

Q		P_f		Velocity		Q		P_f		Velocity	
GPM	L/min	psi/ft	bar/m	fps	m/s	GPM	L/min	psi/ft	bar/m	fps	m/s
500	1892.7	0.0028	0.0006	2.9	0.9	1650	6245.9	0.0257	0.0058	9.6	2.9
525	1987.3	0.0031	0.0007	3.0	0.9	1675	6340.5	0.0264	0.0060	9.7	3.0
550	2082.0	0.0034	0.0008	3.2	1.0	1700	6435.2	0.0272	0.0061	9.9	3.0
575	2176.6	0.0037	0.0008	3.3	1.0	1725	6529.8	0.0279	0.0063	10.0	3.1
600	2271.2	0.0040	0.0009	3.5	1.1	1750	6624.5	0.0287	0.0065	10.2	3.1
625	2365.9	0.0043	0.0010	3.6	1.1	1775	6719.1	0.0294	0.0067	10.3	3.1
650	2460.5	0.0046	0.0010	3.8	1.2	1800	6813.7	0.0302	0.0068	10.5	3.2
675	2555.1	0.0049	0.0011	3.9	1.2	1825	6908.4	0.0310	0.0070	10.6	3.2
700	2649.8	0.0053	0.0012	4.1	1.2	1850	7003.0	0.0318	0.0072	10.7	3.3
725	2744.4	0.0056	0.0013	4.2	1.3	1875	7097.6	0.0326	0.0074	10.9	3.3
750	2839.1	0.0060	0.0014	4.4	1.3	1900	7192.3	0.0334	0.0075	11.0	3.4
775	2933.7	0.0064	0.0014	4.5	1.4	1925	7286.9	0.0342	0.0077	11.2	3.4
800	3028.3	0.0067	0.0015	4.6	1.4	1950	7381.5	0.0350	0.0079	11.3	3.5
825	3123.0	0.0071	0.0016	4.8	1.5	1975	7476.2	0.0358	0.0081	11.5	3.5
850	3217.6	0.0075	0.0017	4.9	1.5	2000	7570.8	0.0367	0.0083	11.6	3.5
875	3312.2	0.0080	0.0018	5.1	1.5	2025	7665.4	0.0375	0.0085	11.8	3.6
900	3406.9	0.0084	0.0019	5.2	1.6	2050	7760.1	0.0384	0.0087	11.9	3.6
925	3501.5	0.0088	0.0020	5.4	1.6	2075	7854.7	0.0393	0.0089	12.1	3.7
950	3596.1	0.0093	0.0021	5.5	1.7	2100	7949.3	0.0402	0.0091	12.2	3.7
975	3690.8	0.0097	0.0022	5.7	1.7	2125	8044.0	0.0410	0.0093	12.3	3.8
1000	3785.4	0.0102	0.0023	5.8	1.8	2150	8138.6	0.0419	0.0095	12.5	3.8
1025	3880.0	0.0107	0.0024	6.0	1.8	2175	8233.2	0.0429	0.0097	12.6	3.8
1050	3974.7	0.0111	0.0025	6.1	1.9	2200	8327.9	0.0438	0.0099	12.8	3.9
1075	4069.3	0.0116	0.0026	6.2	1.9	2225	8422.5	0.0447	0.0101	12.9	3.9
1100	4163.9	0.0121	0.0027	6.4	1.9	2250	8517.2	0.0456	0.0103	13.1	4.0
1125	4258.6	0.0127	0.0029	6.5	2.0	2275	8611.8	0.0466	0.0105	13.2	4.0
1150	4353.2	0.0132	0.0030	6.7	2.0	2300	8706.4	0.0475	0.0107	13.4	4.1
1175	4447.8	0.0137	0.0031	6.8	2.1	2325	8801.1	0.0485	0.0110	13.5	4.1
1200	4542.5	0.0143	0.0032	7.0	2.1	2350	8895.7	0.0494	0.0112	13.6	4.2
1225	4637.1	0.0148	0.0034	7.1	2.2	2375	8990.3	0.0504	0.0114	13.8	4.2
1250	4731.8	0.0154	0.0035	7.3	2.2	2400	9085.0	0.0514	0.0116	13.9	4.2
1275	4826.4	0.0160	0.0036	7.4	2.3	2425	9179.6	0.0524	0.0119	14.1	4.3
1300	4921.0	0.0165	0.0037	7.5	2.3	2450	9274.2	0.0534	0.0121	14.2	4.3
1325	5015.7	0.0171	0.0039	7.7	2.3	2475	9368.9	0.0544	0.0123	14.4	4.4
1350	5110.3	0.0177	0.0040	7.8	2.4	2500	9463.5	0.0554	0.0125	14.5	4.4

(continued)

Table E.10 (continued)

Q		P_f		Velocity		Q		P_f		Velocity	
GPM	L/min	psi/ft	bar/m	fps	m/s	GPM	L/min	psi/ft	bar/m	fps	m/s
1375	5204.9	0.0183	0.0041	8.0	2.4	2525	9558.1	0.0565	0.0128	14.7	4.5
1400	5299.6	0.0190	0.0043	8.1	2.5	2550	9652.8	0.0575	0.0130	14.8	4.5
1425	5394.2	0.0196	0.0044	8.3	2.5	2575	9747.4	0.0586	0.0132	15.0	4.6
1450	5488.8	0.0202	0.0046	8.4	2.6	2600	9842.0	0.0596	0.0135	15.1	4.6
1475	5583.5	0.0209	0.0047	8.6	2.6	2625	9936.7	0.0607	0.0137	15.2	4.6
1500	5678.1	0.0216	0.0049	8.7	2.7	2650	10,031.3	0.0618	0.0140	15.4	4.7
1525	5772.7	0.0222	0.0050	8.9	2.7	2675	10,125.9	0.0628	0.0142	15.5	4.7
1550	5867.4	0.0229	0.0052	9.0	2.7	2700	10,220.6	0.0639	0.0145	15.7	4.8
1575	5962.0	0.0236	0.0053	9.1	2.8	2725	10,315.2	0.0650	0.0147	15.8	4.8
1600	6056.6	0.0243	0.0055	9.3	2.8	2750	10,409.9	0.0661	0.0150	16.0	4.9
1625	6151.3	0.0250	0.0057	9.4	2.9	2775	10,504.5	0.0673	0.0152	16.1	4.9

Table E.11 Water flowing in 8-in. (200 mm) cement lined class 51 ductile iron pipe $C = 140$ I.D. = 8.33 in. (211.6 mm)

Q		P_f		Velocity		Q		P_f		Velocity	
GPM	L/min	psi/ft	bar/m	fps	m/s	GPM	L/min	psi/ft	bar/m	fps	m/s
500	1892.7	0.0016	0.0004	2.9	0.9	1650	6245.9	0.0142	0.0032	9.7	3.0
525	1987.3	0.0017	0.0004	3.1	0.9	1675	6340.5	0.0146	0.0033	9.9	3.0
550	2082.0	0.0019	0.0004	3.2	1.0	1700	6435.2	0.0151	0.0034	10.0	3.0
575	2176.6	0.0020	0.0005	3.4	1.0	1725	6529.8	0.0155	0.0035	10.2	3.1
600	2271.2	0.0022	0.0005	3.5	1.1	1750	6624.5	0.0159	0.0036	10.3	3.1
625	2365.9	0.0024	0.0005	3.7	1.1	1775	6719.1	0.0163	0.0037	10.4	3.2
650	2460.5	0.0025	0.0006	3.8	1.2	1800	6813.7	0.0167	0.0038	10.6	3.2
675	2555.1	0.0027	0.0006	4.0	1.2	1825	6908.4	0.0172	0.0039	10.7	3.3
700	2649.8	0.0029	0.0007	4.1	1.3	1850	7003.0	0.0176	0.0040	10.9	3.3
725	2744.4	0.0031	0.0007	4.3	1.3	1875	7097.6	0.0180	0.0041	11.0	3.4
750	2839.1	0.0033	0.0007	4.4	1.3	1900	7192.3	0.0185	0.0042	11.2	3.4
775	2933.7	0.0035	0.0008	4.6	1.4	1925	7286.9	0.0189	0.0043	11.3	3.5
800	3028.3	0.0037	0.0008	4.7	1.4	1950	7381.5	0.0194	0.0044	11.5	3.5
825	3123.0	0.0040	0.0009	4.9	1.5	1975	7476.2	0.0199	0.0045	11.6	3.5
850	3217.6	0.0042	0.0009	5.0	1.5	2000	7570.8	0.0203	0.0046	11.8	3.6
875	3312.2	0.0044	0.0010	5.1	1.6	2025	7665.4	0.0208	0.0047	11.9	3.6
900	3406.9	0.0046	0.0010	5.3	1.6	2050	7760.1	0.0213	0.0048	12.1	3.7
925	3501.5	0.0049	0.0011	5.4	1.7	2075	7854.7	0.0218	0.0049	12.2	3.7
950	3596.1	0.0051	0.0012	5.6	1.7	2100	7949.3	0.0223	0.0050	12.4	3.8
975	3690.8	0.0054	0.0012	5.7	1.7	2125	8044.0	0.0227	0.0051	12.5	3.8
1000	3785.4	0.0056	0.0013	5.9	1.8	2150	8138.6	0.0232	0.0053	12.7	3.9
1025	3880.0	0.0059	0.0013	6.0	1.8	2175	8233.2	0.0237	0.0054	12.8	3.9
1050	3974.7	0.0062	0.0014	6.2	1.9	2200	8327.9	0.0243	0.0055	12.9	3.9
1075	4069.3	0.0064	0.0015	6.3	1.9	2225	8422.5	0.0248	0.0056	13.1	4.0
1100	4163.9	0.0067	0.0015	6.5	2.0	2250	8517.2	0.0253	0.0057	13.2	4.0
1125	4258.6	0.0070	0.0016	6.6	2.0	2275	8611.8	0.0258	0.0058	13.4	4.1
1150	4353.2	0.0073	0.0017	6.8	2.1	2300	8706.4	0.0263	0.0060	13.5	4.1
1175	4447.8	0.0076	0.0017	6.9	2.1	2325	8801.1	0.0269	0.0061	13.7	4.2
1200	4542.5	0.0079	0.0018	7.1	2.2	2350	8895.7	0.0274	0.0062	13.8	4.2
1225	4637.1	0.0082	0.0019	7.2	2.2	2375	8990.3	0.0279	0.0063	14.0	4.3
1250	4731.8	0.0085	0.0019	7.4	2.2	2400	9085.0	0.0285	0.0064	14.1	4.3
1275	4826.4	0.0088	0.0020	7.5	2.3	2425	9179.6	0.0290	0.0066	14.3	4.3
1300	4921.0	0.0092	0.0021	7.6	2.3	2450	9274.2	0.0296	0.0067	14.4	4.4
1325	5015.7	0.0095	0.0021	7.8	2.4	2475	9368.9	0.0302	0.0068	14.6	4.4
1350	5110.3	0.0098	0.0022	7.9	2.4	2500	9463.5	0.0307	0.0069	14.7	4.5

(continued)

Table E.11 (continued)

Q		P_f		Velocity		Q		P_f		Velocity	
GPM	L/min	psi/ft	bar/m	fps	m/s	GPM	L/min	psi/ft	bar/m	fps	m/s
1375	5204.9	0.0102	0.0023	8.1	2.5	2525	9558.1	0.0313	0.0071	14.9	4.5
1400	5299.6	0.0105	0.0024	8.2	2.5	2550	9652.8	0.0319	0.0072	15.0	4.6
1425	5394.2	0.0109	0.0025	8.4	2.6	2575	9747.4	0.0324	0.0073	15.2	4.6
1450	5488.8	0.0112	0.0025	8.5	2.6	2600	9842.0	0.0330	0.0075	15.3	4.7
1475	5583.5	0.0116	0.0026	8.7	2.6	2625	9936.7	0.0336	0.0076	15.4	4.7
1500	5678.1	0.0119	0.0027	8.8	2.7	2650	10,031.3	0.0342	0.0077	15.6	4.8
1525	5772.7	0.0123	0.0028	9.0	2.7	2675	10,125.9	0.0348	0.0079	15.7	4.8
1550	5867.4	0.0127	0.0029	9.1	2.8	2700	10,220.6	0.0354	0.0080	15.9	4.8
1575	5962.0	0.0131	0.0030	9.3	2.8	2725	10,315.2	0.0360	0.0081	16.0	4.9
1600	6056.6	0.0135	0.0030	9.4	2.9	2750	10,409.9	0.0366	0.0083	16.2	4.9
1625	6151.3	0.0138	0.0031	9.6	2.9	2775	10,504.5	0.0373	0.0084	16.3	5.0

Table E.12 Water flowing in 8-in. (200 mm) cement lined class 52 ductile iron pipe $C = 140$ I.D. = 8.27 in. (210.1 mm)

Q		P_f		Velocity		Q		P_f		Velocity	
GPM	L/min	psi/ft	bar/m	fps	m/s	GPM	L/min	psi/ft	bar/m	fps	m/s
500	1892.7	0.0016	0.0004	3.0	0.9	1650	6245.9	0.0148	0.0033	9.9	3.0
525	1987.3	0.0018	0.0004	3.1	1.0	1675	6340.5	0.0152	0.0034	10.0	3.0
550	2082.0	0.0019	0.0004	3.3	1.0	1700	6435.2	0.0156	0.0035	10.1	3.1
575	2176.6	0.0021	0.0005	3.4	1.0	1725	6529.8	0.0160	0.0036	10.3	3.1
600	2271.2	0.0023	0.0005	3.6	1.1	1750	6624.5	0.0165	0.0037	10.4	3.2
625	2365.9	0.0024	0.0006	3.7	1.1	1775	6719.1	0.0169	0.0038	10.6	3.2
650	2460.5	0.0026	0.0006	3.9	1.2	1800	6813.7	0.0173	0.0039	10.7	3.3
675	2555.1	0.0028	0.0006	4.0	1.2	1825	6908.4	0.0178	0.0040	10.9	3.3
700	2649.8	0.0030	0.0007	4.2	1.3	1850	7003.0	0.0182	0.0041	11.0	3.4
725	2744.4	0.0032	0.0007	4.3	1.3	1875	7097.6	0.0187	0.0042	11.2	3.4
750	2839.1	0.0034	0.0008	4.5	1.4	1900	7192.3	0.0192	0.0043	11.3	3.5
775	2933.7	0.0036	0.0008	4.6	1.4	1925	7286.9	0.0196	0.0044	11.5	3.5
800	3028.3	0.0039	0.0009	4.8	1.5	1950	7381.5	0.0201	0.0045	11.6	3.5
825	3123.0	0.0041	0.0009	4.9	1.5	1975	7476.2	0.0206	0.0047	11.8	3.6
850	3217.6	0.0043	0.0010	5.1	1.5	2000	7570.8	0.0211	0.0048	11.9	3.6
875	3312.2	0.0046	0.0010	5.2	1.6	2025	7665.4	0.0215	0.0049	12.1	3.7
900	3406.9	0.0048	0.0011	5.4	1.6	2050	7760.1	0.0220	0.0050	12.2	3.7
925	3501.5	0.0051	0.0011	5.5	1.7	2075	7854.7	0.0225	0.0051	12.4	3.8
950	3596.1	0.0053	0.0012	5.7	1.7	2100	7949.3	0.0230	0.0052	12.5	3.8
975	3690.8	0.0056	0.0013	5.8	1.8	2125	8044.0	0.0236	0.0053	12.7	3.9
1000	3785.4	0.0058	0.0013	6.0	1.8	2150	8138.6	0.0241	0.0054	12.8	3.9
1025	3880.0	0.0061	0.0014	6.1	1.9	2175	8233.2	0.0246	0.0056	13.0	4.0
1050	3974.7	0.0064	0.0014	6.3	1.9	2200	8327.9	0.0251	0.0057	13.1	4.0
1075	4069.3	0.0067	0.0015	6.4	2.0	2225	8422.5	0.0257	0.0058	13.3	4.0
1100	4163.9	0.0070	0.0016	6.6	2.0	2250	8517.2	0.0262	0.0059	13.4	4.1
1125	4258.6	0.0073	0.0016	6.7	2.0	2275	8611.8	0.0267	0.0060	13.6	4.1
1150	4353.2	0.0076	0.0017	6.9	2.1	2300	8706.4	0.0273	0.0062	13.7	4.2
1175	4447.8	0.0079	0.0018	7.0	2.1	2325	8801.1	0.0278	0.0063	13.9	4.2
1200	4542.5	0.0082	0.0019	7.2	2.2	2350	8895.7	0.0284	0.0064	14.0	4.3
1225	4637.1	0.0085	0.0019	7.3	2.2	2375	8990.3	0.0289	0.0065	14.2	4.3
1250	4731.8	0.0088	0.0020	7.5	2.3	2400	9085.0	0.0295	0.0067	14.3	4.4
1275	4826.4	0.0092	0.0021	7.6	2.3	2425	9179.6	0.0301	0.0068	14.5	4.4
1300	4921.0	0.0095	0.0021	7.8	2.4	2450	9274.2	0.0307	0.0069	14.6	4.5
1325	5015.7	0.0098	0.0022	7.9	2.4	2475	9368.9	0.0312	0.0071	14.8	4.5
1350	5110.3	0.0102	0.0023	8.1	2.5	2500	9463.5	0.0318	0.0072	14.9	4.5

(continued)

Table E.12 (continued)

Q		P_f		Velocity		Q		P_f		Velocity	
GPM	L/min	psi/ft	bar/m	fps	m/s	GPM	L/min	psi/ft	bar/m	fps	m/s
1375	5204.9	0.0105	0.0024	8.2	2.5	2525	9558.1	0.0324	0.0073	15.1	4.6
1400	5299.6	0.0109	0.0025	8.4	2.5	2550	9652.8	0.0330	0.0075	15.2	4.6
1425	5394.2	0.0112	0.0025	8.5	2.6	2575	9747.4	0.0336	0.0076	15.4	4.7
1450	5488.8	0.0116	0.0026	8.7	2.6	2600	9842.0	0.0342	0.0077	15.5	4.7
1475	5583.5	0.0120	0.0027	8.8	2.7	2625	9936.7	0.0348	0.0079	15.7	4.8
1500	5678.1	0.0124	0.0028	9.0	2.7	2650	10,031.3	0.0354	0.0080	15.8	4.8
1525	5772.7	0.0128	0.0029	9.1	2.8	2675	10,125.9	0.0361	0.0082	16.0	4.9
1550	5867.4	0.0131	0.0030	9.3	2.8	2700	10,220.6	0.0367	0.0083	16.1	4.9
1575	5962.0	0.0135	0.0031	9.4	2.9	2725	10,315.2	0.0373	0.0084	16.3	5.0
1600	6056.6	0.0139	0.0032	9.6	2.9	2750	10,409.9	0.0380	0.0086	16.4	5.0
1625	6151.3	0.0143	0.0032	9.7	3.0	2775	10,504.5	0.0386	0.0087	16.6	5.0

Appendix F
Friction Loss Tables for Plastic Underground Pipe

P. Brock and V. Charter, *Fire Protection Hydraulics and Water Supply Analysis*, The Society of Fire Protection Engineers Series, https://doi.org/10.1007/978-3-032-09877-1

Table F.1 Water flowing in 4-in. (100 mm) class 150 PVC plastic pipe $C = 150$ I.D. = 4.24 in. (107.7 mm)

Q		P_f		Velocity		Q		P_f		Velocity	
GPM	L/min	psi/ft	bar/m	fps	m/s	GPM	L/min	psi/ft	bar/m	fps	m/s
100	378.5	0.002	0.0004	2.3	0.7	560	2119.8	0.046	0.0103	12.7	3.9
110	416.4	0.002	0.0005	2.5	0.8	570	2157.7	0.047	0.0106	12.9	3.9
120	454.2	0.003	0.0006	2.7	0.8	580	2195.5	0.049	0.0110	13.2	4.0
130	492.1	0.003	0.0007	3.0	0.9	590	2233.4	0.050	0.0113	13.4	4.1
140	530.0	0.004	0.0008	3.2	1.0	600	2271.2	0.052	0.0117	13.6	4.2
150	567.8	0.004	0.0009	3.4	1.0	610	2309.1	0.053	0.0121	13.9	4.2
160	605.7	0.004	0.0010	3.6	1.1	620	2346.9	0.055	0.0124	14.1	4.3
170	643.5	0.005	0.0011	3.9	1.2	630	2384.8	0.057	0.0128	14.3	4.4
180	681.4	0.006	0.0013	4.1	1.2	640	2422.7	0.058	0.0132	14.5	4.4
190	719.2	0.006	0.0014	4.3	1.3	650	2460.5	0.060	0.0136	14.8	4.5
200	757.1	0.007	0.0015	4.5	1.4	660	2498.4	0.062	0.0140	15.0	4.6
210	794.9	0.007	0.0017	4.8	1.5	670	2536.2	0.063	0.0143	15.2	4.6
220	832.8	0.008	0.0018	5.0	1.5	680	2574.1	0.065	0.0147	15.4	4.7
230	870.6	0.009	0.0020	5.2	1.6	690	2611.9	0.067	0.0152	15.7	4.8
240	908.5	0.009	0.0021	5.5	1.7	700	2649.8	0.069	0.0156	15.9	4.8
250	946.4	0.010	0.0023	5.7	1.7	710	2687.6	0.071	0.0160	16.1	4.9
260	984.2	0.011	0.0025	5.9	1.8	720	2725.5	0.072	0.0164	16.4	5.0
270	1022.1	0.012	0.0027	6.1	1.9	730	2763.3	0.074	0.0168	16.6	5.1
280	1059.9	0.013	0.0029	6.4	1.9	740	2801.2	0.076	0.0172	16.8	5.1
290	1097.8	0.013	0.0030	6.6	2.0	750	2839.1	0.078	0.0177	17.0	5.2
300	1135.6	0.014	0.0032	6.8	2.1	760	2876.9	0.080	0.0181	17.3	5.3
310	1173.5	0.015	0.0034	7.0	2.1	770	2914.8	0.082	0.0186	17.5	5.3
320	1211.3	0.016	0.0037	7.3	2.2	780	2952.6	0.084	0.0190	17.7	5.4
330	1249.2	0.017	0.0039	7.5	2.3	790	2990.5	0.086	0.0195	17.9	5.5
340	1287.0	0.018	0.0041	7.7	2.4	800	3028.3	0.088	0.0199	18.2	5.5
350	1324.9	0.019	0.0043	7.9	2.4	810	3066.2	0.090	0.0204	18.4	5.6
360	1362.7	0.020	0.0045	8.2	2.5	820	3104.0	0.092	0.0209	18.6	5.7
370	1400.6	0.021	0.0048	8.4	2.6	830	3141.9	0.094	0.0213	18.9	5.7
380	1438.5	0.022	0.0050	8.6	2.6	840	3179.7	0.096	0.0218	19.1	5.8
390	1476.3	0.023	0.0053	8.9	2.7	850	3217.6	0.099	0.0223	19.3	5.9
400	1514.2	0.024	0.0055	9.1	2.8	860	3255.4	0.101	0.0228	19.5	6.0
410	1552.0	0.026	0.0058	9.3	2.8	870	3293.3	0.103	0.0233	19.8	6.0
420	1589.9	0.027	0.0060	9.5	2.9	880	3331.2	0.105	0.0238	20.0	6.1
430	1627.7	0.028	0.0063	9.8	3.0	890	3369.0	0.107	0.0243	20.2	6.2
440	1665.6	0.029	0.0066	10.0	3.0	900	3406.9	0.110	0.0248	20.4	6.2

(continued)

Table F.1 (continued)

Q		P_f		Velocity		Q		P_f		Velocity	
GPM	L/min	psi/ft	bar/m	fps	m/s	GPM	L/min	psi/ft	bar/m	fps	m/s
450	1703.4	0.030	0.0069	10.2	3.1	910	3444.7	0.112	0.0253	20.7	6.3
460	1741.3	0.032	0.0072	10.4	3.2	920	3482.6	0.114	0.0258	20.9	6.4
470	1779.1	0.033	0.0074	10.7	3.3	930	3520.4	0.116	0.0263	21.1	6.4
480	1817.0	0.034	0.0077	10.9	3.3	940	3558.3	0.119	0.0268	21.3	6.5
490	1854.8	0.036	0.0080	11.1	3.4	950	3596.1	0.121	0.0274	21.6	6.6
500	1892.7	0.037	0.0084	11.4	3.5	960	3634.0	0.123	0.0279	21.8	6.6
510	1930.6	0.038	0.0087	11.6	3.5	970	3671.8	0.126	0.0285	22.0	6.7
520	1968.4	0.040	0.0090	11.8	3.6	980	3709.7	0.128	0.0290	22.3	6.8
530	2006.3	0.041	0.0093	12.0	3.7	990	3747.5	0.131	0.0295	22.5	6.9
540	2044.1	0.043	0.0096	12.3	3.7	1000	3785.4	0.133	0.0301	22.7	6.9
550	2082.0	0.044	0.0100	12.5	3.8	1010	3823.3	0.136	0.0307	22.9	7.0

Table F.2 Water flowing in 4-in. (100 mm) class 200 PVC plastic pipe $C = 150$ I.D. = 4.08 in. (103.6 mm)

Q		P_f		Velocity		Q		P_f		Velocity	
GPM	L/min	psi/ft	bar/m	fps	m/s	GPM	L/min	psi/ft	bar/m	fps	m/s
100	378.5	0.002	0.0005	2.5	0.7	560	2119.8	0.055	0.0124	13.7	4.2
110	416.4	0.003	0.0006	2.7	0.8	570	2157.7	0.057	0.0129	14.0	4.3
120	454.2	0.003	0.0007	2.9	0.9	580	2195.5	0.059	0.0133	14.2	4.3
130	492.1	0.004	0.0008	3.2	1.0	590	2233.4	0.060	0.0137	14.5	4.4
140	530.0	0.004	0.0010	3.4	1.0	600	2271.2	0.062	0.0141	14.7	4.5
150	567.8	0.005	0.0011	3.7	1.1	610	2309.1	0.064	0.0146	15.0	4.6
160	605.7	0.005	0.0012	3.9	1.2	620	2346.9	0.066	0.0150	15.2	4.6
170	643.5	0.006	0.0014	4.2	1.3	630	2384.8	0.068	0.0155	15.5	4.7
180	681.4	0.007	0.0015	4.4	1.3	640	2422.7	0.070	0.0159	15.7	4.8
190	719.2	0.007	0.0017	4.7	1.4	650	2460.5	0.072	0.0164	15.9	4.9
200	757.1	0.008	0.0019	4.9	1.5	660	2498.4	0.074	0.0169	16.2	4.9
210	794.9	0.009	0.0020	5.2	1.6	670	2536.2	0.077	0.0173	16.4	5.0
220	832.8	0.010	0.0022	5.4	1.6	680	2574.1	0.079	0.0178	16.7	5.1
230	870.6	0.011	0.0024	5.6	1.7	690	2611.9	0.081	0.0183	16.9	5.2
240	908.5	0.011	0.0026	5.9	1.8	700	2649.8	0.083	0.0188	17.2	5.2
250	946.4	0.012	0.0028	6.1	1.9	710	2687.6	0.085	0.0193	17.4	5.3
260	984.2	0.013	0.0030	6.4	1.9	720	2725.5	0.087	0.0198	17.7	5.4
270	1022.1	0.014	0.0032	6.6	2.0	730	2763.3	0.090	0.0203	17.9	5.5
280	1059.9	0.015	0.0035	6.9	2.1	740	2801.2	0.092	0.0208	18.2	5.5
290	1097.8	0.016	0.0037	7.1	2.2	750	2839.1	0.094	0.0214	18.4	5.6
300	1135.6	0.017	0.0039	7.4	2.2	760	2876.9	0.097	0.0219	18.6	5.7
310	1173.5	0.018	0.0042	7.6	2.3	770	2914.8	0.099	0.0224	18.9	5.8
320	1211.3	0.019	0.0044	7.8	2.4	780	2952.6	0.101	0.0230	19.1	5.8
330	1249.2	0.021	0.0047	8.1	2.5	790	2990.5	0.104	0.0235	19.4	5.9
340	1287.0	0.022	0.0049	8.3	2.5	800	3028.3	0.106	0.0241	19.6	6.0
350	1324.9	0.023	0.0052	8.6	2.6	810	3066.2	0.109	0.0246	19.9	6.1
360	1362.7	0.024	0.0055	8.8	2.7	820	3104.0	0.111	0.0252	20.1	6.1
370	1400.6	0.026	0.0058	9.1	2.8	830	3141.9	0.114	0.0258	20.4	6.2
380	1438.5	0.027	0.0061	9.3	2.8	840	3179.7	0.116	0.0263	20.6	6.3
390	1476.3	0.028	0.0064	9.6	2.9	850	3217.6	0.119	0.0269	20.8	6.4
400	1514.2	0.029	0.0067	9.8	3.0	860	3255.4	0.121	0.0275	21.1	6.4
410	1552.0	0.031	0.0070	10.1	3.1	870	3293.3	0.124	0.0281	21.3	6.5
420	1589.9	0.032	0.0073	10.3	3.1	880	3331.2	0.127	0.0287	21.6	6.6
430	1627.7	0.034	0.0076	10.5	3.2	890	3369.0	0.129	0.0293	21.8	6.7
440	1665.6	0.035	0.0080	10.8	3.3	900	3406.9	0.132	0.0299	22.1	6.7

(continued)

Table F.2 (continued)

Q		P_f		Velocity		Q		P_f		Velocity	
GPM	L/min	psi/ft	bar/m	fps	m/s	GPM	L/min	psi/ft	bar/m	fps	m/s
450	1703.4	0.037	0.0083	11.0	3.4	910	3444.7	0.135	0.0305	22.3	6.8
460	1741.3	0.038	0.0086	11.3	3.4	920	3482.6	0.138	0.0312	22.6	6.9
470	1779.1	0.040	0.0090	11.5	3.5	930	3520.4	0.140	0.0318	22.8	7.0
480	1817.0	0.041	0.0094	11.8	3.6	940	3558.3	0.143	0.0324	23.1	7.0
490	1854.8	0.043	0.0097	12.0	3.7	950	3596.1	0.146	0.0331	23.3	7.1
500	1892.7	0.045	0.0101	12.3	3.7	960	3634.0	0.149	0.0337	23.5	7.2
510	1930.6	0.046	0.0105	12.5	3.8	970	3671.8	0.152	0.0344	23.8	7.3
520	1968.4	0.048	0.0108	12.8	3.9	980	3709.7	0.155	0.0350	24.0	7.3
530	2006.3	0.050	0.0112	13.0	4.0	990	3747.5	0.158	0.0357	24.3	7.4
540	2044.1	0.051	0.0116	13.2	4.0	1000	3785.4	0.160	0.0364	24.5	7.5
550	2082.0	0.053	0.0120	13.5	4.1	1010	3823.3	0.163	0.0370	24.8	7.6

Table F.3 Water flowing in 6-in. (150 mm) class 150 PVC plastic pipe $C = 150$ I.D. = 6.09 in. (154.7 mm)

Q		P_f		Velocity		Q		P_f		Velocity	
GPM	L/min	psi/ft	bar/m	fps	m/s	GPM	L/min	psi/ft	bar/m	fps	m/s
200	757.1	0.0012	0.0003	2.2	0.7	660	2498.4	0.0106	0.0024	7.3	2.2
210	794.9	0.0013	0.0003	2.3	0.7	670	2536.2	0.0109	0.0025	7.4	2.2
220	832.8	0.0014	0.0003	2.4	0.7	680	2574.1	0.0112	0.0025	7.5	2.3
230	870.6	0.0015	0.0003	2.5	0.8	690	2611.9	0.0115	0.0026	7.6	2.3
240	908.5	0.0016	0.0004	2.6	0.8	700	2649.8	0.0118	0.0027	7.7	2.3
250	946.4	0.0018	0.0004	2.8	0.8	710	2687.6	0.0121	0.0027	7.8	2.4
260	984.2	0.0019	0.0004	2.9	0.9	720	2725.5	0.0124	0.0028	7.9	2.4
270	1022.1	0.0020	0.0005	3.0	0.9	730	2763.3	0.0127	0.0029	8.0	2.4
280	1059.9	0.0022	0.0005	3.1	0.9	740	2801.2	0.0131	0.0030	8.1	2.5
290	1097.8	0.0023	0.0005	3.2	1.0	750	2839.1	0.0134	0.0030	8.3	2.5
300	1135.6	0.0025	0.0006	3.3	1.0	760	2876.9	0.0137	0.0031	8.4	2.6
310	1173.5	0.0026	0.0006	3.4	1.0	770	2914.8	0.0141	0.0032	8.5	2.6
320	1211.3	0.0028	0.0006	3.5	1.1	780	2952.6	0.0144	0.0033	8.6	2.6
330	1249.2	0.0029	0.0007	3.6	1.1	790	2990.5	0.0148	0.0033	8.7	2.7
340	1287.0	0.0031	0.0007	3.7	1.1	800	3028.3	0.0151	0.0034	8.8	2.7
350	1324.9	0.0033	0.0007	3.9	1.2	810	3066.2	0.0155	0.0035	8.9	2.7
360	1362.7	0.0034	0.0008	4.0	1.2	820	3104.0	0.0158	0.0036	9.0	2.8
370	1400.6	0.0036	0.0008	4.1	1.2	830	3141.9	0.0162	0.0037	9.1	2.8
380	1438.5	0.0038	0.0009	4.2	1.3	840	3179.7	0.0165	0.0037	9.2	2.8
390	1476.3	0.0040	0.0009	4.3	1.3	850	3217.6	0.0169	0.0038	9.4	2.9
400	1514.2	0.0042	0.0009	4.4	1.3	860	3255.4	0.0173	0.0039	9.5	2.9
410	1552.0	0.0044	0.0010	4.5	1.4	870	3293.3	0.0176	0.0040	9.6	2.9
420	1589.9	0.0046	0.0010	4.6	1.4	880	3331.2	0.0180	0.0041	9.7	3.0
430	1627.7	0.0048	0.0011	4.7	1.4	890	3369.0	0.0184	0.0042	9.8	3.0
440	1665.6	0.0050	0.0011	4.8	1.5	900	3406.9	0.0188	0.0042	9.9	3.0
450	1703.4	0.0052	0.0012	5.0	1.5	910	3444.7	0.0192	0.0043	10.0	3.1
460	1741.3	0.0054	0.0012	5.1	1.5	920	3482.6	0.0196	0.0044	10.1	3.1
470	1779.1	0.0056	0.0013	5.2	1.6	930	3520.4	0.0200	0.0045	10.2	3.1
480	1817.0	0.0059	0.0013	5.3	1.6	940	3558.3	0.0204	0.0046	10.3	3.2
490	1854.8	0.0061	0.0014	5.4	1.6	950	3596.1	0.0208	0.0047	10.5	3.2
500	1892.7	0.0063	0.0014	5.5	1.7	960	3634.0	0.0212	0.0048	10.6	3.2
510	1930.6	0.0066	0.0015	5.6	1.7	970	3671.8	0.0216	0.0049	10.7	3.3
520	1968.4	0.0068	0.0015	5.7	1.7	980	3709.7	0.0220	0.0050	10.8	3.3
530	2006.3	0.0071	0.0016	5.8	1.8	990	3747.5	0.0224	0.0051	10.9	3.3
540	2044.1	0.0073	0.0017	5.9	1.8	1000	3785.4	0.0228	0.0052	11.0	3.4

(continued)

Table F.3 (continued)

Q		P_f		Velocity		Q		P_f		Velocity	
GPM	L/min	psi/ft	bar/m	fps	m/s	GPM	L/min	psi/ft	bar/m	fps	m/s
550	2082.0	0.0076	0.0017	6.1	1.8	1010	3823.3	0.0232	0.0053	11.1	3.4
560	2119.8	0.0078	0.0018	6.2	1.9	1020	3861.1	0.0237	0.0054	11.2	3.4
570	2157.7	0.0081	0.0018	6.3	1.9	1030	3899.0	0.0241	0.0055	11.3	3.5
580	2195.5	0.0083	0.0019	6.4	1.9	1040	3936.8	0.0245	0.0055	11.4	3.5
590	2233.4	0.0086	0.0019	6.5	2.0	1050	3974.7	0.0250	0.0056	11.6	3.5
600	2271.2	0.0089	0.0020	6.6	2.0	1060	4012.5	0.0254	0.0057	11.7	3.6
610	2309.1	0.0091	0.0021	6.7	2.0	1070	4050.4	0.0259	0.0058	11.8	3.6
620	2346.9	0.0094	0.0021	6.8	2.1	1080	4088.2	0.0263	0.0059	11.9	3.6
630	2384.8	0.0097	0.0022	6.9	2.1	1090	4126.1	0.0268	0.0061	12.0	3.7
640	2422.7	0.0100	0.0023	7.0	2.1	1100	4163.9	0.0272	0.0062	12.1	3.7
650	2460.5	0.0103	0.0023	7.2	2.2	1110	4201.8	0.0277	0.0063	12.2	3.7

Table F.4 Water flowing in 6-in. (150 mm) class 200 PVC plastic pipe $C = 150$ I.D. = 5.86 in. (148.8 mm)

Q		P_f		Velocity		Q		P_f		Velocity	
GPM	L/min	psi/ft	bar/m	fps	m/s	GPM	L/min	psi/ft	bar/m	fps	m/s
200	757.1	0.0014	0.0003	2.4	0.7	660	2498.4	0.0128	0.0029	7.8	2.4
210	794.9	0.0015	0.0003	2.5	0.8	670	2536.2	0.0131	0.0030	8.0	2.4
220	832.8	0.0017	0.0004	2.6	0.8	680	2574.1	0.0135	0.0031	8.1	2.5
230	870.6	0.0018	0.0004	2.7	0.8	690	2611.9	0.0139	0.0031	8.2	2.5
240	908.5	0.0020	0.0004	2.9	0.9	700	2649.8	0.0142	0.0032	8.3	2.5
250	946.4	0.0021	0.0005	3.0	0.9	710	2687.6	0.0146	0.0033	8.4	2.6
260	984.2	0.0023	0.0005	3.1	0.9	720	2725.5	0.0150	0.0034	8.6	2.6
270	1022.1	0.0024	0.0006	3.2	1.0	730	2763.3	0.0154	0.0035	8.7	2.6
280	1059.9	0.0026	0.0006	3.3	1.0	740	2801.2	0.0158	0.0036	8.8	2.7
290	1097.8	0.0028	0.0006	3.4	1.1	750	2839.1	0.0162	0.0037	8.9	2.7
300	1135.6	0.0030	0.0007	3.6	1.1	760	2876.9	0.0166	0.0038	9.0	2.8
310	1173.5	0.0032	0.0007	3.7	1.1	770	2914.8	0.0170	0.0038	9.2	2.8
320	1211.3	0.0033	0.0008	3.8	1.2	780	2952.6	0.0174	0.0039	9.3	2.8
330	1249.2	0.0035	0.0008	3.9	1.2	790	2990.5	0.0178	0.0040	9.4	2.9
340	1287.0	0.0037	0.0008	4.0	1.2	800	3028.3	0.0182	0.0041	9.5	2.9
350	1324.9	0.0039	0.0009	4.2	1.3	810	3066.2	0.0186	0.0042	9.6	2.9
360	1362.7	0.0042	0.0009	4.3	1.3	820	3104.0	0.0191	0.0043	9.7	3.0
370	1400.6	0.0044	0.0010	4.4	1.3	830	3141.9	0.0195	0.0044	9.9	3.0
380	1438.5	0.0046	0.0010	4.5	1.4	840	3179.7	0.0199	0.0045	10.0	3.0
390	1476.3	0.0048	0.0011	4.6	1.4	850	3217.6	0.0204	0.0046	10.1	3.1
400	1514.2	0.0051	0.0011	4.8	1.4	860	3255.4	0.0208	0.0047	10.2	3.1
410	1552.0	0.0053	0.0012	4.9	1.5	870	3293.3	0.0213	0.0048	10.3	3.2
420	1589.9	0.0055	0.0013	5.0	1.5	880	3331.2	0.0217	0.0049	10.5	3.2
430	1627.7	0.0058	0.0013	5.1	1.6	890	3369.0	0.0222	0.0050	10.6	3.2
440	1665.6	0.0060	0.0014	5.2	1.6	900	3406.9	0.0226	0.0051	10.7	3.3
450	1703.4	0.0063	0.0014	5.4	1.6	910	3444.7	0.0231	0.0052	10.8	3.3
460	1741.3	0.0065	0.0015	5.5	1.7	920	3482.6	0.0236	0.0053	10.9	3.3
470	1779.1	0.0068	0.0015	5.6	1.7	930	3520.4	0.0241	0.0055	11.1	3.4
480	1817.0	0.0071	0.0016	5.7	1.7	940	3558.3	0.0245	0.0056	11.2	3.4
490	1854.8	0.0074	0.0017	5.8	1.8	950	3596.1	0.0250	0.0057	11.3	3.4
500	1892.7	0.0076	0.0017	5.9	1.8	960	3634.0	0.0255	0.0058	11.4	3.5
510	1930.6	0.0079	0.0018	6.1	1.8	970	3671.8	0.0260	0.0059	11.5	3.5
520	1968.4	0.0082	0.0019	6.2	1.9	980	3709.7	0.0265	0.0060	11.7	3.6
530	2006.3	0.0085	0.0019	6.3	1.9	990	3747.5	0.0270	0.0061	11.8	3.6
540	2044.1	0.0088	0.0020	6.4	2.0	1000	3785.4	0.0275	0.0062	11.9	3.6

(continued)

Table F.4 (continued)

Q		P_f		Velocity		Q		P_f		Velocity	
GPM	L/min	psi/ft	bar/m	fps	m/s	GPM	L/min	psi/ft	bar/m	fps	m/s
550	2082.0	0.0091	0.0021	6.5	2.0	1010	3823.3	0.0280	0.0064	12.0	3.7
560	2119.8	0.0094	0.0021	6.7	2.0	1020	3861.1	0.0286	0.0065	12.1	3.7
570	2157.7	0.0097	0.0022	6.8	2.1	1030	3899.0	0.0291	0.0066	12.2	3.7
580	2195.5	0.0100	0.0023	6.9	2.1	1040	3936.8	0.0296	0.0067	12.4	3.8
590	2233.4	0.0104	0.0023	7.0	2.1	1050	3974.7	0.0301	0.0068	12.5	3.8
600	2271.2	0.0107	0.0024	7.1	2.2	1060	4012.5	0.0307	0.0069	12.6	3.8
610	2309.1	0.0110	0.0025	7.3	2.2	1070	4050.4	0.0312	0.0071	12.7	3.9
620	2346.9	0.0114	0.0026	7.4	2.2	1080	4088.2	0.0317	0.0072	12.8	3.9
630	2384.8	0.0117	0.0027	7.5	2.3	1090	4126.1	0.0323	0.0073	13.0	4.0
640	2422.7	0.0121	0.0027	7.6	2.3	1100	4163.9	0.0328	0.0074	13.1	4.0
650	2460.5	0.0124	0.0028	7.7	2.4	1110	4201.8	0.0334	0.0076	13.2	4.0

Table F.5 Water flowing in 8-in. (200 mm) class 150 PVC plastic pipe $C = 150$ I.D. = 7.98 in. (202.7 mm)

Q		P_f		Velocity		Q		P_f		Velocity	
GPM	L/min	psi/ft	bar/m	fps	m/s	GPM	L/min	psi/ft	bar/m	fps	m/s
500	1892.7	0.0017	0.0004	3.2	1.0	1650	6245.9	0.0155	0.0035	10.6	3.2
525	1987.3	0.0019	0.0004	3.4	1.0	1675	6340.5	0.0159	0.0036	10.7	3.3
550	2082.0	0.0020	0.0005	3.5	1.1	1700	6435.2	0.0163	0.0037	10.9	3.3
575	2176.6	0.0022	0.0005	3.7	1.1	1725	6529.8	0.0168	0.0038	11.1	3.4
600	2271.2	0.0024	0.0005	3.8	1.2	1750	6624.5	0.0172	0.0039	11.2	3.4
625	2365.9	0.0026	0.0006	4.0	1.2	1775	6719.1	0.0177	0.0040	11.4	3.5
650	2460.5	0.0028	0.0006	4.2	1.3	1800	6813.7	0.0181	0.0041	11.5	3.5
675	2555.1	0.0030	0.0007	4.3	1.3	1825	6908.4	0.0186	0.0042	11.7	3.6
700	2649.8	0.0032	0.0007	4.5	1.4	1850	7003.0	0.0191	0.0043	11.9	3.6
725	2744.4	0.0034	0.0008	4.6	1.4	1875	7097.6	0.0196	0.0044	12.0	3.7
750	2839.1	0.0036	0.0008	4.8	1.5	1900	7192.3	0.0201	0.0045	12.2	3.7
775	2933.7	0.0038	0.0009	5.0	1.5	1925	7286.9	0.0206	0.0046	12.3	3.8
800	3028.3	0.0040	0.0009	5.1	1.6	1950	7381.5	0.0210	0.0048	12.5	3.8
825	3123.0	0.0043	0.0010	5.3	1.6	1975	7476.2	0.0215	0.0049	12.7	3.9
850	3217.6	0.0045	0.0010	5.4	1.7	2000	7570.8	0.0221	0.0050	12.8	3.9
875	3312.2	0.0048	0.0011	5.6	1.7	2025	7665.4	0.0226	0.0051	13.0	4.0
900	3406.9	0.0050	0.0011	5.8	1.8	2050	7760.1	0.0231	0.0052	13.1	4.0
925	3501.5	0.0053	0.0012	5.9	1.8	2075	7854.7	0.0236	0.0053	13.3	4.1
950	3596.1	0.0056	0.0013	6.1	1.9	2100	7949.3	0.0241	0.0055	13.5	4.1
975	3690.8	0.0058	0.0013	6.3	1.9	2125	8044.0	0.0247	0.0056	13.6	4.2
1000	3785.4	0.0061	0.0014	6.4	2.0	2150	8138.6	0.0252	0.0057	13.8	4.2
1025	3880.0	0.0064	0.0014	6.6	2.0	2175	8233.2	0.0258	0.0058	13.9	4.3
1050	3974.7	0.0067	0.0015	6.7	2.1	2200	8327.9	0.0263	0.0060	14.1	4.3
1075	4069.3	0.0070	0.0016	6.9	2.1	2225	8422.5	0.0269	0.0061	14.3	4.3
1100	4163.9	0.0073	0.0017	7.1	2.1	2250	8517.2	0.0274	0.0062	14.4	4.4
1125	4258.6	0.0076	0.0017	7.2	2.2	2275	8611.8	0.0280	0.0063	14.6	4.4
1150	4353.2	0.0079	0.0018	7.4	2.2	2300	8706.4	0.0286	0.0065	14.7	4.5
1175	4447.8	0.0082	0.0019	7.5	2.3	2325	8801.1	0.0291	0.0066	14.9	4.5
1200	4542.5	0.0086	0.0019	7.7	2.3	2350	8895.7	0.0297	0.0067	15.1	4.6
1225	4637.1	0.0089	0.0020	7.9	2.4	2375	8990.3	0.0303	0.0069	15.2	4.6
1250	4731.8	0.0092	0.0021	8.0	2.4	2400	9085.0	0.0309	0.0070	15.4	4.7
1275	4826.4	0.0096	0.0022	8.2	2.5	2425	9179.6	0.0315	0.0071	15.5	4.7
1300	4921.0	0.0099	0.0022	8.3	2.5	2450	9274.2	0.0321	0.0073	15.7	4.8
1325	5015.7	0.0103	0.0023	8.5	2.6	2475	9368.9	0.0327	0.0074	15.9	4.8
1350	5110.3	0.0107	0.0024	8.7	2.6	2500	9463.5	0.0333	0.0075	16.0	4.9

(continued)

Table F.5 (continued)

Q		P_f		Velocity		Q		P_f		Velocity	
GPM	L/min	psi/ft	bar/m	fps	m/s	GPM	L/min	psi/ft	bar/m	fps	m/s
1375	5204.9	0.0110	0.0025	8.8	2.7	2525	9558.1	0.0339	0.0077	16.2	4.9
1400	5299.6	0.0114	0.0026	9.0	2.7	2550	9652.8	0.0346	0.0078	16.3	5.0
1425	5394.2	0.0118	0.0027	9.1	2.8	2575	9747.4	0.0352	0.0080	16.5	5.0
1450	5488.8	0.0122	0.0028	9.3	2.8	2600	9842.0	0.0358	0.0081	16.7	5.1
1475	5583.5	0.0126	0.0028	9.5	2.9	2625	9936.7	0.0365	0.0083	16.8	5.1
1500	5678.1	0.0130	0.0029	9.6	2.9	2650	10,031.3	0.0371	0.0084	17.0	5.2
1525	5772.7	0.0134	0.0030	9.8	3.0	2675	10,125.9	0.0378	0.0085	17.2	5.2
1550	5867.4	0.0138	0.0031	9.9	3.0	2700	10,220.6	0.0384	0.0087	17.3	5.3
1575	5962.0	0.0142	0.0032	10.1	3.1	2725	10,315.2	0.0391	0.0088	17.5	5.3
1600	6056.6	0.0146	0.0033	10.3	3.1	2750	10,409.9	0.0398	0.0090	17.6	5.4
1625	6151.3	0.0150	0.0034	10.4	3.2	2775	10,504.5	0.0404	0.0091	17.8	5.4

Table F.6 Water flowing in 8-in. (200 mm) Class 200 PVC plastic pipe $C = 150$ I.D. = 7.68 in. (195.1 mm)

Q		P_f		Velocity		Q		P_f		Velocity	
GPM	L/min	psi/ft	bar/m	fps	m/s	GPM	L/min	psi/ft	bar/m	fps	m/s
500	1892.7	0.0020	0.0005	3.5	1.1	1650	6245.9	0.0186	0.0042	11.4	3.5
525	1987.3	0.0022	0.0005	3.6	1.1	1675	6340.5	0.0191	0.0043	11.6	3.5
550	2082.0	0.0024	0.0006	3.8	1.2	1700	6435.2	0.0197	0.0044	11.8	3.6
575	2176.6	0.0026	0.0006	4.0	1.2	1725	6529.8	0.0202	0.0046	11.9	3.6
600	2271.2	0.0029	0.0006	4.2	1.3	1750	6624.5	0.0208	0.0047	12.1	3.7
625	2365.9	0.0031	0.0007	4.3	1.3	1775	6719.1	0.0213	0.0048	12.3	3.7
650	2460.5	0.0033	0.0008	4.5	1.4	1800	6813.7	0.0219	0.0049	12.5	3.8
675	2555.1	0.0036	0.0008	4.7	1.4	1825	6908.4	0.0224	0.0051	12.6	3.9
700	2649.8	0.0038	0.0009	4.8	1.5	1850	7003.0	0.0230	0.0052	12.8	3.9
725	2744.4	0.0041	0.0009	5.0	1.5	1875	7097.6	0.0236	0.0053	13.0	4.0
750	2839.1	0.0043	0.0010	5.2	1.6	1900	7192.3	0.0242	0.0055	13.2	4.0
775	2933.7	0.0046	0.0010	5.4	1.6	1925	7286.9	0.0248	0.0056	13.3	4.1
800	3028.3	0.0049	0.0011	5.5	1.7	1950	7381.5	0.0254	0.0057	13.5	4.1
825	3123.0	0.0052	0.0012	5.7	1.7	1975	7476.2	0.0260	0.0059	13.7	4.2
850	3217.6	0.0055	0.0012	5.9	1.8	2000	7570.8	0.0266	0.0060	13.8	4.2
875	3312.2	0.0058	0.0013	6.1	1.8	2025	7665.4	0.0272	0.0061	14.0	4.3
900	3406.9	0.0061	0.0014	6.2	1.9	2050	7760.1	0.0278	0.0063	14.2	4.3
925	3501.5	0.0064	0.0014	6.4	2.0	2075	7854.7	0.0285	0.0064	14.4	4.4
950	3596.1	0.0067	0.0015	6.6	2.0	2100	7949.3	0.0291	0.0066	14.5	4.4
975	3690.8	0.0070	0.0016	6.7	2.1	2125	8044.0	0.0297	0.0067	14.7	4.5
1000	3785.4	0.0074	0.0017	6.9	2.1	2150	8138.6	0.0304	0.0069	14.9	4.5
1025	3880.0	0.0077	0.0017	7.1	2.2	2175	8233.2	0.0310	0.0070	15.1	4.6
1050	3974.7	0.0081	0.0018	7.3	2.2	2200	8327.9	0.0317	0.0072	15.2	4.6
1075	4069.3	0.0084	0.0019	7.4	2.3	2225	8422.5	0.0324	0.0073	15.4	4.7
1100	4163.9	0.0088	0.0020	7.6	2.3	2250	8517.2	0.0331	0.0075	15.6	4.7
1125	4258.6	0.0092	0.0021	7.8	2.4	2275	8611.8	0.0337	0.0076	15.7	4.8
1150	4353.2	0.0095	0.0022	8.0	2.4	2300	8706.4	0.0344	0.0078	15.9	4.9
1175	4447.8	0.0099	0.0022	8.1	2.5	2325	8801.1	0.0351	0.0079	16.1	4.9
1200	4542.5	0.0103	0.0023	8.3	2.5	2350	8895.7	0.0358	0.0081	16.3	5.0
1225	4637.1	0.0107	0.0024	8.5	2.6	2375	8990.3	0.0365	0.0083	16.4	5.0
1250	4731.8	0.0111	0.0025	8.7	2.6	2400	9085.0	0.0372	0.0084	16.6	5.1
1275	4826.4	0.0116	0.0026	8.8	2.7	2425	9179.6	0.0380	0.0086	16.8	5.1
1300	4921.0	0.0120	0.0027	9.0	2.7	2450	9274.2	0.0387	0.0087	17.0	5.2
1325	5015.7	0.0124	0.0028	9.2	2.8	2475	9368.9	0.0394	0.0089	17.1	5.2
1350	5110.3	0.0128	0.0029	9.3	2.8	2500	9463.5	0.0402	0.0091	17.3	5.3

(continued)

Table F.6 (continued)

Q		P_f		Velocity		Q		P_f		Velocity	
GPM	L/min	psi/ft	bar/m	fps	m/s	GPM	L/min	psi/ft	bar/m	fps	m/s
1375	5204.9	0.0133	0.0030	9.5	2.9	2525	9558.1	0.0409	0.0092	17.5	5.3
1400	5299.6	0.0137	0.0031	9.7	3.0	2550	9652.8	0.0417	0.0094	17.7	5.4
1425	5394.2	0.0142	0.0032	9.9	3.0	2575	9747.4	0.0424	0.0096	17.8	5.4
1450	5488.8	0.0147	0.0033	10.0	3.1	2600	9842.0	0.0432	0.0098	18.0	5.5
1475	5583.5	0.0151	0.0034	10.2	3.1	2625	9936.7	0.0440	0.0099	18.2	5.5
1500	5678.1	0.0156	0.0035	10.4	3.2	2650	10,031.3	0.0447	0.0101	18.3	5.6
1525	5772.7	0.0161	0.0036	10.6	3.2	2675	10,125.9	0.0455	0.0103	18.5	5.6
1550	5867.4	0.0166	0.0038	10.7	3.3	2700	10,220.6	0.0463	0.0105	18.7	5.7
1575	5962.0	0.0171	0.0039	10.9	3.3	2725	10,315.2	0.0471	0.0107	18.9	5.7
1600	6056.6	0.0176	0.0040	11.1	3.4	2750	10,409.9	0.0479	0.0108	19.0	5.8
1625	6151.3	0.0181	0.0041	11.2	3.4	2775	10,504.5	0.0487	0.0110	19.2	5.9

Appendix G
Discharge Tables for Various Nozzles and Orifices

P. Brock and V. Charter, *Fire Protection Hydraulics and Water Supply Analysis*, The Society of Fire Protection Engineers Series, https://doi.org/10.1007/978-3-032-09877-1

Table G.1 Discharge table for circular outlets (pressure measured by pitot gauge) coefficient = 0.80

Pitot pressure		Outlet diameter 2					
		2 1/4 in.	57 mm	1/2 in.	63 mm	2 9/16 in.	65 mm
		Flow rate					
PSI	Bar	GPM	L/min	GPM	L/min	GPM	L/min
5	0.345	270.1	1016.0	333.5	1241.2	350.4	1321.3
6	0.413	295.9	1113.0	365.3	1359.7	383.8	1447.4
7	0.482	319.6	1202.2	394.6	1468.6	414.6	1563.3
8	0.551	341.7	1285.2	421.9	1570.0	443.2	1671.3
9	0.620	362.4	1363.2	447.5	1665.2	470.1	1772.6
10	0.689	382.0	1436.9	471.7	1755.3	495.5	1868.5
11	0.758	400.7	1507.0	494.7	1841.0	519.7	1959.7
12	0.827	418.5	1574.0	516.7	1922.9	542.8	2046.9
13	0.896	435.6	1638.3	537.8	2001.4	565.0	2130.5
14	0.965	452.0	1700.2	558.1	2076.9	586.3	2210.9
15	1.034	467.9	1759.8	577.7	2149.8	606.9	2288.5
16	1.102	483.2	1817.5	596.6	2220.3	626.8	2363.5
17	1.171	498.1	1873.5	615.0	2288.7	646.1	2436.3
18	1.240	512.6	1927.8	632.8	2355.0	664.8	2506.9
19	1.309	526.6	1980.6	650.1	2419.5	683.0	2575.6
20	1.378	540.3	2032.1	667.0	2482.4	700.8	2642.5
21	1.447	553.6	2082.3	683.5	2543.7	718.1	2707.8
22	1.516	566.7	2131.3	699.6	2603.6	735.0	2771.5
23	1.585	579.4	2179.2	715.3	2662.1	751.5	2833.8
24	1.654	591.9	2226.0	730.7	2719.3	767.7	2894.7
25	1.723	604.1	2271.9	745.8	2775.4	783.5	2954.4
26	1.791	616.0	2316.9	760.5	2830.4	799.0	3012.9
27	1.860	627.8	2361.1	775.0	2884.3	814.2	3070.3
28	1.929	639.3	2404.4	789.2	2937.2	829.2	3126.7
29	1.998	650.6	2446.9	803.2	2989.2	843.9	3182.0
30	2.067	661.7	2488.8	816.9	3040.3	858.3	3236.4
31	2.136	672.6	2529.9	830.4	3090.5	872.5	3289.9
32	2.205	683.4	2570.4	843.7	3140.0	886.4	3342.5
33	2.274	694.0	2610.2	856.8	3188.7	900.2	3394.4
34	2.343	704.4	2649.5	869.7	3236.6	913.7	3445.4
35	2.412	714.7	2688.2	882.4	3283.9	927.1	3495.7
36	2.480	724.9	2726.3	894.9	3330.5	940.2	3545.3
37	2.549	734.9	2763.9	907.2	3376.4	953.2	3594.2

(continued)

Table G.1 (continued)

Pitot pressure		Outlet diameter 2					
		2 1/4 in.	57 mm	1/2 in.	63 mm	2 9/16 in.	65 mm
		Flow rate					
PSI	Bar	GPM	L/min	GPM	L/min	GPM	L/min
38	2.618	744.7	2801.0	919.4	3421.7	966.0	3642.4
39	2.687	754.5	2837.6	931.4	3466.5	978.6	3690.1
40	2.756	764.1	2873.8	943.3	3510.6	991.1	3737.1
41	2.825	773.6	2909.5	955.0	3554.2	1003.4	3783.5
42	2.894	782.9	2944.7	966.6	3597.3	1015.5	3829.4
43	2.963	792.2	2979.6	978.0	3639.9	1027.6	3874.7
44	3.032	801.4	3014.0	989.3	3682.0	1039.4	3919.5
45	3.101	810.4	3048.1	1000.5	3723.6	1051.2	3963.8

Table G.2 Discharge table for circular outlets (pressure measured by pitot gauge) coefficient = 0.90

Pitot pressure		Outlet diameter					
		2 1/4 in.	57 mm	2 1/2 in.	63 mm	2 9/16 in.	65 mm
		Flow rate					
PSI	Bar	GPM	L/min	GPM	L/min	GPM	L/min
5	0.345	303.9	1143.0	375.2	1396.3	394.2	1486.4
6	0.413	332.9	1252.1	411.0	1529.6	431.8	1628.3
7	0.482	359.6	1352.5	443.9	1652.2	466.4	1758.7
8	0.551	384.4	1445.8	474.6	1766.3	498.6	1880.2
9	0.620	407.7	1533.5	503.4	1873.4	528.9	1994.2
10	0.689	429.8	1616.5	530.6	1974.7	557.5	2102.1
11	0.758	450.8	1695.4	556.5	2071.1	584.7	2204.7
12	0.827	470.8	1770.8	581.3	2163.2	610.7	2302.7
13	0.896	490.0	1843.1	605.0	2251.5	635.6	2396.8
14	0.965	508.5	1912.7	627.8	2336.5	659.6	2487.2
15	1.034	526.4	1979.8	649.9	2418.5	682.8	2574.5
16	1.102	543.7	2044.7	671.2	2497.9	705.2	2659.0
17	1.171	560.4	2107.7	691.8	2574.7	726.9	2740.8
18	1.240	576.6	2168.8	711.9	2649.4	747.9	2820.3
19	1.309	592.4	2228.2	731.4	2722.0	768.4	2897.5
20	1.378	607.8	2286.1	750.4	2792.7	788.4	2972.8
21	1.447	622.8	2342.5	768.9	2861.7	807.9	3046.2
22	1.516	637.5	2397.7	787.0	2929.0	826.9	3117.9
23	1.585	651.8	2451.5	804.7	2994.8	845.4	3188.0
24	1.654	665.8	2504.3	822.0	3059.2	863.6	3256.6
25	1.723	679.6	2555.9	839.0	3122.3	881.4	3323.7
26	1.791	693.0	2606.5	855.6	3184.2	898.9	3389.5
27	1.860	706.2	2656.2	871.9	3244.8	916.0	3454.1
28	1.929	719.2	2704.9	887.9	3304.4	932.8	3517.5
29	1.998	731.9	2752.8	903.6	3362.8	949.3	3579.7
30	2.067	744.4	2799.9	919.0	3420.3	965.6	3640.9
31	2.136	756.7	2846.1	934.2	3476.9	981.5	3701.1
32	2.205	768.8	2891.7	949.2	3532.5	997.2	3760.3
33	2.274	780.8	2936.5	963.9	3587.3	1012.7	3818.7
34	2.343	792.5	2980.7	978.4	3641.2	1027.9	3876.1
35	2.412	804.1	3024.2	992.7	3694.4	1042.9	3932.7
36	2.480	815.5	3067.1	1006.8	3746.8	1057.7	3988.5
37	2.549	826.7	3109.4	1020.6	3798.5	1072.3	4043.5

(continued)

Table G.2 (continued)

Pitot pressure		Outlet diameter					
		2 1/4 in.	57 mm	2 1/2 in.	63 mm	2 9/16 in.	65 mm
		Flow rate					
PSI	Bar	GPM	L/min	GPM	L/min	GPM	L/min
38	2.618	837.8	3151.1	1034.4	3849.5	1086.7	4097.7
39	2.687	848.8	3192.3	1047.9	3899.8	1100.9	4151.3
40	2.756	859.6	3233.0	1061.2	3949.5	1114.9	4204.2
41	2.825	870.3	3273.2	1074.4	3998.5	1128.8	4256.4
42	2.894	880.8	3312.8	1087.4	4047.0	1142.5	4308.0
43	2.963	891.2	3352.0	1100.3	4094.9	1156.0	4359.0
44	3.032	901.5	3390.8	1113.0	4142.2	1169.4	4409.4
45	3.101	911.7	3429.1	1125.6	4189.0	1182.6	4459.2

Table G.3 Discharge table for circular outlets (pressure measured by pitot gauge) underwriters playpipe, coefficient = 0.97

Pitot pressure		Outlet diameter			
		1 1/8 in.	25 mm	1 3/4 in.	44 mm
		Flow rate			
PSI	Bar	GPM	L/min	GPM	L/min
5	0.345	81.9	237.0	198.1	734.1
6	0.413	89.7	259.6	217.1	804.1
7	0.482	96.9	280.4	234.4	868.6
8	0.551	103.6	299.8	250.6	928.6
9	0.620	109.9	317.9	265.8	984.9
10	0.689	115.8	335.1	280.2	1038.2
11	0.758	121.5	351.5	293.9	1088.8
12	0.827	126.9	367.1	307.0	1137.2
13	0.896	132.0	382.1	319.5	1183.7
14	0.965	137.0	396.6	331.6	1228.4
15	1.034	141.8	410.5	343.2	1271.5
16	1.102	146.5	423.9	354.5	1313.2
17	1.171	151.0	437.0	365.4	1353.6
18	1.240	155.4	449.6	376.0	1392.8
19	1.309	159.6	462.0	386.3	1431.0
20	1.378	163.8	474.0	396.3	1468.2
21	1.447	167.8	485.7	406.1	1504.4
22	1.516	171.8	497.1	415.6	1539.8
23	1.585	175.6	508.3	425.0	1574.4
24	1.654	179.4	519.2	434.1	1608.3
25	1.723	183.1	529.9	443.1	1641.5
26	1.791	186.7	540.4	451.8	1674.0
27	1.860	190.3	550.7	460.5	1705.9
28	1.929	193.8	560.8	468.9	1737.2
29	1.998	197.2	570.7	477.2	1767.9
30	2.067	200.6	580.5	485.4	1798.1
31	2.136	203.9	590.1	493.4	1827.9
32	2.205	207.2	599.5	501.3	1857.1
33	2.274	210.4	608.8	509.0	1885.9
34	2.343	213.5	618.0	516.7	1914.3
35	2.412	216.7	627.0	524.2	1942.2
36	2.480	219.7	635.9	531.7	1969.8
37	2.549	222.8	644.7	539.0	1996.9

(continued)

Table G.3 (continued)

Pitot pressure		Outlet diameter			
		1 1/8 in.	25 mm	1 3/4 in.	44 mm
		Flow rate			
PSI	Bar	GPM	L/min	GPM	L/min
38	2.618	225.7	653.3	546.3	2023.7
39	2.687	228.7	661.9	553.4	2050.2
40	2.756	231.6	670.3	560.4	2076.3
41	2.825	234.5	678.6	567.4	2102.1
42	2.894	237.3	686.8	574.3	2127.6
43	2.963	240.1	695.0	581.1	2152.8
44	3.032	242.9	703.0	587.8	2177.6
45	3.101	245.7	711.0	594.4	2202.3

Appendix H
Coefficients for Hydrant Pumper Outlets

Table H.1 Coefficients for pumper outlets

Velocity pressure	Coefficient
2 psi	0.97
3 psi	0.92
4 psi	0.89
5 psi	0.86
6 psi	0.84
7 psi and over	0.83

P. Brock and V. Charter, *Fire Protection Hydraulics and Water Supply Analysis*, The Society of Fire Protection Engineers Series, https://doi.org/10.1007/978-3-032-09877-1

Appendix I
K-Factors for Automatic Sprinklers

Table I.1 Nominal K-Factors for automatic sprinklers

Orifice diameter		*K*-factor	
Inches	mm	English	S.I.U
1/4	6.4	1.4	20
5/16	8.0	1.9	27
3/8	9.5	2.8	40
7/16	11.0	4.2	57
1/2	12.7	5.6	80
17/32	13.5	8.0	115
5/8	15.9	11.2	160
3/4	19.0	14.0	200

P. Brock and V. Charter, *Fire Protection Hydraulics and Water Supply Analysis*, The Society of Fire Protection Engineers Series, https://doi.org/10.1007/978-3-032-09877-1

Appendix J
Sample Worksheets and Graph Paper

Flow analysis worksheet

P. Brock and V. Charter, *Fire Protection Hydraulics and Water Supply Analysis*, The Society of Fire Protection Engineers Series, https://doi.org/10.1007/978-3-032-09877-1

Loop No.	Line No.	Size	C	Length	Q	P_f	$\frac{1.85P_f}{Q}$	Correc-tion	Q	P_f	$\frac{1.85P_f}{Q}$	Correc-tion	Q	P_f	$\frac{1.85P_f}{Q}$	Correc-tion

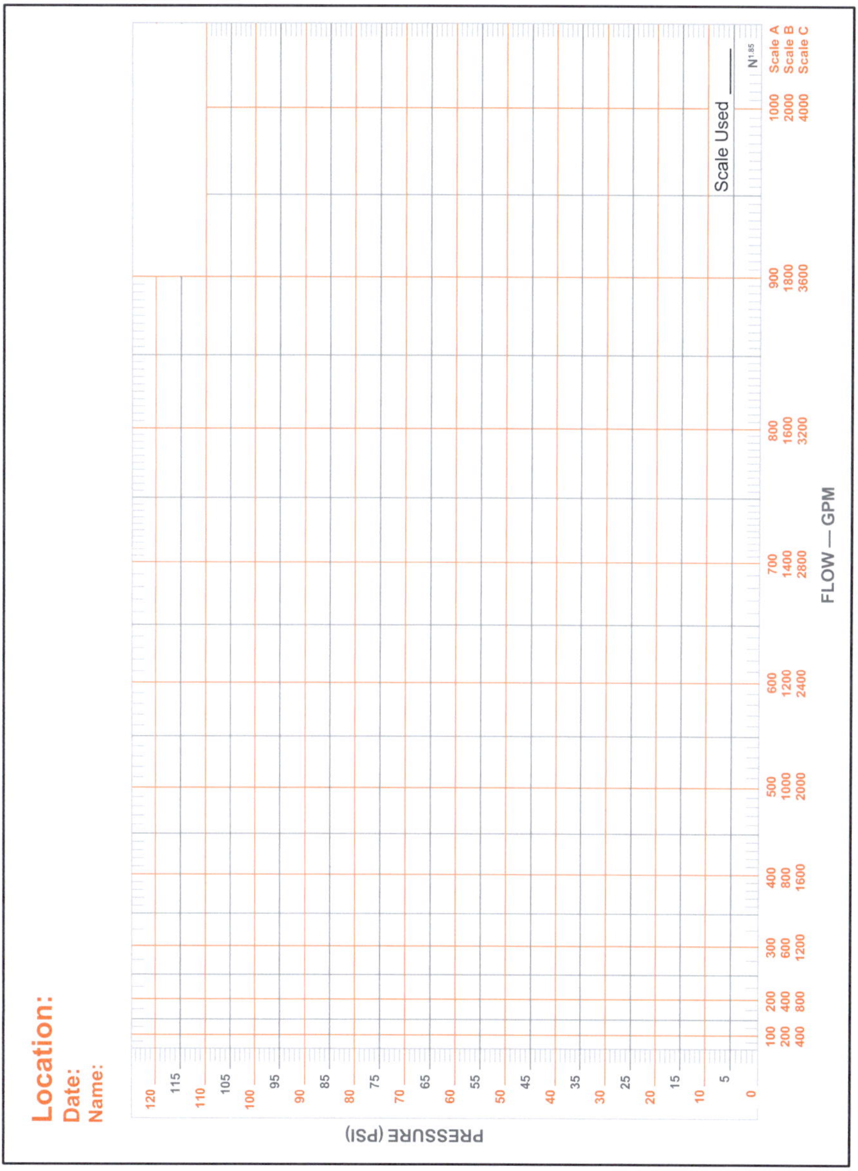
Location:
Date:
Name:
PRESSURE (PSI)
0 5 10 15 20 25 30 35 40 45 50 55 60 65 70 75 80 85 90 95 100 105 110 115 120
FLOW — GPM
Scale A 100 200 300 400 500 600 700 800 900 1000
Scale B 200 400 600 800 1000 1200 1400 1600 1800 2000
Scale C 400 800 1200 1600 2000 2400 2800 3200 3600 4000
$N^{1.85}$
Scale Used ____

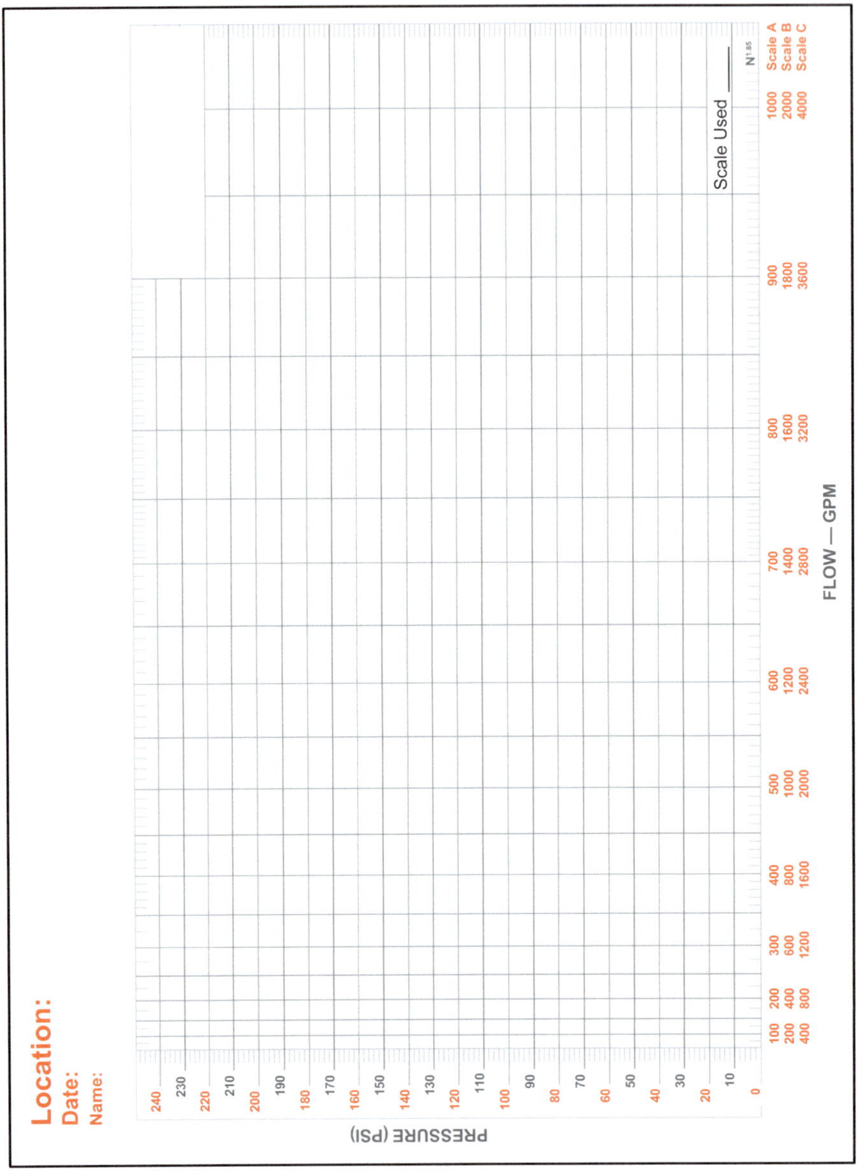
Location:
Date:
Name:
PRESSURE (PSI)
0 10 20 30 40 50 60 70 80 90 100 110 120 130 140 150 160 170 180 190 200 210 220 230 240
FLOW — GPM
100 200 300 400 500 600 700 800 900 1000 Scale A
200 400 600 800 1000 1200 1400 1600 1800 2000 Scale B
400 800 1200 1600 2000 2400 2800 3200 3600 4000 Scale C
$N^{1.85}$
Scale Used ____

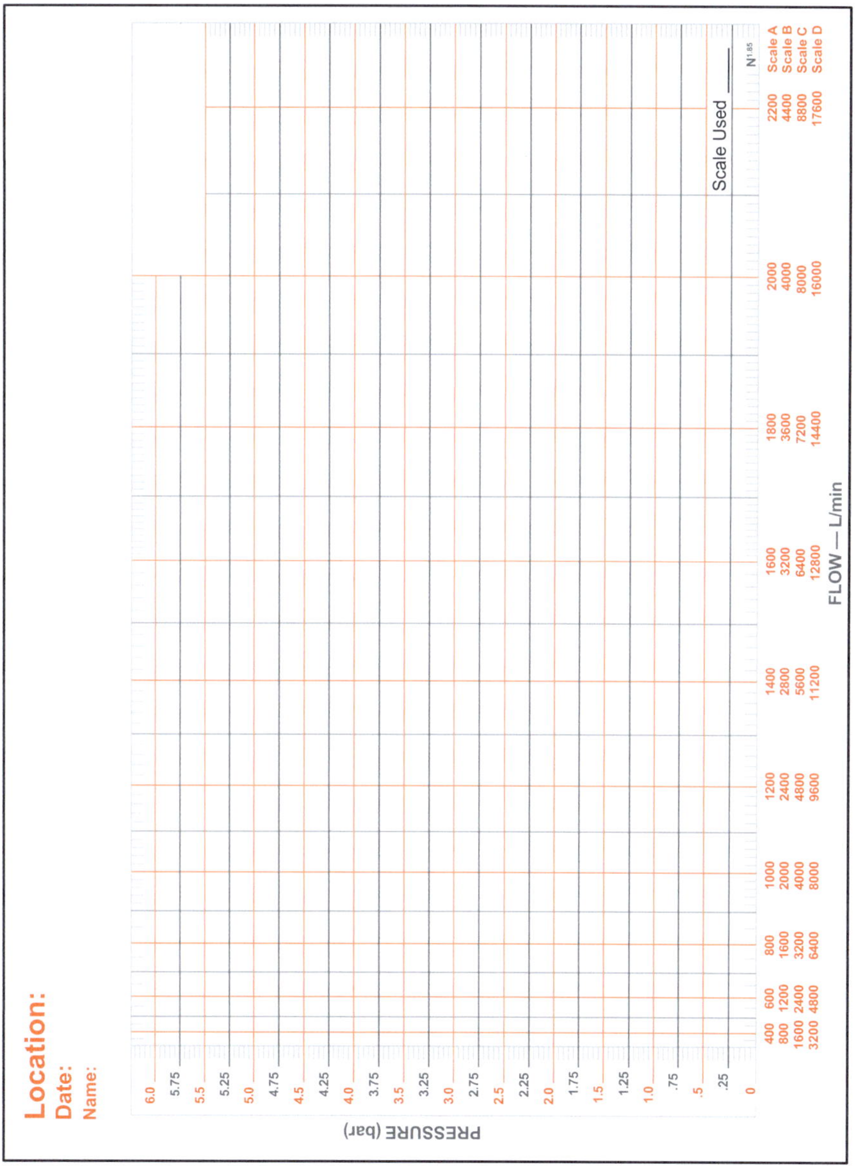
Location:
Date:
Name:
PRESSURE (bar)
6.0 5.75 5.5 5.25 5.0 4.75 4.5 4.25 4.0 3.75 3.5 3.25 3.0 2.75 2.5 2.25 2.0 1.75 1.5 1.25 1.0 .75 .5 .25 0
Scale Used ____
N1.85
400 600 800 1000 1200 1400 1600 1800 2000 2200 Scale A
800 1200 1600 2000 2400 2800 3200 3600 4000 4400 Scale B
1600 2400 3200 4000 4800 5600 6400 7200 8000 8800 Scale C
3200 4800 6400 8000 9600 11200 12800 14400 16000 17600 Scale D
FLOW — L/min

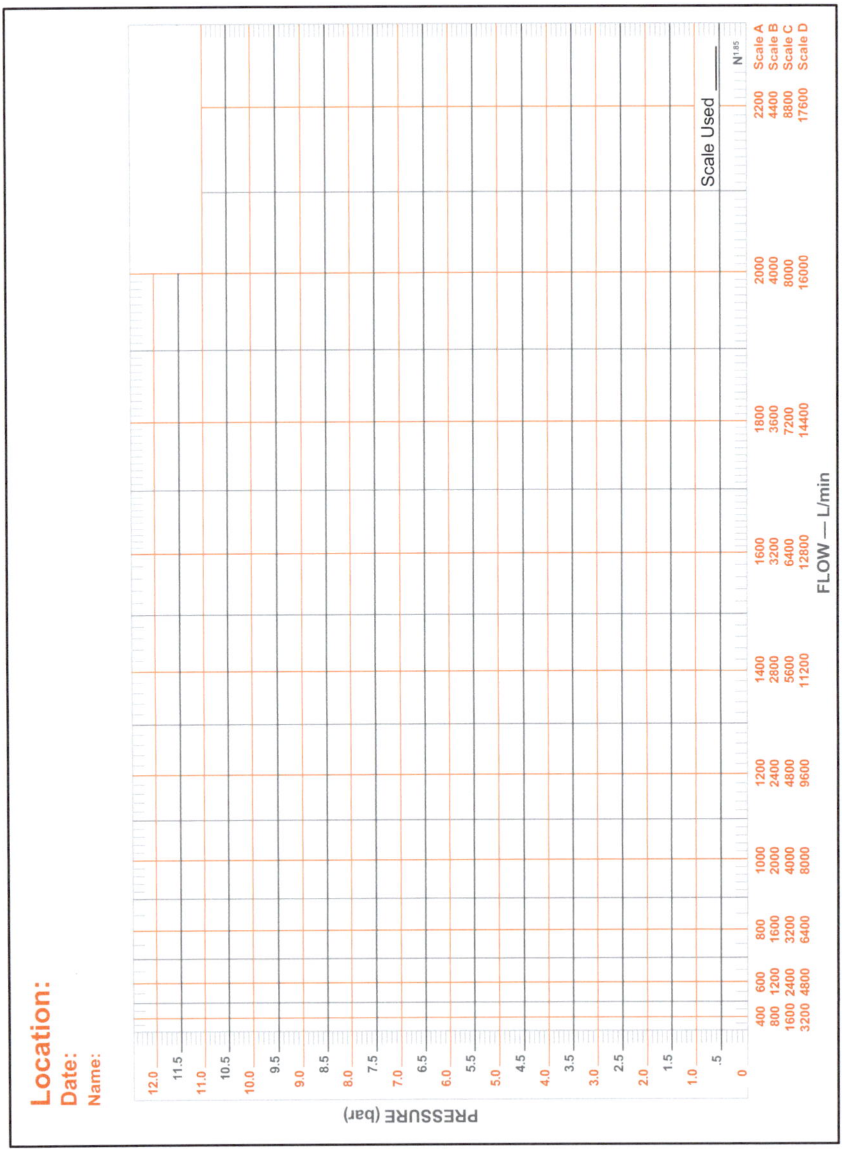
Location:
Date:
Name:
PRESSURE (bar)
12.0
11.5
11.0
10.5
10.0
9.5
9.0
8.5
8.0
7.5
7.0
6.5
6.0
5.5
5.0
4.5
4.0
3.5
3.0
2.5
2.0
1.5
1.0
.5
0
Scale Used ____
N1.85
Scale A 400 600 800 1000 1200 1400 1600 1800 2000 2200
Scale B 800 1200 1600 2000 2400 2800 3200 3600 4000 4400
Scale C 1600 2400 3200 4000 4800 5600 6400 7200 8000 8800
Scale D 3200 4800 6400 8000 9600 11200 12800 14400 16000 17600
FLOW — L/min

Hydraulic Sprinkler Calculations

HYDRAULIC SPRINKLER CALCULATIONS

SHEET ______ OF ______

STEP NO.	NOZZLE IDENT. AND LOCATION	FLOW IN GPM	PIPE SIZE	PIPE FITTINGS AND DEVICES	EQUIV. PIPE LENGTH	FRICTION LOSS PSI/FOOT	PRESSURE SUMMARY	NORMAL PRESSURE	NOTES	REF. STEP
		q			L		Pt	Pt		
					F		Pe	Pv		
		Q			T		Pf	Pn		
		q			L		Pt	Pt		
					F		Pe	Pv		
		Q			T		Pf	Pn		
		q			L		Pt	Pt		
					F		Pe	Pv		
		Q			T		Pf	Pn		
		q			L		Pt	Pt		
					F		Pe	Pv		
		Q			T		Pf	Pn		
		q			L		Pt	Pt		
					F		Pe	Pv		
		Q			T		Pf	Pn		
		q			L		Pt	Pt		
					F		Pe	Pv		
		Q			T		Pf	Pn		
		q			L		Pt	Pt		
					F		Pe	Pv		
		Q			T		Pf	Pn		
		q			L		Pt	Pt		
					F		Pe	Pv		
		Q			T		Pf	Pn		
		q			L		Pt	Pt		
					F		Pe	Pv		
		Q			T		Pf	Pn		
		q			L		Pt	Pt		
					F		Pe	Pv		
		Q			T		Pf	Pn		
		q			L		Pt	Pt		
					F		Pe	Pv		
		Q			T		Pf	Pn		
		q			L		Pt	Pt		
					F		Pe	Pv		
		Q			T		Pf	Pn		

Hydraulic Sprinkler Calculations

HYDRAULIC SPRINKLER CALCULATIONS

SHEET ______ OF ______

STEP NO.	NOZZLE IDENT. AND LOCATION	FLOW IN L/MIN	PIPE SIZE	PIPE FITTINGS AND DEVICES	EQUIV. PIPE LENGTH	FRICTION LOSS BAR/M	PRESSURE SUMMARY	NORMAL PRESSURE	NOTES	REF. STEP
		q			L		Pt	Pt		
					F		Pe	Pv		
		Q			T		Pf	Pn		
		q			L		Pt	Pt		
					F		Pe	Pv		
		Q			T		Pf	Pn		
		q			L		Pt	Pt		
					F		Pe	Pv		
		Q			T		Pf	Pn		
		q			L		Pt	Pt		
					F		Pe	Pv		
		Q			T		Pf	Pn		
		q			L		Pt	Pt		
					F		Pe	Pv		
		Q			T		Pf	Pn		
		q			L		Pt	Pt		
					F		Pe	Pv		
		Q			T		Pf	Pn		
		q			L		Pt	Pt		
					F		Pe	Pv		
		Q			T		Pf	Pn		
		q			L		Pt	Pt		
					F		Pe	Pv		
		Q			T		Pf	Pn		
		q			L		Pt	Pt		
					F		Pe	Pv		
		Q			T		Pf	Pn		
		q			L		Pt	Pt		
					F		Pe	Pv		
		Q			T		Pf	Pn		
		q			L		Pt	Pt		
					F		Pe	Pv		
		Q			T		Pf	Pn		
		q			L		Pt	Pt		
					F		Pe	Pv		
		Q			T		Pf	Pn		

Appendix K
Answers to Selected Problems

Chapter 1

1.3 107,712 gallons (480,000 L)
1.5 106.4 ft (26.5 m)
1.9 153.6 pounds (80.6 kg)
1.11 58.9 psi (433 kg/cm^2 or 4.25 bar)
1.13 27,777 pounds (13,355 kg)
1.15 26,704,960 ft-lb (2,880,000 m-kg)
1.17 3141.6 pounds (1502 kg)
1.19 35.9 gal (133.4 L); 299.6 pounds (133.4 kg)

Chapter 2

2.2 45.9 psi (3.14 bar)
2.4 Pressure will be the same
2.6 80.2 psi (5.77 bar)
2.8 10 psi (0.71 bar)
2.10 17.5 psi (1.2 bar)
2.12 28.4 pounds (12.44 kg)
2.14 31.9 in. (82 cm)
2.16 29,920,000 pounds (13,583,340 kg)
2.18 208 pounds (94 kg)
2.20 375 ft (114 m) at all four points

Chapter 3

3.1 263.8 ft-lb (35.7 kg-m)
3.3 7,343,864 ft-lb (1,014,648 kg-m)
3.5 3.83 ft/s (1.2 m/s)
3.7

a. 176.3 gpm (636.2 L/min)
b. 18 ft/s (5.4 m/s)

P. Brock and V. Charter, *Fire Protection Hydraulics and Water Supply Analysis*, The Society of Fire Protection Engineers Series, https://doi.org/10.1007/978-3-032-09877-1

c. 176.3 gpm (636.2 L/min)

3.9 31 psi (2.1 bar)

3.11

a. 66.1 gpm (238.6 L/min)
b. 66.1 gpm (238.6 L/min)
c. 6.75 ft/s (2.205 m/s)

3.13 37 psi (2.59 bar)

3.15 57.2 psi (3.94 bar)

3.17 21.9 psi (1.52 bar)

3.19 0.1134 psi/ft (0.024 bar/m)

Chapter 4

4.2

a. 677 gpm (2566 L/min)
b. 970 gpm (3679 L/min)
c. 1256 gpm (4761 L/min)

4.4 1.25 in. (31.7 mm)

4.6 32 psi (2.2 bar)

4.8 1338 gpm (4899 L/min)

4.10 0.7

4.12 321 gpm (1432 L/min)

4.14 15.8 ft/s (4.8 m/s)

4.16 10.8 in. (19.6 cm) of mercury

4.18 2351 gpm (8539 L/min)

4.20 505 gpm (1852 L/min)

Chapter 5

5.1 Re = 60,057 (60,612); Turbulent

5.3 Re = 67,152 (67,120); Turbulent

5.5 Re = 72,317 (72,510); Turbulent

5.7 .

a. Approx. 6.4 psi (0.4 bar)
b. Approx. 11.6 psi (0.8 bar)
c. Approx. 7.6 psi (0.6 bar)
d. Approx. 3.4 psi (0.25 bar)
e. Approx. 1.1 psi (0.08 bar)

5.9 Approx. 6.2 psi (0.47 bar)

5.11 6-in. (150 mm)

5.13 .

a. 9.5 psi (0.7 bar)
b. 16.3 psi (1.2 bar)

c. 6.9 psi (0.5 bar)
d. 4.3 psi (0.3 bar)
e. 1.1 psi (0.08 bar)

5.17 1.63 ft (0.56 m)
5.19 86

Chapter 6

6.2

a. 118 gpm (448 L/min), 32 gpm (122 L/min), 2.6 psi (0.18 bar)
b. 321 gpm (1219 L/min), 54 gpm (201 L/min), 4.4 psi (0.3 bar)
c. 1025 gpm (3881 L/min), 475 gpm (1799 L/min), 13.1 psi (0.9 bar)
d. 120 gpm (458 L/min), 80 gpm (302 L/min), 3.6 psi (0.26 bar)
e. 642 gpm (2432 L/min), 508 gpm (1918 L/min), 18.4 psi (1.27 bar)
f. 1062 gpm (4017 L/min), 788 gpm (2983 L/min) 29.9 psi (2.06 bar)

6.4 Ft (3.69 m); Pf = 1.1 psi (0.08 bar) @ 500 gpm (1840 L/min), 4.0 psi (0.28 bar) @ 1000 gpm (3780 L/min), 8.4 psi (0.59 bar) @ 1500 gpm (5670 L/min)

6.6

a. Approx. 10.7 psi (0.82 bar)
b. Approx. 8.3 psi (0.63 bar)
c. Approx. 29.1 psi (2.2 bar)

Chapter 7

7.1

a. 0.72 MGD (2.72 MLD)
b. 1.8 MGD (6.77 MLD)
c. 2.88 MGD (10.89 MLD)

7.3

a. 5069 gpm (19,187 L/min)
b. 6736 gpm (25,500 L/min)
c. 12,778 gpm (48,368 L/min)

7.5 4.5 MGD (17.1 MLD)
7.7 65.0 MGD (245 MLD)
7.9 8333 gpm (31,667 L/min)
7.11 333 gpm (1250 L/min)
7.13 2.27 MGD (9.944 MLD)
7.15 47.28 MGD (178.974 MLD)
7.17 800 gpm (3012 L/min)

Chapter 8

8.4

a. 552 gpm; 0 psi (2004 L/min; 0 bar)
b. 3799 gpm; 84 psi (14,892 L/min; 5.8 bar)
c. 3634 gpm; 45 psi (13,880 L/min; 3.1 bar)
d. 1413 gpm; 42 psi (5300 L/min; 2.9 bar)

8.6 No
8.8 Approx. 585 gpm (2240 L/min)
8.10 No

Chapter 9

9.1 One
9.3 OS&Y
9.5 When the water supply is from a nonpressurized source such as a well or pond.
9.7 40 psi (2.7 bar)
9.9 Suction and Discharge: 10 in. (250 mm)
9.11 Pressure drop on the fire system
9.13 Electric motor, diesel engine, steam turbine
9.15 A compound gauge if suction pressures are less than 20 psi (1.4 bar)
9.17 Must comply with NFPA 70
9.19 A temperature complying with manufacturer's instructions.

Chapter 10

10.2

a. Approx. 915 gpm (3430 L/min)
b. Approx. 3050 gpm (11,657 L/min)
c. Approx. 1095 gpm (4160 L/min)

10.4

a. 500 gpm (1892 L/min)
b. 2000 gpm (7570 L/min)
c. 500 gpm (1892 L/min)

10.14 500 gpm (1892 L/min) @ 40 psi (2.7 bar)
10.16 1996 gpm (7442 L/min); 142 psi (9.9 bar)
10.18 1500 gpm (5677 L/min) @ 140 psi (9.6 bar)

2000 gpm (7570 L/min) @ 110 psi (7.6 bar)

10.20 Approx. 43 psi (3.0 bar)

Chapter 11

11.1 173 ft (53 m)
11.3 37 psi (2.54 bar)

11.5 40 psi (2.74 bar)

11.9a. 167,919 gal (697,000 L)

b. 45 psi (3.14 bar) at both hydrants
c. 82.2 ft/s (25 m/s)

Chapter 12

12.2

a. ISU: 180 gpm (650 L/min); NFA: 600 gpm (2203 L/min); ISO: 1250 gpm (4000 L/min); IFC: 1500 gpm (5677 L/min)
b. ISU: 1152 gpm (4306 L/min); NFA: 3200 gpm (11,946 L/min); ISO: 1750 gpm (7000 L/min); IFC: 2000 gpm (7570 L/min)
c. ISU: 11,520 gpm (42,984 L/min); NFA: 32,000 gpm (143,355 L/min); ISO: 3250 gpm (14,000 L/min); IFC: 3250 gpm (12,301 L/min)

12.4 750 gpm (1839 L/min); 161 psi (11.1 bar)

12.6 100 gpm (378 L/min); 96 psi (6.66 bar)

12.8

a. 610 gpm (2297 L/min)
b. 280 gpm (1063 L/min)
c. 1400 gpm (5289 L/min)

12.10

a. A. ISU: 21,600 gal (78,000 L); NFA: 72,000 gal (264,360 L); ISO: 150,000 gal; (480,000 L); IFC: 180,000 gal (681,240 L)
b. ISU: 138,240 gal (516,720 L); NFA: 576,000 gal (2,150,280 L); ISO: 210,000 gal (480,000 L); IFC: 240,000 gal (908,400 L)
c. ISU: 2,764,800 gal (10,316,160 L); NFA: 7,680,000 gal (34,405,200 L); ISO: 585,000 gal (2,520,000 L); IFC: 585,000 gal (2,214,180 L)

12.12

a. 32,400 gal (121,920 L)
b. 6300 gal (23,940 L)
c. 119,700 gal (452,160 L)

Chapter 13

13.1 600 gpm @ 42 psi (2214 L/min @ 2.73 bar)

13.3 596 gpm @ 42 psi (2194 L/min @ 2.68 bar)

Glossary

Ammeter A device for measuring electrical current in a wiring system.
Bar The units of pressure in Standard International Units.
Barometer An instrument used for the measurement of atmospheric pressure.
Bernoulli's equation A mathematical expression of the principle of conservation of energy.
Bourdon gauge The most common device used to measure system pressures.
Centrifugal pump Modern industrial fire pumps are centrifugal pumps. Pressure is added to the water by the centrifugal force created by a rotating wheel (impeller) or the rotating vanes of a turbine.
Certified shop test curves Before a pump is shipped by the manufacturer, it will be tested in the shop. The results of this test will be plotted on graph paper.
Circulation relief valve A small relief valve that opens up and provides enough water flow into and out of the pump to prevent the pump from overheating when it is operating at churn against a closed system.
Complex loop A piping system or "grid" that is characterized by one or more of the following: more than one inflow point, more than one outflow point, and/or more than two paths between inflow and outflow points.
Controller The electric control panel used to switch a pump on and off and to control its operation.
Darcy-Weisbach method Technique used to establish the pressure lost to friction in a piping system.
Differential manometer A device whose primary application is to reflect the differences in pressures between two points in a system.
Drivers The engine or motor used to turn a pump.
Flow hydrant The hydrant from which the water is discharged during a hydrant flow test.
Flow test Tests conducted to establish the capabilities of water supply systems and referred to as *flow tests* because they involve flowing fire hydrants. The objective of a flow test is to establish quantity (gallons per minute or liters per minute) and pressures available at a specific location on a particular water supply system.

P. Brock and V. Charter, *Fire Protection Hydraulics and Water Supply Analysis*, The Society of Fire Protection Engineers Series, https://doi.org/10.1007/978-3-032-09877-1

Fluid Any substance that can flow; a substance which has definite mass and volume at constant temperature and pressure, but no definite shape; and with the inability to sustain shear stresses.

Fluid mechanics A branch of physics dealing with the behavior of fluids, particularly with respect to their reaction to forces applied to them.

Force In the general terms of physics, force is that which causes motion.

Gridded piping system See complex loop.

Hardy cross method An iterative technique used for solving the complicated problems involving gridded water supply systems.

Hazen-Williams formula An empirical formula for calculating friction loss in water systems that is the fire protection industry standard. To comply with most nationally recognized standards, the Hazen-Williams formula must be used.

Head Pressure expressed in units of feet or meters of water.

Horizontal split-case pump A centrifugal pump with the impeller shaft installed horizontally and often referred to as a split-case pump. This is because the case in which the shaft and impeller rotates is split in the middle and can be separated exposing the shaft, bearings, and impeller.

Hydraulics The branch of fluid mechanics dealing with the mechanical properties of liquids (in this text water) and the application of these properties in engineering.

Hydrokinetics A branch of hydraulics having to do with liquids (water) in motion, particularly in relation to forces created by or applied to the liquid in motion.

Hydrostatics A branch of hydraulics dealing with the properties of liquids (water) at rest, particularly in relation to pressures resulting from or applied to the static liquid.

Jockey pump A jockey pump is a small-capacity, high-pressure pump used to maintain constant pressures on the fire protection system. A jockey pump is often used to prevent the main pump from starting unnecessarily.

Kinematic viscosity The kinematic viscosity of a fluid is the ratio of its absolute viscosity (lb s/ft^2
or kg s/m^2) to its mass density (lb s^2/ft^4
or kg s^2/m^4).

Kinetic energy The energy which a body possesses because of its motion.

Laminar flow A fluid is in the state of laminar flow if its Reynolds number is 2100 or less; laminar flow is related to very low liquid velocities.

Liquid A fluid having a definite volume, unlike gases, which will expand to fill the vessel containing it.

Moody diagram A diagram used with the Darcy-Weisbach friction loss computation technique to relate the Reynolds number, pipe size, and roughness to a friction factor.

Net pressure The net pressure is the pressure added to the system by the pump.

Orifice plate meter An orifice plate meter is a device used for measuring water flow and is similar in principle to a Venturi meter. The change of water velocity is accomplished by using a plate with an orifice that is smaller than the diameter of the pipe in which it is placed.

Pascal's law Principle 1, known as Pascal's Law, points out that pressure acts in all directions and not simply downward.

Piezometer tube This device uses the heights of liquid columns to illustrate the pressures existing in hydraulic systems.

Pitot tube Common device used to measure velocity pressure and thus fluid velocity. The pitot tube consists of a small diameter tube, usually about one- sixteenth inch in internal diameter which is connected to a pressure gauge.

Potential energy Stored energy which has the ability to perform work once released.

Pressure Force per unit of area.

Psi In fire protection, pressure is most often dealt with in units of pounds per square inch (psi) in the English System of units.

Psig Fire protection pressures are also most often gauge pressures (psig) which are deviations from atmospheric pressure.

Relief valve The relief valve is provided to open up and discharge water to a drain should the pressure become excessive. This valve is located between the pump and the discharge check valve and is required with pumps driven by variable speed drivers.

Residual hydrant The fire hydrant where the static and residual pressures are measured during a flow test; also called the test hydrant.

Residual pressure The pressure at the test hydrant while water is flowing. It represents the pressure remaining in the system while the test water is flowing.

Reynolds number The state of flow (laminar or turbulent) of a fluid has traditionally been established mathematically through the calculation of a factor called the Reynolds number (RE).

Simple loop A loop in which there is exactly one inflow point, one outflow point, and exactly two paths between the inflow and outflow points.

Specific gravity The specific gravity (Sg) of a substance may be defined generally as the ratio of the weight density of the substance to the weight density of another substance, usually water.

Static pressure The normal pressure existing on a system before the flow hydrant is opened.

Test hydrant The fire hydrant where the static and residual pressures are measured during a flow test; also called the residual hydrant.

Total energy The total energy (TE) at any point in a system might be defined as the sum of the potential energy and kinetic energy at that point.

Turbulent state Fluid flow is in the turbulent state at higher velocities where there is no definite pattern to the direction of the water particles. Turbulent flow is reflected by a calculated Reynolds number in excess of 2100.

Venturi meter When coupled with a differential manometer, a venturi meter may be used to measure water velocity. The device consists essentially of a piece of pipe in which the cross-sectional area has been constricted.

Vertical-shaft turbine pump This fire pump was originally designed to pump water from wells. Presently, they still have application where the water supply is from a nonpressurized source. Vertical-shaft pumps ordinarily have more than one impeller and are therefore multi-stage pumps.

Vertically mounted split-case pump A centrifugal pump similar to the horizontal split-case, except that the shaft is oriented vertically and the driver is mounted on top of the pump.

Voltmeter A device used for measuring the voltage existing on an electrical system.

Water hammer A violent increase in pressure which can be large enough to rupture the piping. Stopping any flowing stream too rapidly can cause this phenomenon.

Index

P. Brock and V. Charter, *Fire Protection Hydraulics and Water Supply Analysis*, The Society of Fire Protection Engineers Series, https://doi.org/10.1007/978-3-032-09877-1

D

E

F

MIX
Papier aus verantwortungsvollen Quellen
Paper from responsible sources
FSC® C105338

If you have any concerns about our products,
you can contact us on
ProductSafety@springernature.com

In case Publisher is established outside the EU,
the EU authorized representative is:
Springer Nature Customer Service Center GmbH
Europaplatz 3, 69115 Heidelberg, Germany

Printed by Libri Plureos GmbH
in Hamburg, Germany